Ceramics Processing
in Microtechnology

Edited by

Dr. H.-J. Ritzhaupt-Kleissl

Forschungszentrum Karlsruhe GmbH (FZK)

Institute for Materials Research (IMF III), Karlsruhe, Germany

Assoc. Professor Per Johander

Senior Researcher, Swerea IVF, Mölndal, Sweden

Whittles Publishing

Published by
Whittles Publishing,
Dunbeath,
Caithness KW6 6EY,
Scotland, UK
www.whittlespublishing.com

Distributed in North America by
CRC Press LLC,
Taylor and Francis Group,
6000 Broken Sound Parkway NW, Suite 300,
Boca Raton, FL 33487, USA

© 2009   H-J Ritzhaupt-Kleissl and P Johander

ISBN 978-1904445-84-5
USA ISBN 978-1-4398-0868-9

Typeset by Thomson Digital

Printed by Athenaeum Press Ltd.

Contents

Foreword

Ceramic is a fascinating material. On the one hand it is the earliest technical substance which was used to form pottery and may be considered to be as old as mankind itself. On the other hand ceramics today are the material basis for a wide variety of technical products with a great potential for future applications. During the last few years the use of ceramics has been extended into many novel applications such as: highly stressed components in engines and turbines, in communication, and finally in everyday consumer products. In many cases ceramics can replace metals due to their wear resistance, high temperature stability, and chemical inertness. The electronic properties of special ceramics distinguish the material for application in all sorts of sensors and other electronic components.

For many years the development of microelectromagnetic systems (MEMS) was dominated by semiconductor technology. With the proliferation of microsystems other materials such as metals, polymers, and finally ceramics were used for better adaptations to the applications. Today ceramic is a recognized member in the circle of materials for current and future microsystems.

This book was produced within the frame of the 4M Network of Excellence financed by the European Commission. The 4M Network is a knowledge community in Multi-material micro-manufacture to develop micro- and nanotechnology for the manufacture of micro-components and systems in a variety of materials for future products.

The Network comprises more than 150 researchers from 30 partner institutions, supported by over 60 industrial affiliates, and organized into eight research divisions: polymer processing, processing of metals, processing of ceramics, metrology, assembly and packaging, sensors and actuators, microfluidics, and microoptics.

The activities of the divisions include among others: cross-divisional interdisciplinary projects, researcher training and exchanges, workshops in micromanufacture, and technical publications.

The 4M Network organizes the 4M Annual Conference, runs a website with information and publications and maintains an advisory service to industry.

This book is an impressive result of the activities of the 4M Network and can be considered to be a unique collection of special knowledge in ceramics and ceramic technology combined with MEMS, metrology, and related technologies. It is highly recommended to scientists, engineers and everybody else who is interested in the manifold aspects of MEMS and 4M.

Prof. Dr. Wolfgang Menz
Chairman of the 4M Network
University of Freiburg, Germany

Preface

Due to their outstanding properties ceramics are a class of materials that have been used by mankind for millennia. Ceramic materials are still of great interest and importance in modern technology. Generally, they show high hardness, wear resistance and corrosion stability. Modern ceramics can cover a large range of physical properties, ranging from electrical resistivity from metallic-like conductivity to excellent insulation behaviour. The same is true for thermal conductivity. Additionally, there are some special properties, like dielectric, ferroelectric, piezoelectric or pyroelectric properties, which are only possessed by ceramic materials. This may be the reason why a great variety of ceramic materials is used for high-tech applications worldwide. This holds even true for microsystems technology, one of the most important and growing technologies since the end of the 20th century.

So the intention of this volume *Ceramics Processing in Microtechnology* is to present an overview of the current status and future prospects for the processing of ceramic materials in microtechnology.

The contributions to this book are mainly written by scientific partners in the Ceramics Division of the European Network of Excellence Multi-Material-Micro-Manufacturing (4M), but also include contributions by researchers outside the network who also perform R&D in developing and processing ceramic microcomponents.

Funded by the European Commission, the 4M Network seeks to develop micro- and nanotechnology for the batch-manufacture of microcomponents and devices in a variety of materials (except silicon) for future microsystems products. The mission of the network is to establish a knowledge community in 4M. Thus it acts as a knowledge resource for the research community and to industry in the development of microsystems devices that provide new, enhanced, and manifold functionality in tiny packages, integrating micro- and nanoscale features and properties into products and systems.

The Network comprises over 150 researchers from 30 partner institutions, supported by more than 60 industrial affiliates, and organised into eight research divisions: polymer processing, metrology, assembly and packaging, processing of metals, processing of ceramics, microoptics, Microfluidics, sensors and actuators.

The special aspect of 4M technologies covered in this book is related to ceramic micro-manufacturing and favourable applications of ceramic materials.

Comprehensive information is given for materials researchers, for processing engineers and for developers in the field of microsystem technology as well as for students. The objective is to give an impression of the overall development and processing chain from design questions, which are very special for microsystems, via material development (synthesis of microtechnology adequate materials), powder preparation and conditioning, microforming by various techniques, thermal processing up to metrology strategies. This also includes characterization and implementation of adequate quality control with respect to factors such as dimensional stability, reproducibility and the investigation of the dependence of the parts' final properties from the process parameters and from the intrinsic microstructure of the ceramic components.

After an introduction regarding the background and offering a short overview of the 4M Network and the Ceramic Division an overview on status, challenges, requirements, chances, perspectives, and applications for microceramic components is given. As there is a broad palette of ceramic materials with interesting structural or functional properties, various fields of application can be covered. Their potential for application in microsystems will be presented within this chapter.

Another chapter covers modelling and design aspects of microcomponents and microsystems. This chapter is especially important for product development, because establishing design rules for manufacture will prove helpful in assessing the suitability of the process. The second part of this chapter covers modelling and simulation: the possibility of modelling and simulating the production process or the functionality of the whole system is of great importance for the product development process.

The following chapters deal with materials development and various shaping and production processes for ceramic micromanufacturing. It can be seen that materials characteristics and adequate production processes have to be considered together in order to reach an optimal ceramic component. Different synthesis methods, especially regarding the micro-technological requirements, are described, such as: wet chemical synthesis of multinary oxide ceramics, synthesis of nanopowders, a special material development for shrinkage-free sintering ceramics as well as synthesis methods for ferroelectrics or for dielectrics with tuneable permittivity. Processing, typical properties and applications for ceramic–polymer composites are also considered.

Processing technologies are one of the main features of this book. As the most common technologies for the fabrication of ceramic microcomponents are replication techniques, tooling and mould fabrication are of great importance and will be discussed. Besides tooling special topics that will be covered in this chapter are replication techniques such as tape casting, embossing, electrophoretic deposition, high pressure injection moulding, low pressure injection moulding/hot moulding as well as special aspects of prototyping.

Other technologies are subtractive and additive process technologies. The subtractive technologies mentioned here are micromilling, laser milling, and electrodischarge machining. As additive processes layer manufacturing and direct printing of ceramics are described. Hybrid manufacturing processes and relevant applications are also considered. This comprises mainly applications and processing of low temperature confired ceramic materials. Finally, contributions on quality assurance and metrology are given. These contributions include test methods, suitable equipment for testing microcomponents as well as considerations on comparability and reproducibility.

By covering most of the important aspects of modelling, design, materials development, processing and quality control the authors hope not only to present a useful guide for students and readers, who look for a comprehensive overview on nonsilicon microtechnology, but they also wish to give an overview of the technologies available and on the status of R&D in this field as it is performed within the European 4M Network.

Hans-Joachim Ritzhaupt-Kleissl and Per Johander
(Editors)

The Ceramic Division within the 4M Network

Cardiff University, UK
Manufacturing Engineering Centre and Multidisciplinary Micro-technology Centre

IVF Industrial Research and Development Corporation, Sweden

Karlsruhe Research Centre (FZK), Germany
Programme Nano- and Microsystem Technology (NanoMikro)

University of Cranfield, UK
School of Industrial and Manufacturing Science

Vienna University of Technology (ISAS), Austria
Institute of Sensor and Actuator Systems

Royal Institute of Technology, (KTH), Sweden
Stockholm
KTH/KI Nano and Micro Technology Center
KTH Microsystem Technology Lab
KTH ELAB Cleanroom facilities

ROYAL INSTITUTE
OF TECHNOLOGY

FUNDACION TEKNIKER (TMM), Spain

Commissariat à l'énergie Atomique (CEA), France
DTEN, Grenoble

The Institute of Microelectronics (IMEGO), Sweden
Gothenburg

Technical University of Denmark (DTU), Denmark
Department of Manufacturing Engineering and Management

National Institute for R&D in Microtechnologies (IMT), Romania

Katholieke Universiteit Leuven, Belgium
Micro- and Precision Engineering Research Group

Within the European 4M (Multi-Material Micro Manufacture) Network of Excellence the Ceramic Division is represented by the above mentioned 12 research organisations, which contribute with their knowledge and their skills to an inter-European exchange of science and technology in the field of ceramic processing for microtechnical applications. The main objective of the Ceramic Division is to foster nonsilicon microtechnologies and to support the European microsystems industry by offering them their knowledge base and their experimental, analytical and technical experience in order to maintain a well-based international competitiveness.

Within these general targets the Ceramic Division claims to achieve the following objectives:

- To create the needed prerequisites for long-term integration of resources and expertise in microstructuring of ceramic materials.
- To find solutions to problems hindering the application of ceramic materials in microsystem-based products, based on the partners' experience. In particular these problems are:
  o The limitation of existing platforms for manufacture of products requiring integration of several functions as flow channels, 3D electrical interconnects, heat generation, heat transport, optical interconnects etc.
  o Lack of "design for manufacture" rules for microcomponents and products in ceramics.
  o The fragmented research in development of manufacturing platforms for new advanced applications requiring ceramic components.
  o To establish a European working group on "Processing of Ceramic Microcomponents".
  o To perform joint R&D projects funded by national or European institutions.

The documentation and the transfer of knowledge, which has been collected in the last four years (the funding period of the 4M network), is an important task for keeping the generated knowledge sustainable and accessible for ongoing research, development and fabrication in the area of ceramic microdevices, especially in form of inter-European cooperation. The present book contributes to meeting these objectives.

H.-J. Ritzhaupt-Kleissl, P. Johander

Abbreviations

3DP	3D printing
4M	(Multi-Material-Micro-Manufacture)
AES	Auger electron spectroscopy
AFM	atomic force microscopy
BET	Brunauer–Emmett–Teller (method for surface analysis)
BSCCO	bismuth strontium calcium cuprate
BST	barium-strontium-titanate ($Ba_{1-x}Sr_xTiO_3$)
BT	barium titanate ($BaTiO_3$)
C/SiC	carbon fibre reinforced silicon carbide
CAD	computer-aided design
CAM	computer-aided manufacturing
CEA	Commissariat à L'Energie Atomic, Grenoble, France
CIC	cold isostatic compaction
CIM	ceramic injection moulding
CIP	cold isostatic pressing
CM	conformal mapping
CMCs	ceramic matrix composites
CMM	coordinate measurement machine
CNC	computer numerically controlled
CPW	coplanar waveguide
CSCW	computer-supported cooperative work
CTE	coefficient of thermal expansion
CVI	chemical vapour infiltration
CVS	chemical vapour synthesis
DCC	direct coagulation casting
DCIJP	direct ceramic ink-jet printing
DLVO	Derjaguin, Landau, Verwey and Overbeek
DMD	digital micromirror device
DNA	deoxyribonucleic acid
DoD	drop-on-demand
DOE	design of experiments
DPC	deviation from a perfect cylinder
DPD	dissipative particle dynamics
DPDE	DPD with energy conservation
DPS	direct photoshaping
ECM	electrochemical machining
EDM	electrical discharge machining
EHDAP	electrohydrodynamic atomization printing
EPD	electrophoretic deposition
EPI	electrophoretic impregnation

FEM	finite element method
FFF	freeform fabrication
FFM	friction force microscopy
FGM	functionally graded material
FRAM	ferroelectric random access memories
FWHM	full width half maxima
FZK	Karlsruhe Research Centre
HARMS	high aspect ratio microstructure
HDDA	hexane diol diacrylate
HLB	hydrophilic–lipophilic balance
HPIM	high-pressure injection moulding
HRTEM	high resolution transmission electron microscopy
HTCC	high temperature cofired ceramics
HTSC	high temperature superconductors
IDC	interdigital capacitor
IGC	inert gas condensation
ISO	International Standards Organization
ITO	indium tin oxide
KBE	knowledge-based engineering
KMPP	Karlsruhe microwave plasma process
LBM	laser beam machining
LCD	liquid crystal display
LIGA	Lithography, electroplating and replicating (German acronym)
LPI	liquid polymer infiltration
LPIM	low-pressure injection moulding
LSI	liquid silicon infiltrated
LTCC	low temperature cofired ceramic
MEMS	microelectromechanical system
MicroCIM	microceramic injection moulding
MicroEDM	micro electrical discharge machining
MicroPIM	micropowder injection moulding
MIM	metal–insulator–metal
MIT	Massachusetts Institute of Technology
MMA	methylmethacrylic acid
MPE	maximum permissible error
MST	micosystems technology
MyBoK	micro book of knowledge
OCT	optical coherence tomography
PCB	printed circuit board
PEG	polyethylene glycol
PEM	proton exchange membrane (fuel cells)
PET	polyethylene terephtalate

PBG   photonic bandgap
PIM   powder injection moulding
PMMA   polymethylmethacrylate
PN   lead metaniobate
PTC   positive temperature coefficient of electrical sensitivity
PVD   physical vapour deposition
PVS   physical vapour synthesis
PZT   lead zirconate titanate
R&D   Research and Development
RBAO   reaction-bonded aluminium oxide
RBM   reaction bonding of mullite
RBSC   reaction-bonded silicon carbide
RBSN   reaction-bonded silicon nitride
RF   radio frequency
RPPC   rapid prototyping process chain
RSA   residual stress analysis
SEM   scanning electron microscope
SFF   solid freeform fabrication
SiSiC   siliconized silicon carbide
SLS   selective laser sintering
SMEs   small and medium enterprises
SOFC   solid oxide fuel cell
SPH   smoothed particle hydrodynamics
SPM   scanning probe microscope
SRBSN   sintered reaction-bonded silicon nitride
STEM   scanning transmission electron microscopy
STEP   standard for the exchange of product model data
STL   stereolithography
TEC   triethylcitrate
TEM   transmission electron microscopy
TEOS   tetraethyl orthosilicate
th.d.   theoretical density
UV   ultraviolet
XPS   X-ray photoelectron spectroscopy
XRD   X-ray diffraction

List of contributing authors' affiliations

Prof. Albert ALBERS
IPEK – Institute of Product Development,
Universität Karlsruhe (TH), Kaiserstr. 10, D-76128
Karlsruhe, Germany
email: albers@ipek.uka.de

Dr. Werner BAUER
Institute for Materials Research III, Forschungszentrum Karlsruhe GmbH, P.O. Box 3640,
D-76021 Karlsruhe, Germany
email: Werner.Bauer@imf.fzk.de

Dr. Joachim R. BINDER
Institute for Materials Research III, Forschungszentrum Karlsruhe GmbH, P.O. Box 3640,
76021 Karlsruhe, Germany
email: joachim.binder@imf.fzk.de

Dr. Giuliano BISACCO
Assistant professor, Department of Mechanics
and Innovation (DIMEG), University of Padova,
via Venezia 1, 35131 Padova, Italy
email: giuliano.bissacco@unipd.it

Sylvia VOGEL (BONNAS)
Robert Bosch GmbH, PO Box 13 42, D-72762
Reutlingen, Germany
email: Sylvia.Vogel@de.bosch.com

Francesca BORTOLANI
Microsystems and Nanotechnology Centre,
Materials Department, Cranfield University,
Cranfield, Bedfordshire, UK, MK43 0AL
email: f.bortolani@cranfield.ac.uk

Dr. Michael BRUNS
Institute for Materials Research III, Forschungszentrum Karlsruhe GmbH, P.O. Box 3640,
D-76021 Karlsruhe, Germany
email: michael.bruns@imf.fzk.de

Dr. Elis CARLSTRÖM
Swerea IVF AB, P.O. Box 104, SE-431 22
Mölndal, Sweden
email: elis.carlstrom@swerea.se

Tobias DEIGENDESCH
IPEK - Institute of Product Development,
Universität Karlsruhe (TH), Kaiserstr. 10,
D-76128 Karlsruhe, Germany
email: deigendesch@ipek.uka.de

Adam-Mwanga DIECKMANN
Institute of Production Science wbk, Universität
Karlsruhe (TH), Kaiserstr. 12, D-76131 Karlsruhe,
Germany
email: dieckmann@wbk.uka.de

Dr. Robert DOREY
Microsystems & Nanotechnology Centre, Materials Department, Cranfield University, Cranfield,
Bedfordshire, UK, MK43 0AL
email: r.dorey@cranfield.ac.uk

Hendrik ELSENHEIMER
Institute of Microsystems Engineering (IMTEK),
Laboratory for Materials Processing
University of Freiburg, Georges-Koehler-Allee
102, D-79110 Freiburg, Germany
email: hendrik.elsenheimer@imtek.uni-freiburg.de

Dr. Luc FEDERZONI
Department of Transport and Hydrogen. Bat.
E.231, Commissariat à l'Energie Atomique, 17,
Rue des Martyrs, 38054 Grenoble cedex 9, France
email: luc.federzoni@cea.fr

Dr. Eleonora FERRARIS
Division PMA, Department of Mechanical
Engineering, Katholieke Universiteit Leuven,
Celestijnenlaan 300B, BE-3001, Leuven, Belgium
email: Eleonora.Ferraris@mech.kuleuven.be

Dr. Holger GESSWEIN
Institute for Materials Research III, Forschungszentrum Karlsruhe GmbH, P.O. Box 3640,
D-76021 Karlsruhe, Germany
email: Holger.Gesswein@imf.fzk.de

Dr. Andre GIERE
Astyx GmbH, Lise-Meitner-Str. 2a, D-85521
Ottobrunn, Germany
email: a.giere@astyx.de

Dr. Thomas GIETZELT
Institute for Micro Process Engineering, Forschungszentrum Karlsruhe GmbH, PO Box 3640,
76021 Karlsruhe, Germany
email: thomas.gietzelt@imvt.fzk.de

Dr. Andreas GREINER
Department of Microsystems Engineering
(IMTEK), Laboratory for Simulation,

University of Freiburg, Georges-Koehler-Allee
103, D-79110 Freiburg, Germany
email: greiner@imtek.de

Elmar GÜNTHER
Institute for Materials Research III, Forschung-
szentrum Karlsruhe GmbH, P.O. Box 3640,
D-76021 Karlsruhe, Germany

Dr. Thomas HANEMANN
Institute for Materials Research III, Forschung-
szentrum Karlsruhe GmbH, P.O. Box 3640,
D-76021 Karlsruhe, Germany
email: thomas.hanemann@imf.fzk.de
and
Department of Microsystems Engineering,
Albert-Ludwigs-University Freiburg, Georges-
Köhler-Allee, D-79110 Freiburg, Germany
email: hanemann@imtek.uni-freiburg.de

Dr. Stefan HAUSER
IPEK – Institute of Product Development,
Universität Karlsruhe (TH), Kaiserstr. 10,
D-76128 Karlsruhe, Germany
email: hauser@ipek.uka.de

Dr. Richard HELDELE
Robert Bosch GmbH, P.O. Box 30 02 40,
D-70442 Stuttgart, Germany
email: Richard.Heldele@de.bosch.com

Dr. Kirsten HONNEF
Institute of Microsystems Engineering (IMTEK),
Laboratory for Materials Processing, University of
Freiburg, Georges-Koehler-Allee 102, D-79110
Freiburg, Germany
email: kirsten.honnef@imtek.uni-freiburg.de

Dr. Per JOHANDER
Assistant Professor, Swerea IVF AB, P.O. Box
104, SE-431 22 Mölndal, Sweden
email: per.johander@swerea.se

Dr. David KAUZLARIC
Department of Microsystems Engineering
(IMTEK), Laboratory for Simulation,
University of Freiburg, Georges-Koehler-Allee
103, D-79110 Freiburg,Germany
email: kauzlari@imtek.uni-freiburg.de

Dr. Regina KNITTER
Institute for Materials Research III, Forschungszentrum Karlsruhe GmbH, P.O. Box 3640,

D-76021 Karlsruhe, Germany
email: Regina.Knitter@imf.fzk.de

Prof. Jan G. KORVINK
Department of Microsystems Engineering
(IMTEK), Laboratory for Simulation, Univer-
sity of Freiburg, Georges-Koehler-Allee 103,
D-79110 Freiburg, Germany
email: korvink@imtek.uni-freiburg.de

Dr. Jan KOTSCHENREUTHER
Institute of Production Science wbk, Kaiserstr.
12, D-76131 Karlsruhe, Germany
email: sekretariat@wbk.uka.de

Prof. Bert LAUWERS
Division PMA, Department of Mechanical
Engineering, Katholieke Universiteit Leuven,
Celestijnenlaan 300B, BE-3001, Leuven,
Belgium
email: Bert.Lauwers@mech.kuleuven.be

Karin LINDQVIST
Swerea IVF AB, P.O. Box 104, SE-431 22
Mölndal, Sweden
email: karin.lindquist@swerea.se

Kun LIU
Division PMA, Department of Mechanical
Engineering, Katholieke Universiteit Leuven,
Celestijnenlaan 300B, BE-3001, Leuven,
Belgium
email: Kun.Liu@mech.kuleuven.be

Prof. Lars MATTSSON
Industrial Metrology & Optics, Department
of Production Engineering, KTH – The Royal
Institute of Technology, Brinellv. 68, SE-10044
Stockholm, Sweden
email: Larsm@iip.kth.se

Prof. Wolfgang MENZ
Institute of Microsystems Engineering (IMTEK),
Laboratory for Materials Processing, University of
Freiburg, Georges-Koehler-Allee 102, D-79110
Freiburg, Germany
email: wolfgang.menz@imtek.uni-freiburg.de

Dr. Marcus MÜLLER
Institute for Materials Research III, Forschungszentrum Karlsruhe GmbH, P.O. Box 3640,
D-76021 Karlsruhe, Germany
email: Marcus.Mueller@imf.fzk.de

Dr. Rolf OCHS
Institute for Materials Research III, Forschungszentrum Karlsruhe GmbH, P.O. Box 3640,
D-76021 Karlsruhe, Germany
email: rolf.ochs@imf.fzk.de

Lisa PALMQVIST
Swerea IVF AB, P.O. Box 104, SE-431 22
Mölndal, Sweden
email: lisa.palmquist@swerea.se

Dr. Florian PAUL
Institute of Microsystems Engineering (IMTEK),
Laboratory for Materials Processing, University of
Freiburg, Georges-Koehler-Allee 102, D-79110
Freiburg,Germany
email: florian.paul@imtek.uni-freiburg.de

Dr. Jan PEIRS
Division PMA, Department of Mechanical
Engineering, Katholieke Universiteit Leuven,
Celestijnenlaan 300B, BE-3001, Leuven,
Belgium
email: Jan.Peirs@mech.kuleuven.be

Dr. Wilhelm PFLEGING
Institute for Materials Research I, Forschungszentrum Karlsruhe GmbH, P.O. Box 3640,
D-76021 Karlsruhe, Germany
email: wilhelm.pfleging@imf.fzk.de

Dr. Volker PIOTTER
Institute for Materials Research III, Forschungszentrum Karlsruhe GmbH, P.O. Box 3640,
D-76021 Karlsruhe, Germany
email: Volker.Piotter@imf.fzk.de

Prof. Dominiek REYNAERTS
Division PMA, Department of Mechanical
Engineering, Katholieke Universiteit Leuven,
Celestijnenlaan 300B, BE-3001, Leuven,
Belgium
email: Dominiek.Reynaerts@mech.kuleuven.be

Dr. Hans-Joachim RITZHAUPT-KLEISSL
Institute for Materials Research III, Forschungszentrum Karlsruhe GmbH, P.O. Box 3640,
D-76021 Karlsruhe, Germany
email: ritzhaupt-kleissl@imf.fzk.de

Dr. Sophie ROCKS
Microsystems & Nanotechnology Centre,
Materials Department, Cranfield University,

Cranfield, Bedfordshire, UK, MK43 0AL
email: s.rocks@cranfield.ac.uk

Dr. Magnus ROHDE
Institute for Materials Research I, Forschungszentrum Karlsruhe GmbH,
P.O. Box 3640, D-76021 Karlsruhe, Germany
email: magnus.rohde@imf.fzk.de

Dr. Sabine SCHLABACH
Institute for Materials Research III, Forschungszentrum Karlsruhe GmbH,
P.O. Box 3640, D-76021 Karlsruhe, Germany
email: sabine.schlabach@imf.fzk.de

Dr. Johannes SCHNEIDER
Universität Karlsruhe (TH), Institute for
Reliability of Components and Systems,
c/o Forschungszentrum Karlsruhe, P.O.Box 3640,
D-76021 Karlsruhe, Germany
email: Johannes.Schneider@kit.edu

Prof. Volker SCHULZE
Institut für Produktionstechnik (wbk) und Institut für Werkstoffkunde, Universität Karlsruhe
(TH), Kaiserstr. 12, 76131 Karlsruhe, Germany
email: volker.schulze@iwk1.uka.de

Benedikt SCHUMACHER
Institute for Materials Research III, Forschungszentrum Karlsruhe GmbH, P.O. Box 3640,
D-76021 Karlsruhe, Germany
email: benedikt.schumacher@imf.fzk.de

Dr. Walter SMETANA
Institute of Sensor and Actuator Systems, Vienna
University of Technology, Gusshausstraße 27-29,
A-1040 Vienna, Austria
email: walter.smetana@tuwien.ac.at

Dr. Dorothée Vinga SZABÓ
Institute for Materials Research, Forschungszentrum Karlsruhe GmbH, P.O. Box 3640, D-76021
Karlsruhe, Germany
email: dorothee.szabo@imf.fzk.de

Dr. Dazhi WANG
Microsystems & Nanotechnology Centre,
Materials Department, Cranfield University,
Cranfield, Bedfordshire, UK, MK43 0AL
email: d.wang@cranfield.ac.uk

1

Ceramics in microtechnology: status, requirements and challenges

H.-J. Ritzhaupt-Kleissl and R.A. Dorey

Besides polymers and metals ceramics play an important role in microtechnology because of their outstanding properties. They are not only characterized by high hardness, wear resistance, high temperature stability and chemical inertness, but ceramics also show special functionalities, e.g. sensor and actuator properties, ferroelectricity or electric conductivity ranging from insulators via semiconductors to metal-like conductors. Nevertheless, special materials need to be developed to meet the requirements of micro-technology. This is especially true for the synthesis of multinary oxides as functional ceramics or for new ceramic materials with properties tailored for special applications. Ceramic processing also faces challenges in order to meet the microstructural, dimensional and accuracy requirements of microtechnology. Standard shaping processes have to be focused to the fabrication of microparts, or new processes may have to be established. A further aspect is the design of ceramic microcomponents, which require approaches quite different from those needed to design polymer or metal microparts. Thus, the development of ceramic microcomponents requires research and development in the field of materials research as well as in processing and designing the components because these areas are strongly interconnected. Another aspect which has to be taken into account is the integration of ceramic microcomponents into complete microsystems, i.e. suitable assembly and packaging methods must be available for integration.

1.1 Introduction

Besides polymers and metals, ceramic materials are increasingly important in modern technology. This is because of their outstanding properties, such as hardness, high-temperature applicability, abrasion resistance and chemical inertness. But their outstanding physical properties such as di-, ferro- and piezoelectricity as well as conductivity and sensor properties open a wide field of application for modern ceramic materials. Taking into account the need for miniaturization of technical devices and the growth of microsystem technology (MST) there is an increasing requirement for ceramic microcomponents like microparts or microstructured parts.

Examples of ceramic microcomponents are high precision parts for medical and dental application with tolerances in the micrometre range, microstructured piezocomponents, microgas sensors, tunable dielectrics for high-frequency application, transparent or translucent ceramic microcomponents, parts for microreaction technology etc.

Ceramic microcomponents like those mentioned above have in common, that as a prerequisite for their realization and successful development one must take into consideration the strong interdependencies between:

- material;
- processing;
- shape and accuracy; and
- application.

So for example the realization of a ceramic microcomponent for a defined application needs the availability of the appropriate material, which must be processed by an adequate processing route depending again on the material characteristics and also on the envisaged shape, size, accuracy and on the functionality of the final component. The fulfilment of these interdependent requirements is generally accompanied by several challenges along the development path resulting in a functioning ceramic microcomponent.

To meet the requirements and challenges it is often necessary to develop new materials and to establish new manufacturing routes and so generate ceramic components in micro-dimensions, or with microdetails or with accuracies in the micrometre range. This is still more important as a final shaping or finishing of the microcomponents after the moulding and sintering processes is either very expensive or even technically impossible.

One of the greatest challenges is to meet the microstructural, dimensional and accuracy requirements. Moulding processes have to be developed further for the fabrication of micro-components, or even new processes have to be established. An additional aspect is the design of ceramic microcomponents, which requires approaches quite different from designing polymer or metal microparts. As the development areas are strongly interconnected, it is sometimes necessary to establish an integrated line from material development via processing to component design. Examples for possible development approaches will be given in this chapter but will necessarily be also reflected in the subsequent chapters of this book.

This chapter briefly surveys the various fields of research and development currently performed in the area of microprocessing of ceramic materials. The most important of the mentioned topics are described in detail in the following chapters of this book.

1.2 Material development

1.2.1 Moulds and powders

The most important requirements to be met by ceramic microparts are dimensional accuracy and precision down to the micrometre-range as well as smooth surfaces even in the sub-micrometre range without additional surface finishing.

Dimensional accuracy requires high-precision moulds. There has been considerable progress in the manufacture of microstructured tool inserts for replication techniques using micromachining or the lithography, electroplating and replicating (LIGA) technique. However, the tools control the size of the ceramic microcomponent only in the green or presintered state. Whether the accuracy of the sintered part is sufficient mainly depends on the thermal processing step resulting in densification and shrinkage. It is inevitable that the shrinkage must be uniform and reproducible and that no distortion occurs. Deviations from the final dimensions increase as the amount of sintering shrinkage increases. Therefore, especially in microtechnology, where correction of the final dimensions by final finishing is not applicable, green densities should be as high as possible to prevent this. High green densities, however, require perfect powder packing based on optimized particle size distributions of properly mixed larger and smaller particles, where, considering microparts, the "larger" particles must be in the range of, or even below, 1 μm.

Fine powders are necessary for smooth surfaces. Despite the fact, that in sintered parts the surface quality should be controlled by the grain structure and the resultant porosity after sintering rather than by the size of the starting powder, there seems to be a correlation between surface roughness of the sintered parts and the particle size of the powder used, injection molded from powder feedstocks [1]. This is why for dimensional details in the micrometre range and for a high surface quality micro- and nanopowder technologies for advanced ceramic microcomponents have to be applied.

1.2.2 Advanced synthesis methods

Powder synthesis and processing is of great importance for ceramic technology. Especially in the field of microtechnology where developments tend towards smaller parts and finer details there is an urgent need for still finer and more homogeneous powders. Nanopowders, which are aimed at opening new fields of technology due to their unique properties, are either already commercially available or at various states of development at many research institutions. The most important processes for the production of ceramic nanopowders are: gas phase syntheses, wet chemical or laser ablation processes.

Wet chemical processes such as sol-gel and precipitation techniques lead to nanosized powders, which, however, are often agglomerated. Gas phase processes, including flame pyrolysis and chemical vapour synthesis also tend towards particle agglomeration. There are process conditions, however, where agglomeration can be eliminated almost completely [2]. With the so-called Karlsruhe microwave–plasma process [3] not only agglomeration-free ceramic nanopowders but also nanocomposite powders with interesting physical properties can be synthesized, where the ceramic nanoparticles are coated in situ with a second ceramic or polymer layer. The agglomeration tendency can also be reduced by the laser ablation process as was shown e.g. for ZrO_2 [4, 5].

Another promising synthesis method is the wire-explosion technique [6]. Using this technique not only can oxide ceramics be synthesized [7] but also, depending on the atmosphere in the reaction chamber, nitrides or even nanoscaled metal powders, which then can be further processed, e.g. by a passivating coating or by adjacent reactions in order to generate pure ceramics or ceramic–metal mixtures [8].

1.2.3 Sinter shrinkage and distortion

During sintering of a preshaped, porous and not very strong green body to a dense and strong ceramic, shrinkage occurs. This is one of the main disadvantages of ceramic components, especially in microtechnology. Up to 20% linear shrinkage is typical. To obtain the required component size after sintering, the green body must be formed with correspondingly larger dimensions or there must be some finishing after sintering, i.e. in the sintered, hard state. Although compensation of the shrinkage and thus net-shape sintering are state-of-the-art in large-scale production, for narrow dimensional tolerances, final shaping, e.g. by grinding, is still necessary after sintering. In the case of ceramic microcomponents, with minimal detail dimensions in the range of certain micrometres such finishing is extreme difficult and expensive, if not completely impossible. One potential solution for this problem is provided by reaction bonding. In this process precursors of the required ceramic are transformed into the ceramic by an additional reaction during the sintering process. If the precursors are selected to ensure that the reaction product has a larger volume than the starting compo-nents, it is possible to compensate for the inevitable shrinkage in order to obtain dense and net-shape ceramic microcomponents. It has been shown that reaction bonding techniques such as reaction bonding of alumina (RBAO) or the reaction bonding of mullite (RBM) are interesting alternative ceramic processing methods offering advantages compared to conventional processing routes. One of the greatest advantages is the possibility to generate ceramics with low to zero shrinkage during sintering [9–11].

Intermetallic phases e.g. the silicides of titanium and zirconium offer a great potential as inorganic precursors for reaction bonding as they show, compared to metallic precursors, a rather large volume increase combined with an easily controlled oxidation reaction.

For example, according a summary reaction equation for the oxidation of zirconium disilicide, $ZrSi_2$, yields:

$$ZrSi_2 + 3O_2 \rightarrow ZrO_2 + 2\,SiO_2 \text{ or}$$
$$ZrSi_2 + 3O_2 \rightarrow ZrSiO_4 + SiO_2$$

This reaction results in a volume increase of about 100% from $1\,cm^3$ $ZrSi_2$ to $2\,cm^3$ ($ZrSiO_4 + SiO_2$). So, when starting with a defined green density, which is relatively easy to obtain, e.g. by powder pressing, shrinkage can be compensated for and exact net shape ceramic components can be achieved after sintering.

Other aspects which must be taken into account in the course of ceramic processing are organic pressing aids and binders, which are added to the starting powder in order to improve shaping and processing. Generally, these organics must be burnt out during the sintering process. They do not contribute to densification during sintering. On the contrary, they cause new pores after their burnout also influencing the volume change when they are removed. The use of so-called "low-loss binders" improves the situation as they take part in the reaction and ceramize to a certain extent. For example, a certain polysiloxane as a low-loss binder yields about 80% ceramic, i.e. after oxidation about 80% of this organic is transformed to SiO_2, which again together with ZrO_2 can react to $ZrSiO_4$. So the complete thermal process consists of three steps: pyrolysis of the low-loss binder, oxidation i.e. the expansion of the intermetallic phase; and the sintering step.

Figure 1.1 Mechanically green shaped nut (black) and identical nut (white) sintered shrinkage-free by a reaction bonding process with optimal fit to the screw.

A further beneficial effect of the binder addition is that this material can be mechanically microformed well in the green state either by turning or milling with standard hard-metal tools or by embossing with a (microstructured) die. Intensive studies have been performed to develop a $ZrSiO_4$-based non-shrinking ceramic material together with the appropriate microforming technologies [12–16]. Besides the compliance with the high accuracy zero-shrinkage characteristics in order to meet microtechnical requirements, special emphasis was laid on relevant processing routes. These developments resulted in the microstructuring of green blanks either by embossing by a microstructured die, which is appropriate for components with low aspect ratios, or in mechanically microstructuring the green blanks, e.g. by micromilling. The micromilling technique is very favourable, as it allows the fabrication of real net shape 3D microparts in any shape.

An example is shown in Figure 1.1, demonstrating the advantages of the material with respect to mechanical green machining and shrinkage-free sintering. Here the black nut is the green shaped body and the white nut is the sintered part. Both fit very well on the screw.

Current studies [17] show that also zero-shrinkage mullite–ZrO_2 composites can be synthesized from the appropriate intermetallic phases $ZrAl_3/ZrO_2$ and $ZrAl_3/ZrSi/ZrO_2$. Noticeably higher strengths up to about 630 MPa could be realized with these ceramics.

The possibility to shape these types of ceramics in the green state with high accuracy to the final dimensions opens up interesting fields of application. These, besides universal application in microtechnology, include applications in medical or dental technologies.

1.3 Processing routes for the fabrication of ceramic microcomponents

1.3.1 Processing techniques

There are many techniques for the fabrication of ceramic microparts. They may be divided into three basic groups:

- replication techniques;
- ablating processes; and
- generating techniques.

All of them show certain advantages and challenges; none of them is generally suited to fulfill all requirements to fabricate every ceramic micropart of any shape with reasonable effort. So for each material one has to consider the design of the component, the required accuracy and number of parts to be produced.

Replication processes are the most common. This group comprises ceramic injection moulding (CIM), embossing of ceramic tapes, electrophoretic deposition (EPD), screen printing and also special casting and moulding processes, such as soft moulding, gel casting or slurry casting. Even die pressing can be mentioned here, even if this process is not that widely used for the fabrication of microparts.

There are two principal ways to make small-sized ceramic components by replication techniques: powder-free fabrication routes based on preceramic polymers, which can be moulded or lithographically structured, and powder routes.

Replication techniques generate the ceramic replicas by filling moulds. However, the mould materials and the method by which the moulds or the original master patterns are fabricated may differ. For replication the ceramic material must be in a mouldable state, i.e.:

- It must fill the mould completely down to details in the micrometre range.
- It must be consolidated in the mould.
- It must be demouldable after filling so that the structured details are kept and not destroyed.
- It must be sinterable to the desired density without cracking or fragmentation.

These requirements can be well met by colloidal techniques [18–20]. Various replication techniques have been developed for the fabrication of ceramic microparts [21], e.g. gel casting [22, 23], soft moulding [24], direct coagulation casting (DCC) [25], tape casting [26–29], screen printing [30, 31], electrophoretic deposition (EPD) [32–34], lithographic structuring of preceramic polymers [35–37] etc.

Despite promising developments of powder-free fabrication routes [34–37], powder technology is still, and probably will remain, the method of choice for the manufacturing of ceramic microcomponents. The common moulding processes well known from the manufacturing of large ceramic parts are: die-pressing, ceramic injection moulding (CIM), slurry casting or EPD from ceramic suspensions. All of them are followed by sintering to final shape and properties. Among these techniques microceramic injection moulding (MicroCIM) is of great interest for economic production with available equipment.

Ablating processes are: mechanical machining, laser structuring and electrodischarge machining (EDM). They can all be performed after sintering, i.e. they make a final shaping on an already dense and hard ceramic blank in order to impress the final microstructure into the part. While there is no special precondition for mechanical micromachining (provided the suitable equipment is available) materials for EDM must be electroconductive. Furthermore it is possible to carry out micromachining not only with sintered but also with green or partly sintered blanks. In these cases also the sintering shrinkage has to be taken into account. 3D freeform shapes can be realized from all the ablating processes.

Generating processes are mainly 3D printing [38–45], layered manufacturing such as lamination of structured tapes [29, 46] or laminated object manufacturing [47, 48] and selective laser sintering (SLS) [49–51]. These processes also allow 3D freeform manufacturing of ceramic microparts. Originally used for development or prototyping [52, 53] these techniques also claim to be suitable for series production [38–40].

The greatest advantage of the generating processes is that moulding is not needed. The forming process is executed automatically i.e. it is computer controlled. The basic techniques are available; they are either a computer-controlled laser beam (for SLS), an ink-jet technique comparable to that for office printers or a well-known stamping and layering technique for the lamination techniques. But the adaptation to ceramics processing raises specific problems.

A low viscosity ink is necessary for printing techniques. It is filled with ceramic particles which are as small as possible in order to generate even the smallest detail of the desired micropart. So for ink-jet printing stable suspensions containing agglomerate-free nanoparticles are required. To keep the viscosity low, the solid content of the suspension is rather limited, resulting in a rather slow growth of 3D structures.

Another printing technique requires printing a type of glue into a powder bed and allowing the structure grow layerwise by lowering the powder bed layer by layer. Free flowing powders are essential for this technique, because each new powder layer has to be levelled exactly before the powder particles are glued together. So for this process there is a strong need for free flowing nanopowders in order to generate microdetails.

In general ceramic microparts made by printing techniques show a low green density with a surface quality depending on the powder particle size. This low green density not only results in a large amount of shrinkage during sintering to a dense body but can also show distortions when density gradients exist, induced by the printing process.

Laser sintering of ceramics is complicated by the high melting temperatures of ceramic materials. Due to the localized high temperature of the laser beam high local stresses will be induced, which can lead to distortions. Also, the surface quality does not always fulfill the requirements for microcomponents. These deficiencies can be overcome if a mixture of a ceramic with a glassy phase with low melting point is used.

Most of these processing techniques will be described and discussed in detail elsewhere in this book. Nevertheless some general aspects regarding the possibilities, requirements and possible challenges of ceramics microparts forming and processing will be mentioned here, taking injection molding, mechanical machining and EDM as examples.

1.3.2 Microinjection moulding

One of the reasons for the development of powder injection moulding (PIM) for microparts is the fact that conventional, i.e. macroscopic PIM is a widespread manufacturing technology for metallic and ceramic parts. Another advantage is that this technique can cover the whole field from development studies, prototyping or fabricating a few samples to medium- or large-scale series production. This is enabled by the fact that there are two variants of CIM which are both suitable for the fabrication of ceramic microcomponents: low pressure injection moulding (LPIM), also called hot moulding, and the better known high pressure injection moulding (HPIM). While the entire process chain is well established for macroscopic parts, new challenges arise with respect to the miniaturization of PIM for applications in microtechnology. Generally, injection moulding consists of the following process steps: feedstock preparation, injection moulding, debinding and sintering.

Feedstock preparation is an essential step because very fine ceramic powders are necessary for MicroCIM. With only very few exceptions the powders in commercial feedstocks are too large. Because of the small cross-sections of microparts and their high surface-to-volume-ratio, wall friction during mould filling and demoulding becomes important and requires low viscosity for mould filling, and high strength feedstocks and perfectly smooth tool inserts for demoulding. For a perfect flow of a feedstock during mould filling each particle has to be completely coated with binder. So, with decreasing powder particle size, the maximum powder fraction in the feedstock significantly decreases. As the powder content in the feedstock is identical to that in the green part, a reduction of the particle size inevitably results in a reduced green density and in a reduced density after debinding. Therefore the shrinkage during sintering to full density is increased [54]. Thus, despite the fact that sub-micrometre powders are desirable for MicroCIM the usage of nanopowders with particle sizes below 100 nm seems to be limited.

Besides the ceramic powders feedstocks for MicroPIM consist of a multicomponent binder system, adapted to the size and type of the ceramic filler material. Feedstocks for the mass production (HPIM) of microparts or micropatterned parts contain a significant amount of at least one thermoplastic component providing sufficient green strength for a damage-free demoulding of the green parts. They are often combined with waxes and further additives to form binder systems consisting of at least three components [55]. These binder systems are preferred to realize a stepwise debinding process in order to avoid stress in the green compacts during debinding. The waxes reduce the feedstock's viscosity for a complete mould filling [56] and provide a better debinding behaviour [57, 58]. Improved wetting of the polar surfaces of ceramic powders with the non-polar binder can be achieved by using surfactants.

While for series production HPIM is the preferable technique, LPIM at pressures below 5 bar, where paraffin is used as binder, is a promising technique for small series and for rapid prototyping [59].

Because of the low viscosity of paraffin relatively high amounts of ceramic powders can be added without exceeding the acceptable maximum viscosities. Whereas for polymer-based feedstocks the maximum powder loading is in the range of about 55 vol.%, wax-based feedstocks allow powder loadings of up to 70 vol.%, resulting in a reduced sintering shrinkage.

The low viscosity of the wax-based feedstocks also facilitates the preparation of nanopowder-filled feedstocks with reasonable powder loadings without exceeding the tolerable viscosity limit.

So far, mainly metallic tools made by micromilling or by the LIGA process have been used for microinjection molding. For low viscosity binders such as paraffin and for low-pressure injection moulding even silicone moulds can be used. They are cheap and they can be directly copied from existing specimens, e.g. from polymer parts made by rapid prototyping techniques. The demoulding of the green parts from the silicone is possible even with complicated shapes.

1.3.3 Micromachining of ceramics

For flexible and economic production of microsystems containing ceramic microcomponents an economic as well as flexible production chain for ceramic microparts is necessary.

This is especially true for small-to-medium-series production mainly performed by small and medium enterprises (SMEs). For this purpose, where mass production of hundred thousands or millions parts is not envisaged it is essential to establish a competitive production method, enabling European SMEs to stay at the front edge of MST.

These envisaged lot sizes exclude the purchase of expensive tooling as it is necessary e. g. for microinjection moulding.

Mechanical machining of metals is a well-established technology. Micromachining has also become an established technology. This holds true for the available machinery equipment as well as for processing.

There is an interesting potential for ceramic microcomponents, because ceramic micromachining offers several advantages:

- flexible production scale (from few parts to series production);
- use of standard micromachining equipment (milling, grinding);
- equipment can be used for various materials; and
- fabrication of free-formed 3D parts is possible.

Generally, machining of ceramics is not a new technology but a very old one. Already about 6000 years ago information was stored by engraving it with a pin in clay tables, the well-known cuneiform writing. Also pottery is the more or less mechanical forming of a ceramic raw material into a desired shape. But because of the plastic properties of the clay, this material is predestined for pressing into a mould. So many identical replicas can easily be produced.

This may be one of the reasons why replication and not mechanical machining of ceramics became the leading technology. Nevertheless, as stated above, ceramic machining is still an interesting technology, especially for small or microcomponents.

There are three principal ways for shaping ceramic micro parts by machining:

- machining before sintering in the green state;
- machining partly sintered blanks; and
- hard machining of the sintered ceramic.

All these methods have their advantages and disadvantages:

Hard machining

- Mostly milling, grinding or polishing, the process is time consuming and expensive, generally diamond tools are necessary.

- The advantage is that it machines to final dimensions and shape and to very smooth surfaces.

Green machining

- The material is fragile, sharp edges tend to break.
- Advantage: shorter machining times are required and there is reduced tool wear.
- The biggest disadvantage is the shrinking during adjacent thermal processing. Near net shape can be achieved e.g. by infiltration of the machined green parts without sintering or after partly sintering without or with reduced shrinkage. This disadvantage can also be overcome, when a net-shape sintering material is used, which is also easily machinable in the green state as it is shown e.g. in Section 1.2.3 or in Figure 1.1, respectively.

Machining when partly sintered

- Characteristics between green and hard machining. The partly sintered blanks are less fragile than green blanks. There is still shrinking during final sintering. Near net shape parts can be achieved by infiltration.

Currently ceramic micromachining is well established in the dental industry as part of computer-aided design and computer-aided manufacturing (CAD/CAM) systems, where ceramic fixed partial dentures are fabricated by mechanical machining (milling). As mentioned above machining fully dense ceramics results in the final shape, but for machining in the partly sintered state with an adjacent final sintering process, the inevitable shrinkage must be numerically compensated by the CAD/CAM system. To realize this, an exact knowledge of the amount of final shrinkage is necessary in order to meet the required final tolerances and dimensions.

For a broad palette of ceramic materials, which are interesting as 3D components for microsystems further development is necessary in order to make these ceramics micromachinable, to develop improved micromachinable ceramics and to develop and establish optimized processing procedures.

1.3.4 Electrodischarge machining (EDM)

Besides milling, grinding or polishing ceramic (micro) structuring is also possible by EDM. But this process is limited to electroconductive materials. Examples of electroconductive ceramics are the refractory carbides and nitrides of the 4th, 5th and 6th groups of the periodic table of elements. But they can only be densified at high temperatures with additional high external pressure (hot pressing). Usually they are blended with a metal component resulting in ceramic–metal composites, the so-called hard metals, e.g. WC–Co. But it is also possible, to obtain electroconductive ceramics, based on these refractories, that are sintering at ambient pressure. This can be achieved by mixing TiN with Al_2O_3 [60, 61]. This mixture is rather advantageous for several reasons. The thermal expansion coefficients of both these materials are nearly the same, so internal stresses can be avoided. Furthermore, by changing the mixture, the resistivity of the material can be controlled over a wide range. This allows the generation of gradient materials and components, where a heating zone or a hot spot can be defined via the local resistivity by the local material composition and not only by the cross-section. This fact gives interesting aspects especially for the fabrication of electroconductive ceramic microcomponents [62–64]. Detailed information on EDM for the fabrication of ceramic microparts is given in Chapter 13 of this book.

1.4 Component development and fabrication

1.4.1 General considerations

Using the above methods for material development and forming processes the fabrication of microcomponents can easily be realized. But before fabrication starts one has to select the optimal combination of material and processing for the desired components. These considerations include:

- application field of the component;
- selection of the material or of the material combination:
    - using either powders or a powderless process;
    - what type of ceramic powder, synthesized by which process;
- shape of the component, number of parts to be produced;
    - replication processing or freeform shaping;
    - what type of mould;
- properties of the part;
    - dimensional accuracy, surface properties;
    - mechanical or functional properties;
    - resistance against the surrounding environment; and
- assembly, packaging and integration.

When these choices have been made the best combination of material and processing can be chosen.

For some special applications it may be necessary to make special developments in order to fulfill the requirements and to meet actual challenges. This can be true for the ceramic material itself, for the processing, if standard techniques are insufficient, but also in the assembly and integration. The following examples illustrate this.

1.4.2 Materials development for special applications

1.4.2.1 Mesoporous granulates for composites

Agglomerated ceramic nanopowders which are commercially available can also be useful for microtechnical or medical applications. One example is the preparation of mesoporous granulates [65–67]. Starting with sols of pyrogenic silicid acid these mesoporous granulates with tailored properties can be generated by a modified sol-gel process and final spray drying. The types of Aerosil® used in this process define the structure of the open porosity, i.e. dependent on different Aerosils® porous granulates with different pore radii in the region

Figure 1.2 Ceramic granulates with monomodal open porosity in the nanometre range [65].

of about 10–50 nm can be realized (see Figure 1.2). Furthermore, pore diameter, amount of porosity and granulate diameter can be controlled by the processing conditions. These porous granulates can be used as inorganic filling materials in organic matrix composites. Thus the organic matrix material can pour into the open porosity and in this way leads to a mechanical interlocking between the matrix and the filling material. It can be shown that this results in an improved abrasion resistance of the composite. It is also possible to vary the refractive index of the granulates e.g. by adding elements with higher refractive indices to the SiO_2-sol, as it has been realized by addition of defined amounts of zirconia sol, the refractive index of the inorganic component of the composite can be adjusted to that of the organic matrix, giving transparent or translucent ceramic–polymer composites as shown in Figure 1.3.

Composites like these can be structured by the standard microstructuring techniques available for polymer microstructuring. The characteristics and advantages of these composites could be exploited for a special medical application, i.e. for dental restoration. During the development of dental filler materials it was possible to realize composite fillers with an inorganic amount of about 25 vol% which showed the translucent characteristic of natural teeth and also had increased abrasion resistance.

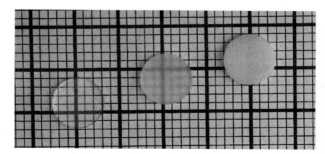

Figure 1.3 Ceramic–polymer composites with adjustable refractive index [65].

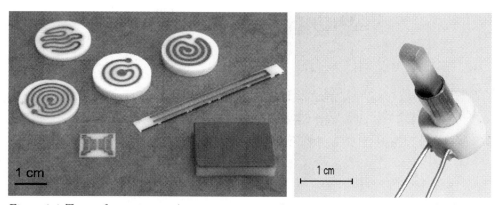

Figure 1.4 Types of ceramic microheaters: screen printed patterns on ceramic substrates (left), heating plug in operation at about 1200°C (right) [69–70].

1.4.2.2 Electroconductive ceramics for ceramic microheaters

A more conventional method is used for the generation of an electroconductive ceramic of the basis of Al_2O_3–TiN (ATN) or Si_3N_4–TiN /SNTN) [68, 69], i.e. the mixture of a metal-like electric conductor (TiN) and an isolator. Here the single constituents are intimately mixed in a planetary ball mill. This type of material has several advantages, especially for microcomponents. The resistivity of the material can be individually adjusted over several orders of magnitude by simply adjusting the conductor/isolator ratio. Gradient materials can easily be realized. This is important e.g. for microheaters, as the heated zone can be locally defined not by changing the cross-section as it is commonly done and which might sometimes be difficult for microparts, but by locally adjusting the composition. A physical advantage is the metal-like characteristic of the resistivity, which is linearly increasing with temperature (positive temperature coefficient of electrical resistivity (PTC) characteristic). This enables a rather easy controlling behaviour. A favorable processing aspect is the sinter-ability of this material. Pure TiN cannot be compacted to reasonable densities by sintering without an additional external pressure. The ATN- and SNTN-mixtures can, however, be sintered though under cover gas atmosphere but at ambient pressure. Further processing can be done by dry pressing of the powder, e.g. for ceramic heaters/igniters for automo-tive application (see Figure 1.4 right), by fabrication of a screen printing paste and screen printing of micropatterns, (Figure 1.4 left) or by setting up a ceramic feedstock for injection moulding [70]. As TiN oxidizes in open air at temperatures higher than about 400°C a protective coating is necessary for high temperature operation. This can be achieved by coating the components with an alumina layer, e.g. by dipping in an alumina suspension or by simply painting with a suspension. Cracking of the protective layer during operation is avoided, because Al_2O_3 and TiN have nearly the same thermal expansion coefficient.

1.4.2.3 Net shape ceramics for dental application

Currently ceramic micromachining is well established in the dental industry as part of CAD/CAM systems, where ceramic fixed partial dentures are fabricated by mechanical

Figure 1.5 Milling of a tooth´s outer surface from a green blank.

machining (milling). As mentioned above machining fully dense ceramics results in the final shape, but for machining in the partly sintered state with an adjacent final sintering process the inevitable shrinkage must be numerically compensated by the CAD/CAM system. To realize this, an exact knowledge of the amount of final shrinkage is necessary in order to meet the required final tolerances and dimensions.

In order to avoid the need for numerical compensation of the sintering shrinkage and dimensional inaccuracies caused by density fluctuations of the green or partly sintered blanks a special material development was performed. The aim of this development was an easily green machinable, net shape sintering material for dental crowns but also for further ceramic microcomponents. This could be realized by a reaction bonding process, starting from inorganic precursors showing a certain volume expansion during the thermal processing and so compensating the shrinkage [15, 71–76], see also Section 1.2.3.

As we can see, Figure 1.5 gives an impression of milling the outer surface of a dental crown from a green blank and Figure 1.6 shows green and sintered dental crowns as well as a perfectly fitting three-unit bridge from this material.

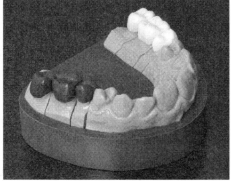

Figure 1.6 Dental crowns milled in green state (left, black) and sintered net shape (left, white), and a three-unit bridge before and after sintering (right).

1.5 Integration of piezoelectric ceramics

Piezoelectric ceramics are able to interact with the surrounding environment to sense, act upon and generate power. Such behaviour makes these ceramics of great technological interest. To make effective use of these materials they need to be combined with other materials to create structures that are able to transmit and amplify motion to create a functioning device that can be applied to do useful work. As with all microelectromechanical system (MEMS) applications, the reduction in size of the devices leads to improvements in terms of sensitivity, efficiency and portability [77, 78].

Lead zirconate titanate (PZT) is one of the most used piezoelectric materials [79] and shares many processing challenges with other piezoelectric materials [80, 81], particularly those that contain lead or other volatile elements, such as Na and K. It is therefore of interest to consider PZT as a "standard" material for purposes of this discussion.

Macroscale devices incorporating PZT are typically manufactured using an assembly type approach where the individual parts are assembled and shaped by mechanical techniques. Such approaches are not suitable for the production of microscale devices due to the difficulty of conducting such shaping operations at the required scale and the difficulty in scaling the thickness of the bond layers with the reduction in the dimensions of the devices.

Layer-by-layer deposition routes can be used to produce small-scale structures and have the advantage that multiple devices can be created simultaneously on a single support wafer. Such processing poses challenges associated with the need to integrate and coprocess multiple materials as well as shaping the materials to create the MEMS devices.

Despite the huge range of piezoelectric MEMS devices available there are a number of features which are common to most devices. The piezoelectric material has at least two integrated electrodes connected to it to allow signals to be applied or detected, and a support structure. In addition to the presence of these three components the material is also shaped/removed from around the active area to produce a structure that is able to sense and actuate.

Each of the components in the MEMS device served a specific function and as such has certain properties. Due to the different properties of the materials present there are constraints (often mutually exclusive) on the processing. Functional ceramic materials are typically brittle and require high temperatures for processing. Bulk PZT, in particular, is generally processed at temperatures in the range 1200–1400°C. The electrodes need to be highly conductive, with metals being favoured for their low electrical resistivity and ease of deposition. However, metals also exhibit high thermal expansion coefficients and poor oxidation/corrosion resistance which pose problems if such materials are to be processed alongside ceramics. For instance, gold, copper and aluminium, melt at 1064°C, 1083°C, and 660°C, respectively and so would not normally be suitable for processing with PZT ceramics at temperatures in the range 1200–1400°C. Support materials are usually passive with a reasonable stiffness and ability to be shaped. They can be metals, polymers, ceramics or semiconductors. Plainly, polymers would not be suitable for coprocessing with ceramics. Some refractory metals, such at Pt or W, can be coprocessed with PZT. Si (a common MST material) melts at 1410°C but undergoes significant oxidation at lower temperatures and so would be unsuitable for coprocessing. While ceramics, such as alumina or zirconia, can readily be used as substrate materials they are significantly more difficult to machine to create delicate features such as cantilevers and diaphragms. In addition, unless the defects within the substrate ceramics are very well-controlled, cracking of the ceramic substrate can occur.

1.5.1 Challenges to integration

The challenges for creating piezoelectric MEMS devices can be grouped under three headings (see Figure 1.7):

Physical changes to materials such as softening, melting or vaporisation. An example of this would be the softening of glass at temperatures above ~700°C which would lead to significant distortion of a device if the wafer were not correctly supported. Melting and vaporisation would occur at higher temperatures or where less thermally stable materials, such as polymers, are present. In these cases the devices would be very likely to be destroyed.

Chemical changes to materials including reaction with neighbouring materials or the atmosphere and changes in the composition through loss of volatile elements. Such challenges are of particular concern for piezoelectric materials with high levels of lead. One example is the formation of a liquid phase at 714°C following the reaction of Pb with SiO_2 [82, 83] This reaction can occur where lead based ceramics are in intimate contact with glass or Si wafers. To prevent this reaction from occurring, special precautions need to be taken when Pb containing piezoelectric materials are integrated with SiO_2-based materials. One approach is to process the material at below 714°C while an alternative approach would be to use a diffusion barrier layer which prevents the Pb diffusing into SiO_2 material. In addition to reactions between two materials, the volatile nature of Pb means that evaporation of Pb at high temperatures can occur, leading to changes in the stoichiometry of the PZT with a consequential change in the properties of the piezoelectric material. Such evaporation of volatile elements can also occur in piezoelectric materials containing Na or K. Loss of volatile compounds affects both bulk and microscale material leading to a depletion zone near the surface. It is most critical in microscale systems as the size of the depletion zone can represent a significant volume fraction of the structure as a while. For this reason, there is a strong drive to minimize the processing temperature of microscale devices.

Mechanical changes to materials/structures such as deformation, fracture or delaminating tend to occur due to the generation of differential dimensional changes between two materials. Sources of such dimensional changes include: shrinkage due to densification, thermal expansion, phase change, crystallographic reorientation, oxidation and chemical reactions. In this way it can be seen that the physical and chemical challenges can have a direct bearing on the mechanical challenges experienced.

The most challenging source of stress is that of shrinkage of the ceramic material during processing. Such shrinkage occurs during drying, pyrolysis and sintering of the ceramic materials and is particularly detrimental as it leads to the generation of a tensile stress [80] within the ceramic film. Ceramics are inherently poor at sustaining tensile stresses and excessive shrinkage can lead to fracturing of the active ceramic material. Along with failure

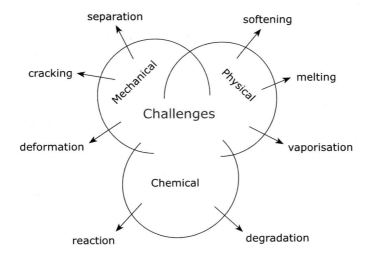

Figure 1.7 Overview of the physical, chemical and mechanical challenges needing to be overcome to successfully integrate ceramics into MEMS.

of the ceramic material, failure of the device at the interface between two materials can also occur due to the build up of tensile or compressive stress.

Thermal expansion mismatch is a major contribution to undesirable deformation of MEMS structures. But this can be minimized through the use of lower processing temperatures and/or the use of alternative materials with better matched thermal expansion coefficients.

1.5.2 Approaches to integration

While the fundamental mechanisms of the challenges to integration are often quite different, they are all interrelated and many of them become more likely and pronounced as the processing temperature increases due to increased reaction rates, overcoming of energy barriers and more extreme changes. Thus, one the primary ways of overcoming the challenges to integrating ceramics into MEMS is to reduce the processing temperatures used during the processing of these materials. Softening, melting, vaporisation and many reactions can be avoided through the use of a lower processing temperature. In addition, the generation of differential shape changes through thermal expansion mismatch is reduced through the use of a lower temperature. Where undesirable reactions can occur between neighbouring materials, even at low temperatures, diffusion barrier materials can be introduced to separate the two materials. It should be noted, however, that adding components will increase the complexity of the system as the new components must also be compatible with the system. The addition of diffusion barriers can allow the processing temperature to be raised slightly if undesirable reactions are the only cause for concern.

While reducing the processing temperature alleviates the challenges it also retards the processing of the functional ceramic material. For this reason many of the developments in integration have focused on developing new low temperature routes for fabricating ceramic materials. Examples include sol-gel, composite sol-gel, chemical vapour deposition and physical vapour deposition [80, 81].

Conclusions

It could be shown that ceramics cover a wide range of application in microtechnology. This holds for structural ceramics (the most widespread materials here are Al_2O_3 and ZrO_2) as well as for functional ceramics. There are many ceramic materials and processing techniques which can fulfill microtechnical requirements. However, for an efficient development of ceramic micro components an integrated concept reaching from the design idea and the selected material to the final microcomponent, especially with respect to the desired functionality, the material characteristics and quality and the suitable processing route is of great importance. These requirements gain even more significance when the shift from micro- to nanotechnology, where markedly smaller details are needed, is borne in mind.

References

[1] Piotter, V., Benzler T., Gietzelt T., Ruprecht R. and Haußelt, J. 2000. Micro powder injection molding. *Advanced Engineering Materials* **2**, 639–642.

[2] Hahn H. 2003. Unique features and properties of nanostructured materials. *Advanced Engineering Materials*, **5**, 277–284.

[3] Vollath D., Szabo D.V. and Fuchs S. 1999. Synthesis and properties of ceramic–polymer composites. *Nano Structured Materials*, **12**, 433–438.

[4] Müller E., Oestreich Ch., Popp U., Michel G., Staupendahl G. and Henneberg K.-H. 1995. Characterization of nanocrystalline oxide powders prepared by CO_2 laser evaporation. *KONA. Powder and Particle*, **13**, 79–90.

[5] Müller E. 2002. Fortschrittsberichte der DKG. *Verfahrenstechnik*, **17**, 18–27.

[6] Kwon Y.-S., Jung Y.-H., Yavorovsky N.A., Illyn A.P. and Kim J.-S. 2001. Ultra-fine powder by wire explosion method. *Scripta Materialia*, **44**, 2247–2251.

[7] Giri V.S., Sarathi R., Chakravarthy S.R. and Venkataseshaiah C. 2004. Studies on production and characterization of nano-Al_2O_2 powder using wire explosion technique. *Materials Letters*, **58**, 1047–1050.

[8] Gromov A. and Vereshchagin V. 2004. Study of aluminium nitride formation by superfine aluminium powder combustion in air. *Journal of the European Ceramic Society*, **24**, 2879–2884.

[9] Claussen N., Le T., Wu S. 1989. Low-shrinkage reaction-bonded alumina. *Journal of the European Ceramic Society*, **5**, 29-35.

[10] Holz, D., Pagel S., Bowen C., Wu S. and Claussen N. 1996. Fabrication of low-to-zero shrinkage reaction-bonded mullite composites. *Journal of the European Ceramic Society*, **16**, 255–260.

[11] She J. H., Schneider H., Inoue T., Suzuki M., Sodeoka S. and Ueno, K. 2001. Fabrication of low-shrinkage reaction-bonded alumina–mullite composites. *Materials Chemistry and Physics*, **68**, 105–109.

[12] Hennige V. D., Ritzhaupt-Kleissl H.-J. and Haußelt J. 1998. Verfahren zur Herstellung schrumpfungsfreier $ZrSiO_4$-Keramiken. *Keramische Zeitschift*, **50**, 262–265.

[13] Hennige V. D., Haußelt J., Ritzhaupt-Kleissl H.-J. and Windmann T. 1999. Shrinkage-free ZrSiO$_4$-ceramics: characterisation and applications. *Journal of the European Ceramic Society*, **19**, 2901–2908.

[14] German Patent DE 195 47 129, 1995. Hennige V.D., Ritzhaupt-Kleissl H.-J. and Haußelt J. Verfahren zur Herstellung eines oxidkeramischen Sinterkörpers und dessen Verwendung. Also European Patent EP 59 603 853.

[15] Binder J. R., Ritzhaupt-Kleissl H.-J. and Haußelt J. 2001. Keramischer Zahnersatz aus einer schwindungsfreien Zirkonkeramik. *Dental Dialogue*, **6**, 684–686.

[16] Ritzhaupt-Kleissl H.-J., Binder J. R., Klose E. and Haußelt J. 2002. Net-shape ceramic microcomponents by reaction bonding. *Ceramic Forum International*, **79**(10), E9–E12.

[17] Geßwein H., Binder J.R., Ritzhaupt-Kleissl H.-J. and Haußelt, J. 2006. Fabrication of net shape reaction bonded oxide ceramics. *Journal of the European Ceramic Society*, **26**, 697–702.

[18] Lewis J.A. 2000. Colloidal processing of ceramics. *Journal of the American Ceramic Society*, **83**, 2341–2359.

[19] Lange F.F. 1989. Powder processing science and technology for increased reliability. *Journal of the American Ceramic Society*, **72**, 3–15.

[20] Martin C.R. and Aksay I.A. 2004. Submicrometer-scale patterning of ceramic thin films. *Journal of Electroceramics* **12**, 53–68.

[21] Ritzhaupt-Kleissl H.J., von Both H., Dauscher M. and Knitter R. 2005. Further ceramic replication techniques. In Baltes H., Brand O., Fedder G.K., Hierold C., Korvink J. and Tabata O. (Eds.) *Advanced Micro- and Nanosystems*, **3**, *Microengineering of Metals and Ceramics*, Wiley VCH.

[22] Omatete O.O., Janney M.A. and Nunn S.D. 1997. Gelcasting: From laboratory development toward industrial production. *Journal of the European Ceramic Society*, **17**, 407–413.

[23] US Patent 4894 194, 1990. Janney M.A. Method for molding ceramic powders.

[24] Xia Y. and Whitesides G.M. 1998. Soft lithography. *Angewandte Chemie International Edition*, **37**, 550–575.

[25] Gauckler L. J., Graule T. and Baader F. 1999. Ceramic forming using enzyme catalyzed reactions. *Materials Chemistry and Physics*, **61**, 78–102.

[26] German Patent DE-PS 43 10 068, 1994. Knitter R., Odemer C., Günther E and Maciejewski U., Verfahren zur Herstellung eines plattenförmigen Mikrostrukturkörpers aus Keramik.

[27] Knitter R., Günther E., Maciejewski U. and Odemer C. 1994. Herstellung keramischer Mikrostrukturen. *Ceramic Forum International*, **71**, 549–556.

[28] Roosen A. 1999. Tape casting of ceramic green tapes for multilayer device processing. In *Ceramic Transactions*, **97**, Jean H.J., Gupta T.K., Nair K.M. and Niwa K. (Eds.) *American Ceramic Society*, Columbus, Ohio, 103–121.

[29] Schindler K. and Roosen A. 2006. Cold low pressure lamination of ceramic green tapes for the manufacture of 3D structures. *Ceramic Forum International*, **83**, 27– 30.

[30] Stolz S. 2004. Siebdruck von elektrisch leitfähigen Keramiken zur Entwicklung heizbarer keramischer Mikrokomponenten. Scientific Report, FZKA-6906, Forschungszentrum Karlsruhe, Karlsruhe, Germany.

[31] Stolz S., Bauer W., Ritzhaupt-Kleissl H.J. and Hausselt J. 2004. Screen printed electroconductive ceramics. *Journal of European Ceramic Society*, **24**, 1087–1090.

[32] Tabellion J. and Clasen R. 2004. Electrophoretic deposition from aqueous suspensions for near-shape manufacturing of advanced ceramics and glasses-applications. *Journal of Materials Science*, **39**, 803–811.

[33] Clasen R. 2004. Nanotechnologie im Saarland–vom Pulver zum Hochleistungswerkstoff. *Ceramic Forum International,* **81(8),** 12–16.

[34] German Patent DE 44 25 978 C1, 1994. Ritzhaupt-Kleissl H.J. and Laubersheimer J., Verfahren zur Herstellung keramischer Mikrostrukturen.

[35] German Patent DE-PS 19 815 978. 2003. Hanemann T. and Hausselt J., Verfahren zur Herstellung von Klein- und Mikroteilen aus Keramik. Method for producing small and micro-sized ceramic parts. Also US Patent US-PS 6 573 020

[36] Hanemann T., Börner M., Göttert J., Heldele R., Motz G., Schulz M., and Hausselt J. 2003. Lithographie an präkeramischen Polymeren. Paper presented at *Materials Week,* 16–18 September 2003, Munich, Germany.

[37] Hanemann T., Ade M., Börner M., Motz G. and Schulz M. 2002. Microstructuring of preceramic polymers. *Advanced Engineering Materials,* **4,** 869–873.

[38] Kaufmann U., Ritzhaupt-Kleissl H.J., Harrysson U. and Johander P. 2006. FASTFAB – a process for the free-form fabrication of 3D ceramic components. *Ceramic Forum International,* **83,** 13–17.

[39] Cawley J.D. 1999. Solid freeform fabrication of ceramics. *Current Opinion in Solid State and Materials Science,* **4,** 483–489.

[40] Johander P., Harrysson U., Kaufmann U. and Ritzhaupt-Kleissl H.J. 2005. Direct manufacture of ceramic micro components with layered manufacturing methods. *4M2005 : Proceedings of the First International Conference on Multi-Material Micro Manufacture,* 29 June–1 July, 2005 Karlsruhe, Germany. Menz, W. (Ed.) Elsevier, Amsterdam, 231–236.

[41] Melcher R., Zhang W., Travitzky N. and Greil P. 2006. 3D-printing of Al_2O_3/Cu composites. *Ceramic Forum International,* **83,** 1–22.

[42] Tay B.Y., Evans J.R.G. and Edirisinghe M.J. 2003. Solid freeform fabrication of ceramics. *International Materials Review,* **48,** 341–370.

[43] Windle J. and Derby J. 1999. Ink jet printing of PZT aqueous ceramic suspensions. *Journal of Materials Science Letters,* **18,** 87–90.

[44] Khalyfa A., Meyer W., Schnabelrauch M., Vogt S. and Richter H.J. 2006. Manufacturing of biocompatible ceramic bone substitutes by 3D-printing. *Ceramic Forum International,* **83,** 23–26.

[45] Deisinger U., Irlinger F., Pelzer R. and Ziegler G. 2006. 3D-printing of HA-scaffolds for the application as bone substitution material. *Ceramic Forum International,* **83,** 75–78.

[46] Roosen A. 2001. New Lamination technique to join ceramic green tapes for the manufacturing of multilayer devices. *Journal of European Ceramic Society,* **21,** 1993–1996.

[47] Weisensel L., Windsheimer H., Travitzky N. and Greil P. 2006. Laminated onject manufacturing (LOM) of SiSiC-composites. *Ceramic Forum International,* **83,** 31–35.

[48] Klostermann D., Chartoff R., Osborne N., Graves G., Lightman A., Han G., Bezeredi A., Rodrigues S. 1999. Development of a curved layer LOM process for monolithic ceramics and ceramic matrix composites. *Rapid Prototyping Journal,* **5,** 61–71.

[49] Exner H., Hartwig L., Streek A., Horn M., Klötzer S., Ebert R. and Regenfuss P. 2006. Laser micro sintering of ceramic materials. *Ceramic Forum International,* **83,** 45–52.

[50] Günster J., Gahler A. and Heinrich J.G. 2006. Rapid prototyping of ceramic components. *Ceramic Forum International,* **83,** 53–56.

[51] Fischer H., Wilkes J., Bergmann Ch., Kuhl I., Meiners W., Wissenbach K., Poprawe R., Telle R. 2006. Bone substitute implants made of TCP/glass composites using selective laser melting technique. *Ceramic Forum International,* **83,** 57–60.

[52]   Gebhardt A. 2000. *Rapid prototyping – Werkzeuge für die schnelle Produktentstehung*, Carl Hanser Verlag, Munich, Germany.

[53]   Gebhardt A. 2006. Vision rapid prototyping. Generative manufacturing of ceramic parts – A survey. *Ceramic Forum International*, **83**, 7–12.

[54]   Haußelt J. 2001. Micro powder injection moulding of metals and ceramics, Status, potential and limits. In *Proceedings of 4th International Workshop on High-aspect-ratio Micro-structure Technology (HARMST 2001)*, 17–19 June, 2001, Baden-Baden, Germany, 125–127.

[55]   German R. M. 1990. Powder injection moulding. Metal Powder Industries Federation, Princeton, New Jersey.

[56]   Song J. H. and Evans J. R. G. 1995. The injection moulding of fine and ultra-fine zirconia powders. *Ceramics International*, **21**, 325–333.

[57]   Song J. H. and Evans J. R. G. 1996. Ultrafine ceramic powder injection moulding: The role of dispersants. *Journal of Rheology*, **40**, 131–152.

[58]   Rhee B. O., Cao M. Y., Zhang H. R., Streicher E. and Chung C. I. 1991. In *Advances in Powder Metallurgy 2: Powder Injection Moulding*, Princeton, New Jersey, 43–58.

[59]   Bauer W. and Knitter R. 2002. Development of a rapid prototyping process chain for the production of ceramic microcomponents. *Journal of Material Sciences*, **37**, 3127–3140.

[60]   Winter V. and Knitter R. 1997. Al_2O_3/TiN as a material for microheaters. Paper presented at *Micro Materials '97*, 16–18 April, 1997, Berlin, Germany.

[61]   Winter V. 1998. Elektrische Heizbarkeit und Mikrostrukturierbarkeit einer Mischkeramik aus Aluminiumoxid und Titannitrid. (Scientific Report, FZKA-6173, October 1998). PhD-thesis, Universität Karlsruhe, Germany.

[62]   Winter V. and Lurk R. 2000. Vollkeramische Heizelemente. Scientific Report, Forschungszentrum Karlsruhe FZKA-6528, 129–34.

[63]   Knitter R., Lurk R., Rohde M., Stolz S. and Winter V. 2001. Heating concepts for ceramic microreactors. In M. Matlosz. (Ed.) *Microreaction Technology: IMRET 5; Proceedings of the Fifth International Conference*, 27–30 April, 2001, Strasbourg, France, Springer, Verlag, Berlin, 2001. 86–93.

[64]   Stolz S., Lurk R., Rohde M. and Winter V. 2001. Heizkonzepte für keramische Mikroreaktoren. Forschungszentrum Karlsruhe Scientific Report, FZKA-6662. 119–23.

[65]   Binder J.R., Ritzhaupt-Kleissl H.J., Rentsch H., Haußelt J., Dermann K. 2000. Preparation of porous fillers for dental composites. Paper presented at *Materials Week 2000*, 25–28 September, 2000, Munich, Germany.

[66]   European Patent No. EP 1119337 B1 1999. Alkemper J., Dermann K., Rentsch H., Ritzhaupt-Kleissl H.J., Haußelt J., Albert P. and Gall C. Dental material with porous glass ceramics.

[67]   European Patent EP 1181924 B1 2001. Alkemper J., Binder J.R., Rentsch H., Ritzhaupt-Kleissl H.J. and Haußelt J. Dental composite material including a hybrid filler and method for producing the same.

[68]   Winter V. and Knitter R. 1997. Al_2O_3/TiN as a material for microheaters. Paper presented at *Micro Materials 1997*, 16–18, April, 1997, Berlin, Germany.

[69]   Kaufmann U. and Stolz S. 2004. Elektrisch leitfähige Keramik auf Basis von Al_2O_3/TiN, Si_3N_4/TiN und Si_3N_4/MoSi$_2$ - Eigenschaften und Anwendung. *Annual Meeting of the German Ceramic Society 2004*, 11–13, October, 2004, Karlsruhe, Germany.

[70]   Stolz S., Bauer W., Ritzhaupt-Kleissl H.J. and Haußelt J. 2004. Screen printed electro-conductive ceramics. *Journal of European Ceramic Society*, **24**, 1087–1090.

[71] Hennige V. D. 1998. *Verfahrens- und Werkstoffentwicklung zur Herstellung oxidkeramischer Mikroformteile mit minimiertem Sinterschrumpf.* Shaker-Verlag, Aachen, Germany.

[72] Hennige V. D. and Ritzhaupt-Kleissl H.J. 1998. Verfahren zur Herstellung schrumpfungsfreier ZrSiO4-Keramiken. *Keramische Zeitschrift*, **50**, 262–265.

[73] Hennige V. D., Haußelt J., Ritzhaupt-Kleissl H.J. and Windmann T. 1999. Shrinkage-free $ZrSiO_4$-ceramics: characterisation and applications. *Journal of European Ceramic Society*, **19**, 2901–2908.

[74] German Patent DE 195 47 129, 1995. Hennige V.D., Ritzhaupt-Kleissl H.J. and Haußelt J. Verfahren zur Herstellung eines oxidkeramischen Sinterkörpers und dessen Verwendung. Also European Patent EP 0779 259, 1999.

[75] Ritzhaupt-Kleissl H.J., Binder J. R., Klose E. and Haußelt J. 2002. Net-shape ceramic microcomponents by reaction bonding. *Ceramic Forum International* **79**, E9–E12.

[76] Geßwein H., Binder J.R., Ritzhaupt-Kleissl H.J. and Haußelt J. 2006. Fabrication of net shape reaction bonded oxide ceramics. *Journal of European Ceramic Society*, **26**, 697–702.

[77] Jeon Y., Seo Y.G, Kim S.J. and No K. 2000. Low temperature sintering of screen-printed Pb(ZrTi)O3 thick films. *Integrated Ferroelectrics*, **30**, 91–101.

[78] Simon L., Le Dren S. and Gonnard P. 2001. PZT and PT screen-printed thick films. *Journal of European Ceramic Society*, **21**, 1441–1444.

[79] Koch M., Evans A. and Brunnschweiler A. 2001. Microfluidic Technology and Applications, Research Studies Press, Baldock, United Kingdom.

[80] Dorey R.A. and Whatmore R.W. 2004. Electroceramic thick film fabrication for MEMS. *Journal of Electroceramics*, **12**, 19–32.

[81] Whatmore R.W., Zhang Q., Huang Z. and Dorey R.A. 2003. Ferroelectric thin and thick films for Microsystems. *Materials Science in Semiconductor Processing* **5**, 65–76.

[82] Duval F. F. C., Dorey R. A., Haigh R.H., Whatmore R. W. 2003. Stable TiO_2/Pt electrode structure for lead containing ferroelectric thick films on silicon MEMS structures. *Thin Solid Films*, **444**, 235–240.

[83] Dauchy F. and Dorey R.A. 2007. Patterned crack-free PZT thick films for micro-electromechanical system applications. *International Journal of Advanced Manufacturing Technology*, **33**, 86–94.

2

Design rules for microcomponents

A. Albers, T. Deigendesch and S. Hauser

In development processes for microcomponents made of ceramic materials comprehensive knowledge of subsequent stages in a product life-cycle: process preparation, e.g. mould manufacturing as well as production, e.g. injection moulding or quality assurance. These design restrictions have to be considered mandatorily in order to design a functioning microsystem. A proposed methodological tool is usage of design rules, which represent knowledge from stages subsequent of design.

2.1 Introduction

Today's products follow megatrends such as integration or miniaturization [1]. In this context microsystems are increasingly used. Most successful applications are monolithic silicon-based MEMS such as ink-jet nozzles, accelerometers or micromirror arrays in video projectors. The trend of miniaturization proceeds by spreading to the purely mechanical domain, which tries to miniaturize macroscopically known components and systems.

The possibility of batch processing of MEMS devices on silicon wafers decreases production cost which decides on market success. Another possibility for cost reduction is to use replication techniques, such as injection moulding. Depending on the required material, injection moulding of thermoplastics (thermoplastic injection moulding), metal powder feedstocks (metal injection moulding) or ceramic powder feedstocks (ceramic injection moulding) is possible. However, all injection moulding processes require moulds, the production of which encounters many restrictions. Especially in microtechnology with a restricted number of possible processes these restrictions have to be considered in the early stages of design.

The present contribution briefly describes two ways of producing microcomponents with replication techniques, i.e. applying the lithography, electroplating and replicating (LIGA) process and miniaturised conventional manufacturing methods. Knowing the processes, design rules can be derived. These rules can support designers in different ways of application, e.g. in an integrated knowledge-based engineering (KBE) environment or within an external knowledge representation system, e.g. a wiki-system.

2.2 Derivation of design rules

Simultaneously with the development of the production technology, its impact on the design phase has to be considered. For this reason two different groups from different disciplines have to work closely together: stakeholders from production science (e.g. process preparation, production and quality assurance) and product development. One way to improve communication between them and sustainably secure the acquired information are so-called design rules.

In general, design rules are hints to the designer about how to behave in a certain situation. With the scope of ceramic microcomponents they can describe the link between manufacturing technology and geometrical possibilities in the design. They can be acquired in two ways, either theoretically before any production process starts or empirically by monitoring a running process. The first group includes the more obvious design rules, e.g. if using micromilling, each cavity has to be wider than the diameter of the end mill. In the second group one can formulate the results of tests, e.g. an obtainable surface roughness when a part had been processed by microelectrodischarge-machining.

Design rules have a two-fold character. On the one hand, they are informative. By browsing in a design rule kit, development engineers are enabled to obtain a brief overview on what is producible. This front-loading approach results in a more sophisticated design proposal from the start. On the other hand, design rules may have a diagnostic character, i.e. the design rule can be described formally in a way that is processable by a KBE system. Sections 2.3 and 2.4 describe possible ways of representing design knowledge by design rules. Section 2.3 gives an overview on how design rules can be used to inform the designers on what is producible and how a promising design has to look like. Section 2.4 describes design rules in a formal context as they are embedded in a KBE system.

2.3 Design rules for knowledge representation

When knowledge from production or production preparation has to be used in the early stages of the product life-cycle, i.e. for design, several factors have to be considered. For example, the human factor cannot be disregarded. Knowledge management tools require an enterprise culture, which enables sharing of knowledge to find fertile ground. However, this contribution concentrates on the technical aspects for sharing design and production knowledge. Hurdles for sharing knowledge should be as low as possible. If the knowledge-sharing person needs knowledge about programming languages or has to take special syntax rules into consideration, the hurdle can already be too high. Adding information must be as easy as possible, i.e. a contribution by plain text editing would the easiest to handle. In order to be able to trace back information, the source should be identifiable. Further on, the system should be accessible from several locations, so that persons from different disciplines along the life-cycle chain can easily contribute. This results in a claim for web-accessibility. When being accessible from intranet or internet, different groups of users should have different possibilities of adding new information, editing exiting information or only reading existing information. Thus, there must be a way of handling access restrictions. Finally, acquisition, installation and maintenance of the system should be cost-effective.

During recent years, many authors, especially from the field of microelectronics [2], MEMS or LIGA, proposed design rules as a means of support that were stored in databases [3–7]. Also the design rules being addressed in this chapter were represented within a database system with a customized front-end based on active server pages in the beginning. The experience of the authors led to the conclusion, that storing data in databases is very effective for predictable data. However, especially the early design stages are characterized by a high level of uncertainty. Which type of information is going to be represented is usually not predictable and not even known. The established database system required a high level of maintenance due to the customized front-end and was not flexible enough for sharing all the knowledge required. Instead, only that information, which was intentionally supposed to be included, was added. The effort required to implement new data fields in the database and front-end always was high as it required modifications to the software. There was no possibility of integrating additional information, e.g. drawings, CAD-models, several images etc. Ongoing research on knowledge management led to the conclusion that wiki-systems [8] and semantic wikis [9] seemed to be a very good tool that fulfilled the given requirements and had the necessary flexibility for open integration of different types of information.

A wiki system is an open content management system that was intentionally developed to document design patterns by computer programmer Ward Cunningham. The most famous representative of a wiki is Wikipedia [10], an online encyclopaedia. Wikis are also an established tool for computer-supported cooperative work (CSCW) [11]. Based on the given requirements, the authors decided on a Dokuwiki [12] system for knowledge representation. This wiki system is open source software that is continuously being improved. Thus the software itself is very cost-effective as well as installation- and maintenance-friendly. It features an access control list function, page locking and versioning for secure editing. Other file formats, such as documents, images or CAD models can be integrated by linking or uploading. The content is written into so-called pages, which are stored in a multilevel structure, so-called namespaces that could be described as folders. A sidebar listing all namespaces and pages, a page-specific table of contents and finally full-text search enable different ways of accessing the information. A new page can easily be added by entering the link to that page and clicking it. The wiki displays the name of the last user who edited a page. This enables other users to know who is responsible for the information, e.g. for asking. Just by pressing a button, a user can leave a signature consisting of name and a date/time stamp. Additionally, a user who always needs up-to-date information can enable a subscribe-function, that automatically notifies that person in case of changes to page contents.

Figure 2.1 depicts the customized Dokuwiki system as platform for knowledge representation by design rules. The wiki, called micro book of knowledge (MyBoK), provides a complete design rule kit that supports design of PIM-microcomponents. Of course, knowledge that does not fit into a design rule pattern can easily be added. In another namespace, design project documentation is done within the same system. So, design and production engineers are confronted by only a single tool that enables them to store, to share and to access knowledge on products and processes along the product life-cycle.

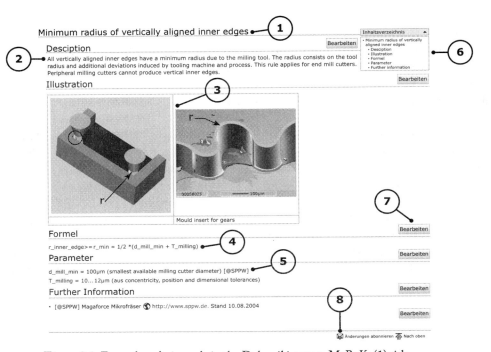

Figure 2.1 Exemplary design rule in the Dokuwiki system MyBoK: (1) title, (2) description, (3) illustration, (4) formal description, (5) formal parameter, (6) table of contents, (7) edit-button, (8) subscription button.

2.4 Design rules in knowledge-based engineering

Design rules can not only give advice to a designer, they can also be checked automatically in a CAD system as long as they refer to geometrical or other information that is stored within the CAD file. Since the rules themselves are universally valid they have to be stored outside the CAD in a separate system so they can be used with any CAD system and any design. Hence the KBE system presented here consists of three parts: the CAD system, which holds all geometrical and other data related to the product; a database with all design rules; and the actual KBE system which can combine both information and derive design proposals.

Most parametric CAD systems use "features". A feature mainly is a (basic) geometrical element together with its dimensions and its placement relative to other features. Today, there are more than 100 different features defined in a CAD system to support the designer. Hence, there are different possibilities to define the same shape. For example one obtains the same resulting geometrical part if extruding a square with a circle in the middle and if using a cube and placing a hole in it. Production related design rules should only refer to something you really produce, that means the shape of the part and not its "CAD history".

Therefore only this shape should be used to describe the design rules. Most commercially available CAD systems provide it as the so-called "boundary representation", which is also defined in the standard for the exchange of product model data (STEP). The main advantage of this approach is that it is nearly independent of the history of the CAD model; it is even possible for imported CAD models from another system. If two models have the same shape, they also have the same boundary elements. It is possible, depending on the CAD system, that a circle can be divided into two concentric half-circles. The boundary elements of real 3D parts are for example faces and edges (the intersection of two adjacent surfaces). For use in a design rule, the geometrical properties rather than the shape element itself are important. Hence algorithms are needed that are able to calculate them. They can be defined for edges (e.g. angle), faces (e.g. height), two faces (e.g. gaps, free distance between the faces) or special groups of faces (e.g. notches).

The design rules are stored in an independent database. They are formulated as abstractly as possible without any special process data. One rule consists of a condition when it has to be applied (e.g. for all micromilling processes), the rule itself (e.g. the mill diameter has to be larger than the diameter of a rounding), the rule is formulated as a mathematical equation for an automatic evaluation and eventually some advice is given if the rule is infringed (e.g. enhance radius).

The process specific data is stored in a separate database. It represents the current state of technology and thus changes over time. The data is stored at two hierarchical levels. The first represents the process (e.g. milling), the second an additional differentiation (e.g. a special end mill with $d = 100\,\mu m$). Parameters can be defined on both levels. If the same parameters are defined multiple times it is possible to add them (e.g. tolerances of a milling machine and an end mill) or use the largest, mean or smallest value (e.g. maximum part dimensions).

2.4.1 Integration in the CAD system

In general, any assistance system will only be used if it means no additional effort and has an obvious advantage to the designer. Thus no new system besides the normal CAD tool should be introduced to the designer, but the rule-based assistance system has to be integrated in the regular CAD system. The realized prototype extends the toolbar of the CAD system with new options to define parameters and perform a rule check. This makes it possible to check all defined rules within the CAD system. The results will be printed on the screen and directly marked in the CAD model.

2.4.2 Scheme of rule check

The rules are evaluated in several steps: first parameters have to be defined. Since many design rules describe the primary shaping process used for ceramic microcomponents the system is able to handle both, the mould and the part, and it is possible to define two subsequent manufacturing processes, e.g. milling of the mould and PIM. These parameters, the two processes and whether it is a mould or a part, have to be defined prior to checking.

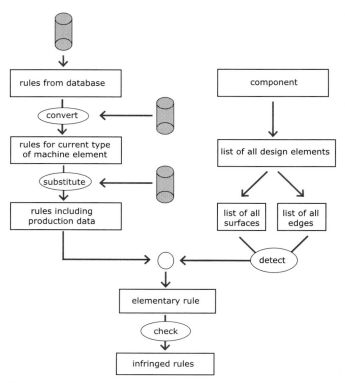

Figure 2.2 Schematic of rule adaptation and evaluation.

2.4.3 Adapting rules

By adapting the general rules of the database to the special geometry, as well as to the currently used processes, elementary rules are obtained. Consequently, they are not generally applicable anymore, but describe the concrete circumstances and hence can be evaluated by the computer. The preparation for this is carried out in two parallel steps: the adaptation and concretization of the rules and the analysis of the geometry. The processing of the rules as provided by the database on to the connection with the geometric parameters is carried out in three steps.

(1)   The rules are loaded from the database and their scope is directly checked. If they do not refer to the process, tools or materials, they are not considered.

(2)   Since rules can be defined for parts and for moulds they have to be converted eventually. Although the part and the mould are different in their shape and dimensions, they are correlated in an exact way. If you can assume a linear shrinkage, all dimensions have a defined ratio and the shape is inverted i.e. a cylindrical shaft becomes a hole with a smaller diameter and length. Thus if a rule is defined for the other part type (model or mould insert), it is "translated" by means of a transformation table, e.g. mould: bore diameter → model: shaft diameter × shrinkage.

(3)   In accordance with the applied process chain the technological data is loaded
from a separate database and placeholders in the rules' formulas are replaced. If no
equivalent for a placeholder can be found in the database, the value "0" is assumed.
E.g. if you have a process without shrinkage, shrinkage will not be defined.

2.4.4 Analysing CAD part geometry

The evaluation of the geometry of the part is carried out in two phases. First, all boundary
elements from the database of the CAD system need to be read; and secondly, the corresponding properties of each of them are to be determined if they are needed by some rule.

Geometric properties can be defined for solid bodies (i.e. all surfaces of one part),
single surfaces, pairs of surfaces, special groups of connected surfaces or edges (identical
to the intersection of two adjacent surfaces). For the determination of geometric properties topologic information about the individual boundary elements are available within the
CAD system from the part database, such as unit (normal) vectors, maximum width and
height etc. Simple geometric parameters can be derived directly from them. For example
two surfaces are parallel if they have identical unit normal vectors.

Other parameters are more complicated to calculate, since there are various influencing factors or interdependencies. One example is the angle of an edge, which is defined
as the angle measured between the adjacent surfaces directly at the edge. It is less than
180° for convex structures and greater than 180° for concave ones. This separation is
possible since the normal vectors of surfaces of solid bodies point always outwards from
the volume (see Figure 2.3).

In particular it can be calculated in the following steps:

(1)   Determine the adjacent surfaces.
(2)   Specify a common point of both surfaces on the edge.
(3)   Determine the unit normal vectors of the two surfaces at this point.
(4)   Calculate the angle between the unit normal vectors:

$$\cos\varphi = n_1 \, n_2$$ (Eqn 2.1)

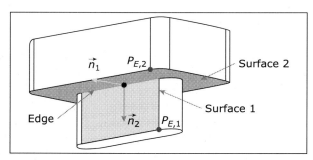

Figure 2.3 Boundary elements of a sample part used to
calculate the angle on the marked edge.

(5) Determine the small angle between the surfaces:

$$\alpha^* = 180° - \varphi \qquad \text{(Eqn 2.2)}$$

(6) If the surfaces are parallel ($\alpha^* = 0°$ or $\alpha^* = 180°$) the angle on this "edge" is $\alpha = \alpha^*$. Otherwise it has to be determined whether it is an "outer" or an "inner" edge.

(7) Check if one of the surfaces exceeds the common edge in the direction of the other normal vector. This is done by determining the extreme point on the surface 1 in the direction of the normal vector of surface 2. To obtain an unambiguous point two further directions are needed, the first normal vector and the cross-product of both vectors are used. This is of no importance to the angle but is needed for the algorithm to work. The second point is determined analogously.

(8) If one of these two points is on the common edge of the surfaces, it is an outer edge; if not, it is an inner edge.

(9) Calculation of the angle: $\alpha = \alpha^*$ (outer edge) or $\alpha = \alpha^* + 180°$ (inner edge), respectively.

2.4.5 Evaluation of the rules

A large set of elementary rules is created by combining all the adapted rules with all the appropriate geometric elements. In this step the last placeholders (for the geometric parameters) are replaced by exact values. Finally these elementary rules are evaluated to a logic result (true or false). If the rule evaluates to "false" the system creates an appropriate message and proposals how to correct the CAD model. In some cases, this optimization can be done automatically.

2.4.6 Realized prototype

The above described system has been realized as a software prototype at the University of Karlsruhe. It is based on the commercially available CAD system "Unigraphics" (now Siemens PLM Software) and its programmer interface "UG/Open API". The handling of the rules and the analysis of the geometry is performed by a program written in Visual C++. The evaluation of the elementary rules to logic results is done by Matlab and its C-interface. Therefore, all of Matlab's algebraic functions can be used inside the rules as well as user-defined functions. To extend the possibilities of the rule-based engineering system, new functions can be defined in Matlab without changing the core program. For example, a set of functions to calculate the notch effect according to the German standard DIN 743 [13] is implemented in this way and allows load-related design rules on shafts, to be used.

The user interfaces for setting up the parameters of a part and checking the rules are accessible by an extended toolbar within the CAD system. Thus the user can work regularly in the CAD system and activate the knowledge based system "by one click" (see Figure 2.4). The result of the rule checking is also presented in a dialog box. All rules that are evaluated to false are marked (see Figure 2.5). If the user selects a marked rule, the corresponding

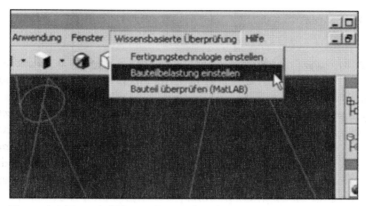

Figure 2.4 Extended Unigraphics toolbar for direct access to knowledge-based system.

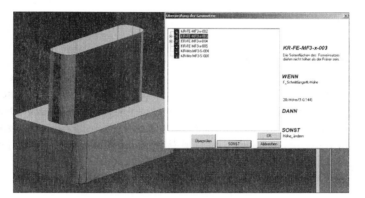

Figure 2.5 Dialogue box with results of rule check. Boundary elements infringing selected design rule are marked.

elements in the CAD model are highlighted. If an automatic correction is defined, the user can apply it [14, 15].

Conclusions

When designing microcomponents from ceramic materials, the subsequent product life-cycle stages have to be considered. Restrictions from mould manufacturing, e.g. micromilling or laser beam machining, and from the production process itself, e.g. PIM or sintering, have to be identified and provided to the designer. Design rules are a successful means of support in other microdomains, such as microelectronics, MEMS or LIGA. Basically, design rules have two functions at different levels of abstraction. First, they give information to the designer. Secondly, they bear information that is computer processable and thus has a more diagnostic character. Promising ways of application are presented for both functions.

For sharing knowledge between production and design, a wiki system is proposed due to its inherent possibilities for open integration, its ease of use, its accessibility and its ease of installation. For design diagnosis, a KBE system integrated in CAD is shown that enables the designer to design rule checks during design. However, all design rule tools are only as good as the knowledge that is represented by them. Thus, for an effective application of these tools, a culture has to be established, that encourages individual production and design engineers to share their knowledge during the product lifecycle.

Acknowledgements

The authors wish to thank the German Research Foundation (DFG) for their support of the Collaborative Research Centre (SFB 499) "Development, Production and Quality Assurance of Primary Shaped Micro Components from Metallic and Ceramic Materials".

References

[1]   Linde H. and Hill B. 1993. Erfolgreich erfinden. Widerspruchsorientierte Innovationsstrategie für Entwickler und Konstrukteure, Hoppenstedt Verlag, Darmstadt, Germany.

[2]   Mead C. and Conway L. 1980. *Introduction to VLSI Systems*, Addison-Wesley, Reading, Massachusetts, USA.

[3]   Scherer K. P., Buchberger P., Eggert H., Stiller P. and Stucky U. 1996. *Knowledge-based Support for Manufacturing of Microstructures. Microsystem Technologies*, **2**, Springer, Heidelberg, Germany, 167–170.

[4]   Görlitz S. and Kiehnscherf R. 1996. Werkzeugentwicklung für den Entwurf von Silizium-Mikrostrukturen. *Proceedings of Workshop Methoden und Werkzeugentwicklung für den Mikrosystementwurf*, Karlsruhe, Germany, 389–416.

[5]   Hansen U., Triltsch U., B?ttgenbach S., Germes C and Franke H.J. 2003. Rule-based validation of processing sequences. *Proceedingss of the International Conference on Modeling and Simulation of Microsystems* (MSM'03), San Francisco, California, USA, 484–487.

[6]   Leßmöllmann C. 1992. *Fertigungsgerechte Gestaltung von Mikrostrukturen für die LIGA - Technik*. Doctoral Thesis, Universität Karlsruhe, Karlsruhe, Germany.

[7]   Hahn K. 1999. *Methoden und Werkzeuge zur fertigungsnahen Entwurfsverifikation in der Mikrotechnik*, Fortschrittberichte VDI Reihe 20, Nr.88, VDI-Verlag, Düsseldorf, Germany.

[8]   Albers A., Deigendesch T., Drammer M., Ellmer C., Meboldt M. and Sauter C. 2007. Wikis as a cooperation and communication platform within product development. *Proceedings of the International Conference on Engineering Design* (ICED'07), 28–31 August, 2007, Paris, France.

[9]   Albers A., Sauter C., Deigendesch T. and Meboldt M. 2007. Semantic Wikis in knowledge management for multidisciplinary product development. *Proceedings DAAAM International Symposium*, 24–27 October, 2007, Zadar, Croatia.

[10] Wikipedia: http://www.wikipedia.org (accessed 07 November, 2008).

[11] Gross T. and Koch M. 2007. *Computer-Supported Cooperative Work*. Oldenbourg, Munich, Germany.

[12] Dokuwiki: http://wiki.splitbrain.org/wiki:dokuwiki (accessed 07 November, 2008).

[13] Deutsches Institut für Normung (Ed.) 2000, *DIN 743: Shafts and Axles, Calculation of Load Capacity*. Beuth, Berlin, Germany.

[14] Albers A., Burkardt N., Hauser S. and Marz J. 2005. *Knowledge-based design environment for primary shaped micro parts. Microsystem Technologies*, **11**, Springer, Heidelberg, Germany, 254–260.

[15] Albers A., Burkardt N., Hauser S. and Marz J. 2005. Computer Aided Evaluation of Design Rules for Micro Parts. *Proceedings of the International Conference on Engineering Design (ICED'05)*, 16–18 August, 2005, Melbourne, Australia.

3

Process simulation with particle methods: micropowder injection moulding and hot embossing

D. Kauzlarić, A. Greiner and J.G. Korvink

A particle-based approach for the predictive simulation of the injection moulding process and the hot embossing process for microcomponents is presented. The approach combines two methods: macroscopic finite interpolation as given by the smoothed particle hydrodynamics method and the mesoscopic approach of dissipative particle dynamics. Both approaches allow us to include internal degrees of freedom in the material particles. This approach is used for nonisothermal problems by adding the internal energy as a degree of freedom. For feedstocks with suspended solid particles the solid fraction is added as the degree of freedom.

3.1 Introduction

Classical MEMS, as they originate from the silicon device manufacturing technology used in microelectronics, are the most common MEMS devices. However, there is a great need for microstructured polymers, metals, and ceramics. The application areas range from DNA analysis instruments in life science over biocompatible materials for medical applications to a variety of electronic devices for sensors and actuators. For the microstructuring of pure polymers hot embossing and injection moulding are favoured for mass production. The PIM process is suitable for polymeric feedstocks with ceramic or metallic particle load. In order to reduce production costs there is a need for predictive process simulation which is able to take the special effects of the microworld into account. An increased surface-to-volume ratio leads to effects which are less important in the macroworld. In special cases, even microscopic thermal fluctuations are of importance for the result of the process. Additionally, embossing and injection moulding lead to large material deformation fields, which require the simulation to span several orders of magnitude in length scale, which is a nontrivial task.

There are a variety of finite element-based simulation approaches [1, 2]. Usually, their disadvantage is that it is difficult to handle the free surfaces and large deformations.

Particle-based approaches are able to deal with free surfaces and large deformations. We present two of those methods to the given problem field: macroscopic finite interpolation methods such as smoothed particle hydrodynamics SPH [3, 4] and the mesoscopic approach of dissipative particle dynamics DPD [5]. The choice of methods is based on the goal of reducing the microscopic effects in the material to the necessary minimum for an accurate and useful simulation for the process engineer. Therefore, we include nonlinear effects such as shear-thinning and yield stresses. Additionally, the DPD method allows for the inclusion of thermal fluctuations. A big advantage of SPH and DPD is that both allow for the implementation of internal degrees of freedom of the material particles. This is applied to nonisothermal problems by adding an internal energy and to feedstocks filled with ceramic or metallic particles by adding a solid fraction.

3.2 Particle methods

A brief, basic description of the applied particle methods is now given. Their basis is the interpolation idea that an arbitrary function $f(x)$ may be expressed by [3, 4]:

$$f(\mathbf{x}) = \int f(\mathbf{x}') \, W_h(\mathbf{x} - \mathbf{x}') \, d\mathbf{x}' \qquad (3.1)$$

where $W_h(\mathbf{x} - \mathbf{x}')$ is an interpolation function of width, h and its volume-integral is normalized to unity. Note that for $h \to 0$ Equation (3.1) is exact.

Given Equation (3.1) we introduce the so-called finite interpolation methods by replacing the integral by a sum over a finite set of points or "particles" distributed in space and $f(\mathbf{x})$ is replaced by the particle-centred value f_i, i.e.:

$$f_i = \sum_j \frac{m_j}{\rho_j} f_j W_{ij} \qquad (3.2)$$

where m_j is the mass of the particle and ρ_j its associated local density. Additionally, $W_{ij} = W_h (\mathbf{x}_i - \mathbf{x}_j)$, and $\mathbf{x}_i$ and $\mathbf{x}_j$ are the positions of particles and i and j, respectively. The range h of the interpolation function is now of finite size.

Equation (3.2) is applied to directly compute the local density ρ_i around a particle, to read:

$$\rho_i = \sum_j m_j W_{ij}$$

A representation of derivatives like $(\nabla f)_i$ is often needed. Therefore we assume f to vary smoothly between particles i and j. From Equation (3.2) we then obtain the expression:

$$(\nabla f)_i = \sum_j \frac{m_j}{\rho_j} f_j \nabla W_{ij}$$

3.2.1 Smoothed particle hydrodynamics

The above interpolations can be directly applied to the Lagrangian form of the conservation equation of momentum:

$$d\mathbf{v}/dt = (1/\rho) \nabla \cdot \sigma$$

to give:

$$d\mathbf{v}_i/dt = -\sum_j m_j \left((\sigma_i/\rho_i^2) + (\sigma_j/\rho_j^2) \right) \cdot \nabla W_{ij} \qquad (3.3)$$

which is one of the possible discretisations often used in SPH, with **v** being the velocity field and σ the total stress tensor [6].

The particle migration model in Section 3.4.2 requires the use of a discretisation of second spatial derivatives. Español and Revenga [7] proposed the expression:

$$(\nabla \cdot [B\,(\mathbf{r})\,\nabla A\,(\mathbf{r})])_i \approx -\sum_j (B_i + B_j)\,(A_i - A_j)\,F_{ij}/\rho_j \qquad (3.4)$$

for two scalar fields $A\,(\mathbf{r})$ and $B\,(\mathbf{r})$. F_{ij} is defined through $\nabla W_{ij} = -\mathbf{r}_{ij} F_{ij}$ and $\mathbf{r}_{ij} = \mathbf{r}_i - \mathbf{r}_j$ is the distance vector between particles i and j.

3.2.2 Dissipative particle dynamics

DPD, as it was first introduced [5], is a particle-based approach with microscopic background. As in SPH, there is a discrete equation of motion of the particle's velocities, which may be written in the general form:

$$\frac{d\mathbf{v}_i}{dt} = \frac{1}{m_i}\sum_j \left(\mathbf{F}^C_{ij} + \mathbf{F}^D_{ij} + \mathbf{F}^R_{ij}\right) \qquad (3.5)$$

where $\mathbf{F}^C_{ij}$ is a conservative force, $\mathbf{F}^D_{ij}$ accounts for dissipative forces, and $\mathbf{F}^R_{ij}$ is a stochastic force. The stochastic force respresents the thermal fluctuations due to the microscopic particles comprised in the DPD particles. All forces act between the pair of particles i and j. The general form of the forces is given by:

$$\mathbf{F}_{ij} = A\omega_{ij}\mathbf{r}_{ij}/r_{ij} \qquad (3.6)$$

where $r_{ij} = |\mathbf{r}_{ij}|$ is the distance between the particles i and j. The factor A may be a constant or may depend on various particle properties. ω_{ij} is an appropriate weighting function. If we choose:

$$\nabla W_{ij} = \omega_{ij}\mathbf{r}_{ij}/r_{ij}$$

the similarities between Equation (3.3) for SPH and Equation (3.5) together with (3.6) for DPD become obvious.

This is why, in [7], the two methods have been merged. We expect that thermal fluctuations will have no influence in standard embossing or high-pressure PIM simulations, since the length scales are very large compared to the structural features of the material and the high viscosity will, in addition, lead to a high damping rate of any type of fluctuation. On the other hand, low-pressure PIM for rapid-prototyping in the micrometre range [8] or the welding of polymeric materials on nanoscales [9] may show effects due to thermal fluctuations.

As a consequence, for both the simulation of embossing and of PIM, we decided to use DPD as a basis to maintain flexibility. A second advantage of this choice is the numerical robustness of DPD, which often needs less stabilization than SPH due to the stabilizing effect of the thermal noise.

3.2.3 Local thermostat for dissipative particle dynamics

Generalizations of DPD to nonisothermal systems have been developed in various approaches. We developed [10] a method similar to existing thermostating schemes that gives an explicit algorithm. It is based on the Peters thermostat for isothermal DPD [11].

The thermostat possesses one adjustable parameter, the friction constant γ_{ij}, for a given pair of particles. For later use, a functional dependency of the effective shear viscosity of the fluid in dependence of the model parameter γ_{ij} and the chosen time step Δt is necessary. It turns out that:

$$\eta \approx c_1 (1 - \exp[-c_2 \gamma_{ij} \Delta t]$$
$$+ \sqrt{1 - \exp[-2c_2 \gamma_{ij} \Delta t]})/\Delta t + c_3 \tag{3.7}$$

fits the behaviour well, where c_1, c_2 and c_3 denote fitting parameters.

3.3 A yield-stress model for polymethylmethacrylate

By increasing the relaxation frequency $1/\Delta t$ in Equation (3.7) we are able to increase the maximum viscosity, and thus increase the accessible viscosity bandwidth. Since we want to simulate a softening process we choose a very small time step, which provides us with a viscosity bandwidth of about two orders of magnitude. In order to simulate micrometre length scales and viscosities of the order of 10^5 Pa s as it is necessary for softening processes, we would need to resolve time steps of 10^{-13} s. This is unacceptable for process times of the order of microseconds to seconds. Therefore we propose to perform an equivalent flow simulation. The underlying idea is to set the ratio of clamping pressure and viscosity to the correct value by simultaneously reducing the pressure and viscosity. The invariant under this transformation is the product $Re\ Eu$, where Re is the Reynolds number and Eu is the Euler number. To describe the softening behaviour of polymethylmethacrylate (PMMA) we introduce a modified Bingham model incorporating both softening and shear thinning. The local viscosity is a function of the temperature T_i and shear rate γ_i, given by:

$$\eta_i(T_i, \dot{\gamma}_i) = \eta_1 \exp[-t_1 \dot{\gamma}_i]$$
$$+ \eta_2 ((1 + \exp[(\dot{\gamma}_i - \dot{\gamma}_c)/t_2])^{-1} \tag{3.8}$$
$$+ (1 + \exp[(T_i - T_c)/(E_a/k_B T_i)])^{-1})/2$$

The viscosity drops when exceeding the critical values $\dot{\gamma}_c$ and T_c for the shear rate and temperature respectively. The parameters t_2 and E_a define the slope of this viscosity drop, while η_1 and η_2 define the maximum viscosity. Shear thinning behaviour is captured by the constant t_1 in the first term.

$\dot{\gamma}_i$ is the second invariant of the shear tensor $\dot{\gamma}_i^{\alpha\beta}$, to read:

$$\dot{\gamma}_i = \sqrt{\sum_{\alpha,\beta} (\dot{\gamma}_i^{\alpha\beta})^2} \tag{3.9}$$

The discretization of the divergence operator can be generalized in order to obtain a discrete form of the shear rate tensor, which is:

$$\dot{\gamma}_i^{\alpha\beta} = \sum_j (r_{ij}^\alpha v_{ij}^\beta + r_{ij}^\beta v_{ij}^\alpha) F_{ij}/(2\rho_i)$$

Walls in the simulations given in this section and Section 3.4 are modelled by frozen particles and a stochastic reflection ensuring no-slip conditions. This simple setup does not guarantee no-slip conditions for every case. But for the embossing simulations we observed quite good results as the applied excess pressure leads to good contact between the fluid particles and the walls.

3.4 A model for powder migration

For micro-PIM detection of the migration of powder particles during the injection process is crucial. The goal is to minimize this effect with the assistance of a predictive simulation. In this section, we first present the chosen continuum model for powder migration. We then apply the SPH-discretization formalism to it.

3.4.1 Continuum model

A phenomenological model for shear-induced powder migration was first presented by Phillips *et al.* [12]. The proposed equation of motion for the volume fraction ϕ of the suspended solid material is:

$$\partial\phi/\partial t = -\nabla \cdot \phi\mathbf{v} - \nabla\cdot(\mathbf{J}_c + \mathbf{J}_\eta) \tag{3.10}$$

where $\mathbf{v}$ is the flow velocity of the solution.

The flux $\mathbf{J}_c$ represents the powder migration due to a spatially varying collision frequency of the solid powder particles. It is assumed to be:

$$\mathbf{J}_c = -D_c a^2 \phi\nabla\dot\gamma\phi = -D_c a^2 \left(\phi^2 \nabla\dot\gamma + \phi\dot\gamma\nabla\phi\right) \tag{3.11}$$

where D_c is a diffusion constant, a is the radius of the spherical powder particles and $\dot\gamma$ is a measure for the shear rate, as given in Equation (3.9). The effect of $\mathbf{J}_c$ is an accumulation of the solid fraction in regions with a low shear rate.

$\mathbf{J}_\eta$ is supposed to be a flux of powder particles towards regions of low viscosity. Its expression is:

$$\mathbf{J}_\eta = -D_\eta\dot\gamma\phi^2 \frac{a^2}{\eta}\frac{d\eta}{d\phi}\nabla\phi \tag{3.12}$$

Usually, the viscosity increases with the solids load. Then, $\mathbf{J}_\eta$ will counteract the effect of $\mathbf{J}_c$. For the dependency of the viscosity on the solid fraction, we use Krieger's model, which is $\eta(\phi) = (1-\phi/\phi_m)^{-c}$. For the simulations presented here, we use a saturation value of $\phi_m = 0.68$ and an exponent of $c = 1.82$ as Phillips *et al.* in their original work [12].

3.4.2 Smoothed particle hydrodynamics discretization

Instead of discretizing equation (Equation 3.10) directly, we use the equation of motion for the volume $V_\phi = \phi/\rho$ occupied by powder particles. The discretized equation for V_ϕ will be antisymmetric under exchange of fluid particles and this leads to the advantage that we have exact volume conservation. We are seeking the discretized equations of motion for V_ϕ on a fluid-particle moving with the flow. Hence we have to switch from the Eulerian to the Lagrangian description. Introducing the substantial derivative $d/dt = \partial/\partial t + \mathbf{v}\cdot\nabla$ transforms Equation (3.10) into:

$$\frac{d\phi}{dt} = -\phi\nabla\cdot\mathbf{v} + \nabla\cdot\left[D_c a^2\phi\nabla\phi\,\dot\gamma + D_\eta\dot\gamma\phi^2\frac{a^2}{\eta}\frac{d\eta}{d\phi}\nabla\phi\right]$$

where we have already inserted the fluxes (Equations (3.11) and (3.12)). Substituting $\phi = \rho V_\phi$ gives:

$$\frac{dV_\phi}{dt} = -V_\phi \frac{\dot{\rho}}{\rho} - V_\phi \nabla \cdot \mathbf{v}$$

$$+ \frac{1}{\rho} \nabla \cdot \left[D_c a^2 \phi \nabla \phi \dot{\gamma} + D_\eta \dot{\gamma} \phi^2 \frac{a^2}{\eta} \frac{d\eta}{d\phi} \nabla \phi \right] \tag{3.13}$$

which is the equation of motion for the occupied volume. The divergence of the velocity field gives the relative volume expansion, i.e. $\nabla \cdot \mathbf{v} = \dot{V}/V = -\dot{\rho}/\rho$. Thus the first two terms on the right-hand side of Equation (3.13) cancel each other out. Now, we apply the rule (4) for the SPH-discretization of second derivatives to the right-hand side of Equation (3.13). This gives:

$$\frac{dV_{\phi,i}}{dt} = -D_c a^2 \sum_j \frac{F_{ij}}{\rho_i \rho_j} (\phi_i + \phi_j)(\phi_i \dot{\gamma}_i - \phi_j \dot{\gamma}_j)$$

$$- D_\eta a^2 \sum_j \frac{F_{ij}}{\rho_i \rho_j} \left(\dot{\gamma}_i \phi_i^2 \left(\frac{d\eta}{\eta d\phi} \right)_i + \dot{\gamma}_i \phi_i^2 \left(\frac{d\eta}{\eta d\phi} \right)_j \right)(\phi_i - \phi_j)$$

The expressions within the sums are antisymmetric under particle exchange. This is not the case for the discretized form of the equation of motion for ϕ.

3.5 Results

We now present simulation results for two topics. First, we present embossing simulations testing our yield–stress model and then we show simulations of PIM where we track the solid fraction of the feedstock.

3.5.1 Segregation through migration

For the following tests, we choose a particle diameter of $a = 0.1$ and rate constants of $D_c = 1$ and $D_\eta = 2$. The velocities and shear rates are given in reduced units. As a first consistency test we consider the particle migration on a fixed grid of fluid-particles in a geometry with periodic boundary conditions in all directions. A sinusoidal velocity profile is superimposed on the discretization particles and an initial homogeneous solids load of $\phi = 0.5$ is assigned to each fluid particle. Figure 3.1 shows the steady-state profile of the volume fraction. The solid particles aggregate at the positions of least shear.

The depletion of powder particles in the regions of high shear, as they are represented by walls or edges in the mould, was detected over time. Figure 3.2 shows the transient behaviour of the relative volume fraction in such a region where the shear rate is highest. We see that the ratio D_c/D_η determines the magnitude of migration. As a more complex application we take an injection moulding test geometry (see Figure 3.3(c). We concentrate on the region marked by the rectangle. The arrow indicates the gate of the mould geometry and the flow direction.

Figure 3.3(a) is a snapshot of an early stage of the injection. At this stage, the velocity field in the volume directly behind the gate is still inhomogeneous, i.e., there is a strong

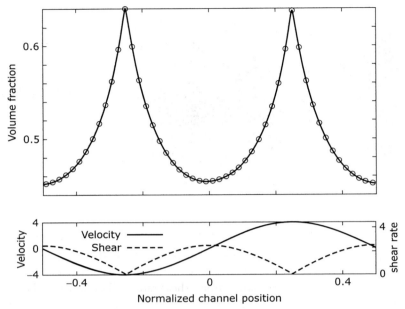

Figure 3.1 Upper figure shows distribution of concentration of powder after migration due to a shear wave in a cuboid simulation domain. Lower figure shows velocity and shear rate distribution through cross-section of the channel. There is a pronounced increase in ceramic powder concentration in low shear rate regions.

shear field. The distribution of the solids load indicates an aggregation at convex corners (i.e. pointing into the mould material) and a decrease at concave corners (i.e. pointing into the cavity). This is intuitively understandable since the shear in the proximity of convex

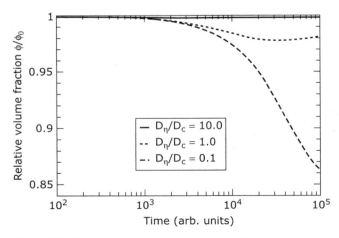

Figure 3.2 Depletion due to shear migration of powder particles. Transient decrease of volume fraction of powder concentration for three different ratios between D_c and D_η.

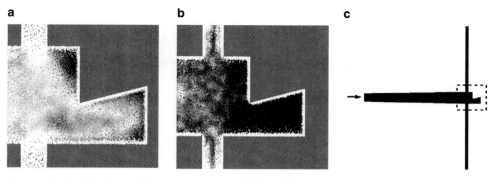

Figure 3.3 Solid load in different stages of mould filling. The solid load of a single particle is grey scale encoded: the higher the load the darker the particle: (a) early stage of mould filling; (b) later stage; (c) test geometry for simulation of injection process.

corners should be lower than in the bulk of the geometry. On the other hand, concave corners are regions which have large shear rates. In later stages of the filling, the flow in the volume close to the gate becomes more homogeneous and directed towards the two arms at the top and the bottom. In these arms strong shear rates will occur since they are very narrow. The effect can be seen in Figure 3.3(b). The concentration in the part of the entrance volume behind the two arms is rather homogeneous and large. The concentration in the front part is a few per cent lower since this is the region where the feedstock is still flowing towards the arms and it undergoes shearing motion. Inside the arms the average concentration is even lower. This indicates that a larger fraction of the solid particles prefers to stay in the entrance volume where shear rates and flow velocities are lower. Additionally, the solids load has a maximum in the centre of the arms, as is expected for flow in a channel.

It should be noticed that, in the presented simulations, we did not consider full or partial solidification of the material. This is the reason why the inhomogeneities from Figure 3.3(a) can dissappear or be replaced by other types of inhomogenities. If the material solidified at least partially in the entrance region, e.g., in the right part behind the arms, the inhomogeneous solids load would be frozen in. In the worst case, all observed inhomogeneities could occur simultaneously, eventually leading to cracks or deformations during the sintering step.

3.5.2 Embossing simulation of polymethylmethacrylate

The simulations are performed for a cut-out of an embossing geometry (see Figure 3.4 top-right). The wall particles are kept at a constant temperature of 25°C to account for the large heat conduction of the brass material. Figure 3.4 shows the simulation and experimental results. As can be seen in the simulations, the highest temperature is about 100°C. This shows that the PMMA softens below the glass transition temperature due to shear. Despite the difficulty to measure local temperatures in the experiments, experimental observations support these simulation results. Additionally, we can see a step in the flow profile close to the corners of the geometry, both experimentally and in the simulation. Apparently, the large heat conduction of the wall prevents a boundary layer of the fluid from heating up to the critical temperature and thus inhibits flow in this region.

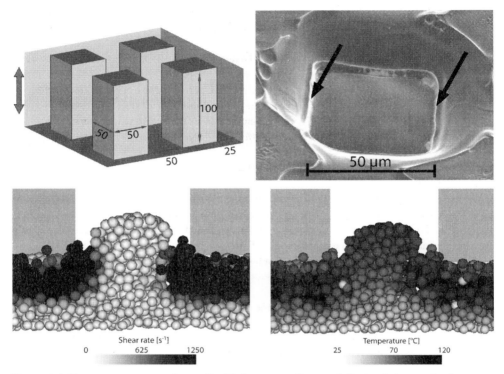

Figure 3.4 Top test geometry and partially filled structure. Bottom-left: simulation results for shear-rate distribution. Bottom-right: simulation results for temperature distribution.

Conclusions

We have shown that with the help of the SPH and DPD particle methods, important information about the microproduction processes micropowder injection moulding and embossing can be gained. These methods were used for the simulation of powder migration in micropowder injection moulding and for the shearing during embossing.

Acknowledgments

The authors wish to express their gratitude to the Deutsche Forschungsgemeinschaft (DFG) for funding of this project within the framework of the Sonderforschungsbereich Mikrourformen (SFB499).

References

[1] Bilovol V. V., Kowalski L. and Duszczyk J. 2000. Numerical simulation of the powder injection moulding process for optimization of mould design and process parameters. *Advanced Engineering Materials.*, **2**(3), 127–131.

[2] Barriere T., Gelin J. C. and Liu B. 2001. Experimental and numerical investigations on properties and parts produced by MIM. *Powder Metallurgy.*, **44**(3), 228–234.

[3] Gingold R. A. and Monaghan J. J. 1997. Smoothed particle hydrodynamics: Theory and application to non-spherical stars. *Monthly Notices of the Royal. Astronomical Society* **181**, 375–389.

[4] Lucy L. B. 1997. A numerical approach to the testing of the fission hypothesis. *Astronomical Journal* **82**(12), 1013–1024.

[5] Hoogerbrugge P. J. and Koelman J. M. V. A. 1992. Simulating microscopic hydrodynamic phenomena with dissipative particle dynamics. *Europhysics Letters* **19**(3), 155–160.

[6] Randles P. W. and Libersky L. D. 1996. Smoothed particle hydrodynamics: Some recent improvements and applications. *Computer Methods in Applied Mechanics and Engineering.*, **139**(1–4), 375–408.

[7] Español P. and Revenga M. 2003. Smoothed dissipative particle dynamics. *Physical Review E*, **67**, 026705.

[8] Knitter R. and Bauer W. 2003. Ceramic microfabrication by rapid prototyping process chains. *Sadhana-Academy Proceedings in Engineering Sciences* **28**(1–2), 307–318.

[9] Kunihiko T. and Hirokazu T. 1998. Molecular movement during welding for engineering plastics using langevin transducer equipped with half-wavelength step horn. *Ultrasonics*, **36**, 75–78.

[10] Pastewka L., Kauzlaric D., Greiner A. and Korvink J.G. 2006. Thermostat with a local heat-bath coupling for exact energy conservation in dissipative particle dynamics. *Physical Review E*, **73**, 037701.

[11] Peters E. A. J. F. 2004. Elimination of time step effects in DPD. *Europhysics Letters* **66**(3), 311–317.

[12] Phillips R. J., Armstrong R. C., Brown R. A., Graham A. L. and Abbott J. R. 1992. A constitutive equation for concentrated suspensions that accounts for shear-induced particle migration. *Physics Fluids A*, **4**(1), 30–40.

4

Finite element modeling of the pressing of ceramic powders to predict the shape of a pellet after die compaction and sintering

Luc Federzoni

We describe the use of finite element method simulation for the die pressing of a ceramic part. Results obtained for a nuclear fuel pellet are presented. There are some advantages related to the use of the simulation in the prediction of the shape of the part. In association with an optimization routine, the simulation can offer the possibility of optimizing the processing route and leads to a better part.

4.1 Introduction

This chapter describes the interest of the die compaction simulation and sintering of ceramics. Using an example, it describes the advantages of using the simulation even for a simple shape such as a pellet.

The example chosen is a nuclear fuel pellet. It is not really a micropart (about several cm^3 size), but the methodology for the simulation of the formation is the same as for microparts and is relevant for this chapter.

4.2 Mechanical modeling of die-pressing

4.2.1 Main steps of ceramics forming by pressing: case of the UO_2

At the beginning of the die pressing step, calibrated amounts of a given mixture of oxide powders, including lubricant and porogen, are poured into the closed die of a press and then compacted by the movement of an upper punch. The die filling density is about 20–30% [1–4] of the theoretical density (th.d) of the material although the green compact density reaches 60–65% th.d [4]. Punches and die are designed by taking into account the shrinkage of the compact during sintering to obtain the specified shape for the pellet, which may include dishings and chamfered edges. The mechanical strength of the green compact is sufficient to allow the pellets to be transported to the sintering furnaces. The sintering step, which includes a heating stage at about 1700°C under specific atmosphere leads to a consolidation of the compact which nearly reaches the theoretical density (95–98% th.d) [4].

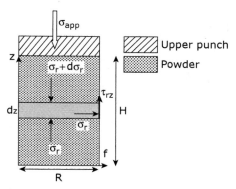

Figure 4.1 Balance of forces acting on a powder
slice during compaction (simple mode with
upper punch movement).

4.2.2 Influence of frictional forces

During closed die pressing, the powder material is submitted to a quasi-biaxial stress state
where σ_z, the axial pressure, is transmitted by the punches and σ_r is the resulting radial
constraint exerted by the cylindrical die (see Figure 4.1). Significant frictional forces develop
between the granular material and inner die surface and have to be taken into account. The
equilibrium of axial forces on a given powder slice (width dz) situated at a distant z from
the lower punch can be written [3]:

$$\pi R^2 d\sigma_z = 2\pi R \tau_{rz} dz \tag{4.1}$$

We assume that:

- the frictional stress obeys to Coulomb's law $|\tau_{rz}| = \mu\sigma_r$ (4.2)
- the radial stress is proportional to the axial component $\sigma_r = \beta\sigma_z$ (4.3)

where μ is the friction coefficient and β is a powder dependent parameter (flow index).

In the simple case where the powder compaction results from the single downward
displacement of the upper punch (single effect compaction), the direction of the frictional
forces is the same along the interface so Equation (4.1) can be integrated as:

$$\sigma_z(z) = \sigma_{app} \exp\left[\frac{-2\mu\beta(H-z)}{R}\right] \tag{4.4}$$

where σ_{app} is the axial pressure applied by the upper punch.

This simple relationship shows that due to friction forces, the compaction is not uniform
along the powder column and the resulting stress and density gradients increase with the
product $\mu\beta$ (so-called friction index).

For more complex cases corresponding to actual shaped pellets and press tool
displacements, the finite element method (FEM) has to be used to calculate the stress
components.

4.2.3 Powder compaction model

The constitutive laws used to determine the powder density for a given stress state are based on a macroscopic elastoplastic formulation (Cam-clay model [4]). Other formulations can be selected, like the Drucker–Prager cap model or Shima [5]. The selection of the formulation is the result of a compromise between the accuracy of the representation of the reality and of the ease of the determination of the input data. It is considered that the Cam-clay models and Drucker–Prager cap model fit these criteria [5].

In the approach of the Cam-clay model, the density, ρ, is treated as an internal variable which allows the effects of previous loadings to be considered. For a given density, the compacted powder obeys Hooke's law at low stresses, when the current stress state belongs to the elastic domain sized by the density. The boundary of this domain in the space of stress tensor components is represented by the expression:

$$f(P,Q) = Q^2 + M^2 (P-P_0)(P-P_1) = 0 \tag{4.5}$$

where Q is the von Mises stress and P the isostatic stress (first and second invariants of stress tensor).

With such a yield criterion, the summation of diagonal terms of the plastic strain tensor corresponds to the increment of powder density according to:

$$d\varepsilon^P_{ii} = \frac{d\rho}{\rho} \tag{4.6}$$

The elastic domain (which is elliptical in the P,Q plane) expands as the parameter P_1 rises according to the following consolidation law:

$$P_1 = \left[\frac{\rho}{\rho_0}\right]^k \tag{4.7}$$

where ρ_0 is the filling density and k is a powder parameter.

An associated flow rule is used to calculate the plastic strain when the yield condition $f(P,Q) = 0$ is fulfilled:

$$d\varepsilon^P_{ij} = d\lambda \cdot \frac{df}{d\sigma_{ij}} \tag{4.8}$$

The term $d\lambda (d\lambda > 0)$ is determined by the consistency condition ($df = 0$):

$$d\lambda = -\frac{\partial f/\partial Q \cdot dQ + \partial f/\partial P \cdot dP}{\rho \cdot \partial f/\partial \rho \cdot \partial f/\partial P} \tag{4.9}$$

The system of equations (4.5–4.9) has to be solved numerically to determine the density increment, $d\rho$.

This mechanical approach allows us to take into account the impact of different stress states on the powder flow. For instance, cold isostatic compaction as well as die pressing processes can be treated by the same formalism. For die pressing, the effect of the shear component which can improve the powder densification by grains gliding is well caught by this model.

Moreover, the elastic expansion of the compact (axial and radial springback) after ejection can be calculated when the appropriate elastic parameters (Young's modulus and Poisson's coefficient) are given.

4.2.4 Calculation of pellet shrinkage after sintering

Finally, the dimensional changes induced by sintering are simply calculated using the hypothesis that the final pellet density is homogeneous. Indeed, the shrinkage of each part of the compact can easily be deduced from the difference between the local green density (previously calculated) and the sintered density by considering the mass balance.

4.3 Identification of model parameters

4.3.1 Methodology

Four material parameters (function of density) have to be determined experimentally:

- plastic flow parameters: $M(\rho)$ and $k(\rho)$
- elastic parameters: $E(\rho)$ and $v(\rho)$

A combination of the elastic parameters can be directly evaluated from the volume expansion corresponding to the springback of the green compact after ejection:

$$\overline{E}\,(\overline{\rho}_e) = \frac{3}{(1 - 2\overline{v})} \frac{\overline{P}}{\ln(\overline{\rho}_p / \overline{\rho}_e)} \tag{4.10}$$

with

$$\overline{P} = \frac{\overline{\sigma}_z + 2\overline{\sigma}_r}{3} \tag{4.11}$$

where $\overline{\rho}_p$ is the mean compact density under loading at the end of die pressing (known by recording the punch displacement) and $\overline{\rho}_e$ is the mean density after pellet removal determined by geometrical measurements.

The mean pressure $\overline{P}$ is evaluated by Equation (4.11) from the experimental values of axial and radial stresses and is averaged over all the pellet volume.

The instrumented press used allows us to measure the applied stress continuously at the upper punch σ_{app}, the transmitted stress to the lower punch σ_{tr} ($\sigma_{app} > \sigma_{tr}$ due to friction) and also the radial forces, σ_r, exerted on the die [5]. Equation (4.4) is used to determine the stresses at different heights.

A careful analysis of the evolution of the stresses, σ_z and σ_r, recorded during compact axial discharge gives a value for Poisson's coefficient. Figure 4.2 compares the calculated and experimental paths that the stress state in the (P,Q) plane follows during compact discharge. It is shown that the evolution of the stress state is closer to the experimental data for $v = 0.1$ than for $v = 0.3$.

The friction coefficient, μ, is then determined from the friction index deduced from single effect compaction tests where Equation (4.4) can be transformed to:

$$\mu\beta = \frac{R}{2H}\ln\left(\frac{\sigma_{app}}{\sigma_{tr}}\right) \tag{4.12}$$

The parameter $\beta(\sigma_r/\sigma_z)$. is obtained by dividing the experimental values of σ_r by values of σ_z (rescaled at the height z_m where the measurement of σ_r is performed).

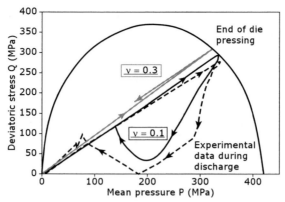

Figure 4.2 Paths of stress states in the (P,Q) plane during a die pressing of a UO_2 powder at 600 MPa (simple mode) and subsequent discharge. The elliptical curve represents the yield surface at the end of die pressing. The path corresponding to the springback and calculated with a Poisson's coefficient $v = 0.1$ is consistent with the experimental data (dotted line).

The flow parameters, M and k, are also determined from die pressing tests by considering two conditions:

- the current stress state always belongs to the yields surface i.e. the following relation is always true:

$$Q_m^2 + M^2 (P_m - P_0)(P_m - P_1) = 0 \tag{4.13}$$

where $P_m = \sigma_2 (1 + 2\beta)/3$ and $Q_m = \sigma_2 (1-\beta)$ \hfill (4.14)

- the radial component of plastic strain can be neglected ($d\varepsilon_r = 0$) so Equations (4.7) and (4.8) give after transformation:

$$\frac{2Q_m}{M^2 [2P_m - P_0 - P_1]} = \frac{2}{3} \tag{4.15}$$

Table 4.1 Values of elastoplastic parameters of the Cam-clay model for a UO_2 powder.

Compact density	Young's modulus ($v = 0.1$)	Cam-clay parameter M	Cam-clay parameter P_1
g/cm³	MPa		MPa
4	Not determined	1.77	35
4.5	Not determined	1.77	50
5	4650	1.77	100
5.5	10450	1.77	180
6	15800	1.77	330
6.24	24750	1.77	420

For low cohesive materials like nuclear ceramics, the traction resistance, P_0, can be neglected and the resolution of Equations (4.13) and (4.15) gives a set of M and P_1 (or k) related to discrete values of density.

4.3.2 Model parameters for a UO_2 powder

Material characterization and parameters identification have been performed for a UO_2 powder (Mimas ADU dry route). The values of model parameters are given in Table 4.1 for powder densities higher than $4\,g/cm^3$. It has been observed that macroscopic mechanical models give poor results for the first stage of densification where the rearrangements of the granules take place (2–$4\,g/cm^3$). A rather good approximation is obtained by considering the powder to have a uniform density of $4\,g/cm^3$ (higher than the filling density) at the beginning of the simulation (the height of the powder at the initial state is rescaled to ensure mass conservation).

4.4 Code performances

4.4.1 Green compact characteristics

Several comparisons between simulation and experimental data have been carried out for this powder. FEM calculations have been performed by considering powder and tools (punches and die) geometries and mechanical behaviour. The experimental value of the friction coefficient (0.15) is also introduced as an input data. Actual tool displacement recorded by the press instrumentation are specified as boundary conditions for the compaction stage, so that the resulting forces can be calculated.

Figure 4.3 shows good agreement between the axial evolution of the green density for a single effect compaction (600 MPa) measured with a gamma densitometry technique and

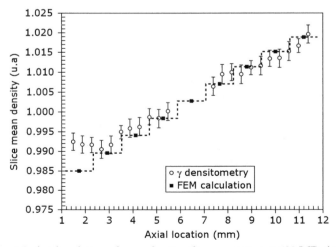

Figure 4.3 Axial evolution of green density after compaction at 600 MPa (single effect) of UO_2 powder. Each point correspond an averaged value of density over an axial slice (10 slices have been considered for FEM posttreatment).

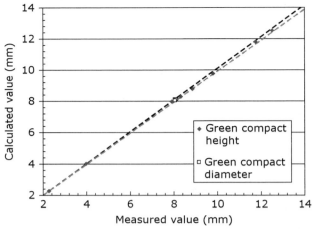

Figure 4.4 Calculated vs. experimental values of green compact dimensions after die pressing and ejection of a UO_2 powder, scattering is less than ±1%.

using the FEM. It can be seen that for this pellet, with a ratio H/D of 1.5, the axial variation of the green density is lower than 4%. Despite this low value, the pellet has the shape of a truncated cone after sintering and the difference between the maximal and minimal radius reaches 45 μm.

Figures 4.4 and 4.5 present comparisons of calculated and measured values of green compact geometry (height and diameter) and forces acting on punches for different H/D (0.25–1.5) pellet ratios and green densities (5–6.5 g/cm³).

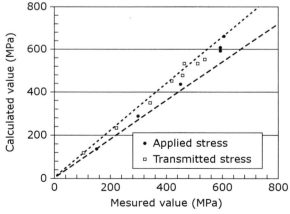

Figure 4.5 Calculated vs. experimental values of mean axial stress exerted on lower and upper punches at the end of a single effect compaction of a UO_2 powder, scattering is less than ±10%.

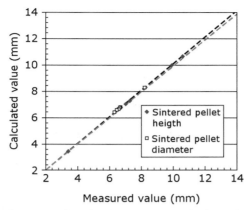

Figure 4.6 Calculated vs. experimental values of sintered pellet dimensions after forming of a UO_2 powder, scattering is less than ±1%.

4.4.2 Sintered pellets characteristics

Experimental and calculated values of sintered pellets dimensions are plotted in Figure 4.6. Again, values are consistent within the range (0.25–1.5) of H/D ratios. Figure 4.7 depicts the values of the deviation from a perfect cylinder (DPC) which is defined as the difference between the maximal and minimal pellet radius.

It should be noticed that the simulation reproduces the experimental sensitivity on H/D but tends to slightly overestimate the values of DPC.

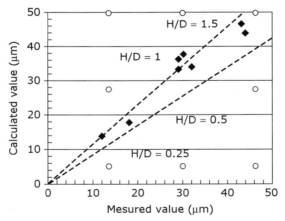

Figure 4.7 Calculated vs. experimental values of deviation from perfect cylinder ($R_{max} - R_{min}$) of sintered pellets, scattering is less than ±15%.

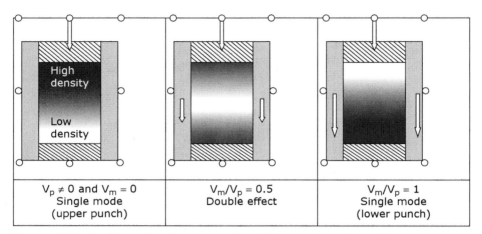

Figure 4.8 Powder density distribution according to die-pressing conditions for a cylindrical part.

This trend may be explained by the poor description of the pellet ejection which is considered to be purely elastic although inelastic strains (dilatancy) could occur during the extraction of the pellet and could change the density distribution.

4.5 Analysis of the effect of tools displacements

Industrial presses used to form complex parts by powder metallurgy include several tools (punches, cores and die) which can move separately in a specific sequence that prevents powder transfers from a high density part to low density part during pressing.

Even for a simple shape like a fuel pellet, the displacement of the upper punch that controls the volume of powder is accompanied by a downward movement of the die which influences the friction forces and density gradient as a consequence. Three main cases can be considered according to the ratio between the speed of the upper punch, V_p, and those of the die, V_m (see Figure 4.8). Single effect modes lead to a monotonic decrease in density, the higher density zone being in contact with the moving punch (relative to the die). The double effect mode ($V_m/V_p = 0.5$) produces a symmetrical density distribution which gives a final "hourglass-like" shape after sintering.

However, the density gradients may differ from these ideal cases when the speed ratio changes during the pressing stage. We have considered the actual sequence where first both the upper punch and die move in a quasidouble effect mode ($V_m/V_p \sim 0.5$) and then, at a given moment, the die stops and the upper punch moves downward until the pressing has been completed.

Simulation shows that the time at which the die stops is a key parameter. An optimized value of this time was numerically determined and then experimentally tested. Comparisons of the profile of the diameters of two pellets obtained after sintering are shown in Figure 4.9, one concerns a reference pressing cycle, the second is related to an optimized cycle. In the first case, the volume of material removed by grinding to obtain a perfect cylinder reaches 0.92% of the pellet volume (0.75% calculated) while it is lowered to 0.29% (0.23% calculated) after

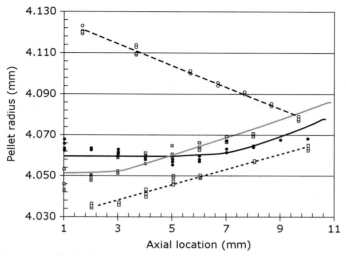

Figure 4.9 Pellet diametrical profiles obtained after different pressing cycles: reference cycle (gray) and optimized cycle (black), symbols denote measured values, continuous lines represent the calculated profiles. Open marks combined with dotted lines correspond to two cases: when die stops early (squares), when die stops after the punch has stopped (circles).

optimization. In these calculations, the diameter of the cylinder is defined as the minimal value obtained on the sintered pellet and it should be noted that this value also depends on the cycle (see Figure 4.9).

Moreover two limiting cases have been reported on Figure 4.9: an experimental profile obtained when the die stops early in the cycle (quasi simple mode) and a profile obtained when the die moves until the end of the cycle and goes on after the upper punch has stopped. In the latest case, a complete inversion of the profile is observed as a result of stress redistribution (at constant volume) due to friction between powder and the moving die. It is thus shown that the variation in the pellet shapes may be strongly correlated with the variation of a simple press parameter.

Conclusions

FEM simulations and experimental results clearly show that the profile of the diameter of a fuel pellet oxide depends on the die-pressing conditions as a consequence of the frictional forces between the die and the powder. Since friction depends on relative movements, the pressing cycle parameters (speed ratio and changes) may be adapted to minimize the density gradient and to lower the geometrical defects of the sintered pellet. In this study, the volume of material that has to be removed to obtain a perfect cylinder has been reduced by a factor of 3, just by fitting a single press parameter.

FEM simulation of the shaping process constitutes an interesting tool for performing one or even multiparameter optimizations (a specific algorithm has been implemented in the PreCAD® software). This code may be also used for design purposes (tools, cycles etc.)

or to perform sensitivity analysis (powder properties, friction coefficient etc.). It clearly shows that die compaction and sintering simulations of ceramics can be used to solve some technical problems.

References

[1] Baccino R. and Moret F. 2000. Numerical modeling of powder metallurgy processes. *Materials and Design*, **21**, 359–364.
[2] Dellis C., Abondance D., Baccino R., Bernier F., Moret F., 1996. PreCAD®, a computer assisted design and modelling tool for powder precision moulding. *Proceedings of the International Conference on Hot Isostatic Pressing*, (HIP' 96), 20–22 May 1996, Andover, Massachusetts, USA, 75–78.
[3] Pavier E. 1998. Characterisation of the behaviour of an iron powder for die compaction process. PhD thesis, Institut National Polytechnique de Grenoble, France.
[4] Roure S., Bouvard D., Donemus P. and Pavier E. 1999. Analysis of die compaction of tungsten carbide and cobalt powder mixtures, *Powder Metallurgy*, **42**, (4), 345–352.
[5] PM Modnet Methods and Measurement Group. 2000. Measurement of friction for powder compaction modeling–comparison with laboratories. *Powder Metallurgy*, **43**(4), 364–374.

5

Wet chemical synthesis for multinary oxide ceramics

E. Guenther and R. A. Dorey

High performance ceramics for functional application often have a great many constituent atoms arranged in specific configurations. To maintain the functional properties of these materials it is of great importance that these constituents are homogeneously distributed within the material. Promising methods to achieve this are wet chemical synthesise routes including sol-gel and related processing, coprecipitation and hydrothermal synthesis. Through these processes it is possible to synthesise homogeneous multinary ceramic powders, which are then available for further forming and processing (e.g. die pressing, tape casting, injection moulding etc.), or to directly generate layered sol-gel microstructures, which can be converted to a ceramic at a later stage to produce microstructured thin films.

This chapter gives an overview of the sol-gel technology for the formation of thin and thick ceramic films as well as on the sol-gel related two-step process for the production of multinary ceramic oxide powders. Examples of the synthesis of different functional ceramic materials and films are also given.

5.1 Introduction

Functional ceramics are classified as ceramics which possess properties in addition to their structural properties. Such properties may include: dielectric, piezoelectric, pyroelectric, ferroelectric, electronic and ionic conductivity and photonic characteristics. By making use of such functional properties, these ceramic materials can be integrated into engineering systems to produce devices that can act upon or sense the surrounding environment. Integrating these functional ceramics with microscale systems brings additional advantages of small size, low weight, increased robustness, lower power consumption and reduced cost.

To produce high-performance functional ceramics of high quality, synthesis of the mostly multinary, oxide starting powders requires a highly homogeneous distribution of the heterometallic components and in particular of the small fractions of dopant elements. For this purpose, modern production methods are increasingly applied as alternatives to traditional mixing and milling routes. Examples of these alternative processing techniques include the thermal two-step process [1–3], the coprecipitation method [4–6], hydrothermal synthesis [7–15] and the sol-gel process [16–27].

The traditional mixing and milling process is associated with contamination caused by abrasion of the container and milling media due to the long milling time and a potentially inhomogeneous element distribution as a result of solid–solid mixing, which does not allow for the ideal mixing of the individual particles. Moreover, solid-state reactions are diffusion-dependent and require high temperatures and long reaction times to ensure complete reactions.

Atomic level mixing and rapid syntheses at low temperatures may be achieved readily in liquids. This requires that the constituent elements be available in liquid or dissolved form. Such wet chemical methods [28–30] are often based on soluble salt or metal–organic compounds. The advantage of wet chemical synthesis routes is that the chemical compositions, including small fractions of additives, can be adjusted precisely to give a high degree of control over the resultant properties of the ceramics. Another advantage lies in the fact that finely dispersed and reactive powders with favourable sintering properties are obtained. The processes reported here, describe such wet chemical routes for the fabrication of a range of functional ceramic materials.

5.2 Thermal two-step process

The thermal two-step process results in powder with high chemical purity, microscopic homogeneity, and the fine scale required for the production of miniaturized parts e.g. for microsystems technology. In addition, production of such ceramic powders is characterised by a relatively short duration of the process. The two-step process consists of:

- the synthesis of a precursor solution consisting of heterometallic organocomplexes or salt solutions and their conversion into a powder; and
- direct thermochemical conversion of this precursor substance into a sinterable ceramic powder suited for the production of compacts in particular for ceramic parts.

The technical process consists of two thermal process steps to produce a powder: first the precursor solution (a mixture of carboxylates and alkoxides or certain inorganic salts) is spray dried. In the resultant powder, the individual components are distributed homogeneously. The second step consists of calcining the solid product, in a high-temperature fluidized bed, to yield the final ceramic. Here, complete conversion into a multinary oxide takes place, with the decomposition products being separated. After calcination, no further conditioning is required in most cases. A ceramic powder with good moulding and sintering properties is obtained.

5.3 Spray drying

Production of a solid precursor substance from the solution mixture of inorganic and organic salts or from the solution of heterometallic oligomers takes place by means of a drying process in a spray drying facility, mostly under inert gas. Under constant stirring, the solution is fed into a spraying wheel driven by nitrogen and split up into fine droplets due to the high rotating speed of this centrifugal sprayer. In the drying column, these droplets are passed by a hot, humidity-unsaturated gas as a drying agent. As a result of the large, free liquid surface area, the solvent is removed in a very short time and a finely dispersed powder is obtained. This powder mainly consists of spherical granules that are made up of nanoscaled primary particles.

5.4 Calcination

In a high-temperature fluidized bed, the precursor powder that has been isolated by spray drying is converted thermochemically into the oxide ceramic powder. In this facility [31], thermal processes are carried out at temperatures of up to 1200°C with gases of variable composition and variable solid quantities. Due to the favourable heat transfer properties of the turbulent bulk material in the fluidized bed, isothermal powder annealing and chemical reactions may take place in the bulk material and between it and the fluid. In such a facility, calcination of the precursor substance is carried out. The quartz glass reaction tube is located in a vertically arranged high-temperature furnace (Figure 5.1). Preheated synthetic air enters the tube from

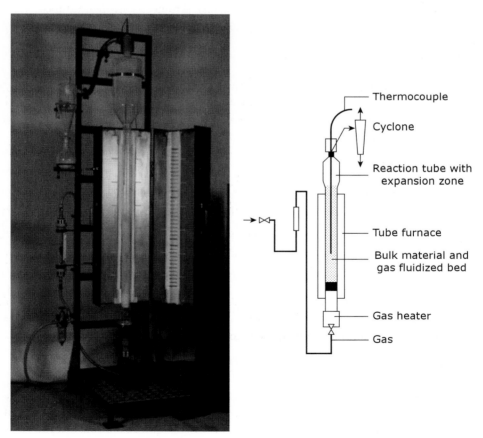

Figure 5.1 High-temperature fluidized bed calcination equipment.

Figure 5.2 Spherical powder particles produced via the two-step process: (a) PZT powder; (b) lead niobate (PN) powder.

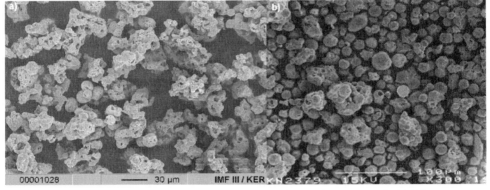

Figure 5.3 Agglomerated powder particles produced via the two-step process: (a) indium tin oxide (ITO) powder, a base material for electroconductive transparent thin films; (b) Pb-doped bismuth strontium calcium cuprate (BSCCO) powder, a base material for high-temperature superconductors.

below via a frit and keeps the bulk material suspended as a fluidized bed in the thermal zone of the furnace. Due to the geometry of the transition area between the reaction tube and an expansion part at the upper end, the flow rate of the fluid is reduced rapidly, such that the fine fraction of the bulk material entering this area may flow back to the fluidized bed and material discharge is prevented. When the gas flow rate is increased significantly, the material can be discharged pneumatically and the reacted product separated in a cyclone. Particle morphology of the powders obtained is largely spherical with a coarsely structured surface area (Figure 5.2). Such particles sometimes agglomerate in the fluidized bed (Figure 5.3).

5.5 Examples of powders produced

In general, the method described can be used to produce a number of multicomponent oxide ceramic powders, provided that the initial substances react chemically with each other and/ or form a common solution. The production of four electroceramic materials is described:

Ceramics based on PZT and lead niobate are mainly used as piezoceramic sensors, actuators and transducers.

5.5.1 Lead zirconate titanate

- To synthesize a neodymium-doped PZT powder of the stoichiometric composition of $Pb(Zr_xTi_{1-x})O_3 + yNd_2O_3$, the corresponding amounts of titanium ethylate $Ti(OC_2H_5)_4$, zirconium propylate $(Zr(OC_3H_7)_4)$, lead acetate $(Pb(CH_3COO)_4)$, and neodymium acetate $(Nd(CH_3COO)_3)$ in a methanol solution react with each other by reflux cooking in the presence of H^+ ions. Apart from ethyl acetate and propyl acetate, complex compounds with alkoxide ligands and central atoms of lead, zirconium and titanium bonded by oxygen atoms are formed [31, 32]. Distillation of the esters causes the reaction equilibrium to shift, as a result of which conversion is promoted. The solid precursor substance is obtained from the methanol solution of these oxoalkoxides by spray drying under inert gas. The powder thus produced is subjected to thermochemical conversion in a high-temperature fluidized bed at temperatures in the range 700–800°C with air as reaction gas. Mostly spherical PZT particles of perovskite crystal structure are produced (Figure 5.2a).
- The compaction behaviour of the PZT powder is very good: compacts were produced by sintering the spray granules for one hour at 1150°C. It was possible to achieve ceramics with 97% of the theoretical density of 7.99 g/cm³.
- Lead metaniobate (PN).
- For the synthesis of an undoped PN powder of the stoichiometric composition of $PbNb_2O_6$, the corresponding amounts of niobium ethylate $(Nb(OC_2H_5)_5)$ and lead acetate $(Pb(CH_3COO)_2)$ are dissolved in ethanol under nitrogen. This results in the formation of a solution mixture of both educts. The solid precursor substance is obtained from the ethanol mixture by spray drying under inert gas. The powder generated is converted thermochemically in a high-temperature fluidized bed at temperatures of 700°C for two hours. Air is the reaction gas. Mainly spherical PN particles of rhombohedral crystal structure are produced (Figure 5.2b).
- At room temperature, the ferroelectric, polymorphous PN is metastable. Above 1200°C, it possesses a tetragonal tungsten bronze structure. When slowly cooled down to below 1200°C, it assumes the rhombohedral structure which is paraelectric at room temperature. Rapid cooling from at least 1200°C to 600°C results in a stable tetragonal structure. Its Curie point is 560°C. Below that temperature, the tetragonal lattice is subjected to an orthorhombic distortion and the PN becomes ferroelectric [33]. In practice, this means that powders converted into stoichiometric $PbNb_2O_6$ in the fluidized bed and subjected to an annealing treatment for one hour at 1250°C have to be cooled down to room temperature rather rapidly. In the course of this process, the paraelectric francombite phase turns into the desired ferroelectric goodmanite phase [34]. To increase the compactability of PN powder treated in such a way, it is milled for a short while under slight mechanical loading. The resulting press granules can be compacted to a high degree even without the addition of compaction additives. The compacts can be sintered for two hours at 1280°C to densities greater than 90% of the theoretical density of 6.66 g/cm³.

5.5.2 Indium tin oxide

- Indium tin oxide (ITO) powders are used for the production of thin electrically conducting layers on glass substrates. These transparent and electrically conducting oxide films are also highly permeable to wavelengths in the visible range and reflect the infrared part well. Amongst other applications, the thin layers can be used for electrochromic and liquid crystal displays [35, 36].

- For the synthesis of an ITO powder of the stoichiometric composition of $(In_{2-X} Sn_X)O_3$, the initial substances used in the form of a hydroxide and acetate have to be dissolved by chemical conversion. For this purpose, the required amount of indium hydroxide $(In(OH)_3)$ is made to react with concentrated acetic acid. Together with the reaction water arising, the resultant indium acetate forms hydroxoacetates of variable composition. A colloidal solution is produced, which is mixed with an already prepared solution of tin acetate $(Sn(CH_3COO)_4)$ in glacial acetic acid. This sol is subjected to spray drying as rapidly as possible in order to prevent slow hydrolysis of the acetates. The precursor powder obtained is pyrolysed in the high-temperature fluidized bed at about 600°C under air.

- Reaction annealing in the fluidized bed is completed by a short-term temperature increase to 1000°C. It ensures that a pure oxide is generated in the form of a mixed crystal of cubic structure (Figure 5.3(a)). Cold compaction of the produced ITO powder still remains to be optimized. Although it is hardly agglomerated, it should be milled under a small mechanical load for a short time. The granules are highly compactable even without compaction additives. Compacts produced from such granules can be sintered for five hours at 1350°C to densities in excess of 96% of a theoretical density of 7.16 g/cm³. They are used as sputter targets to coat glass surfaces.

5.5.3 Bismuth strontium calcium cuprate

- Oxide ceramics that become superconducting when cooled down below a critical temperature (jump temperature), i.e. the electric resistance of which becomes nearly zero, include bismuth cuprates with earth-alkali dopants that may be substituted. The major representative of this class of high temperature superconductors (HTSC) is $Bi_2Sr_2Ca_{n-1}Cu_nO_{2n+4}$ with characteristic BiO_x layers and central CuO_2 planes in the crystal lattice. When the number of these planes is increased, a jump temperature of 110 K is reached for the so-called "triple layer", i.e. the 2223 phase of the bismuth strontium calcium cuprate (BSCCO) compounds. These HTSC powders are particularly suitable for the production of silver-coated superconducting wires or bands [38].

There are various options for the synthesis of a Pb-doped BSCCO powder of the stoichiometric composition of $(Bi,Pb)_2Sr_2Ca_2Cu_3O_{10+X}$. Starting from a pure acetate synthesis, the acetates of the elements involved may be substituted successively by their nitrates for pure nitrate synthesis. Alternatively, the hydroxides of these elements, which are soluble, may be used. They are mixed with the acetates and nitrates of the remaining elements.

The solvents for all synthesis paths are aqueous acetic acid or nitric acid solutions. The resultant salt solutions are spray-dried under air. The X-ray amorphous precursor powder is dried under synthetic air in the high-temperature fluidized bed for ten hours at temperatures of up to 200°C. The powder is then converted thermochemically into cuprates at 600°C and annealed at 800°C to form crystalline phases. Depending on the temperature, a phase mixture is obtained, which corresponds to a homolog cuprate series of the general formula $(Bi, Pb)_2Sr_2Ca_{n-1}Cu_nO_{2n+4}$, where n may be 1, 2, or 3. Preferably, the 2212 phase is formed. In the resultant calcinate that consists of agglomerated powder particles, the fraction of the 2223 phase may be increased largely by thermomechanical cycling for relatively short periods of time [37].

This reaction takes place rather rapidly, because the diffusion paths of the individual cations are short due to the highly homogeneous element distribution in the precursor powder. Spray granules obtained from such a powder have a good flowability and a high compactability (Figure 5.3b).

5.6 Sol-gel film formation

Sol-gel chemistry differs from that of a simple solution of metal organic complexes or metal salts in that clusters of reacted metal organic species are created where the ratio of metal and oxygen atoms is fixed by the reactions that have taken place [38]. In this way a homogeneous distribution of the constituent atoms is ensured unlike in simple solutions where some metal organics or salts may decompose preferentially to others resulting in local chemical inhomogeneities. Through careful processing and heat treatment of these sol-gel systems it is possible to create powders [39] of oxide ceramics where the size and morphology of the powders is controlled by the concentration of the sol-gel, the sol aging times and the thermal treatment temperatures.

Sol-gel processing makes use of reactive metal organic compounds such as metal acetates, propoxides, and butoxides. In the sol-gel process two mechanisms occur: alcoxide hydrolysis and condensation [40]. During the hydrolysis process the metal alcoxides react with water to become partially hydrolized through the replacement of one or more of the alcoxide side groups. The condensation reactions take place between the partially hydrated metal alcoxides and leads to their polymerization where atomic networks consisting of metal–oxygen–metal connections are formed and water or an alcohol are released. Such reactions ensure that there is atomic level mixing of the constituent atoms that make up the ceramic material. Sol-gel processing works for simple ceramics, such as ZrO_2 or TiO_2, but comes into its element for complex ceramics (particularly functional ceramics) where multiple metal atom species are required to form the required material. A prime example of this is PZT which contains Pb, Ti and Zr along with a number of dopants to modify the properties [41]. The processing of doped PZT clearly demonstrates the flexibility of the process, which allows compositions to be varied with ease.

The sol-gel systems are converted to an oxide ceramic through a thermal treatment termed pyrolysis. During heating a number of changes occur to the sol-gel system. Initially the solvents within the sol-gel system are evaporated leaving behind an amorphous

Figure 5.4 Example of sol-gel derived PZT powder.

metal–oxide–organic network of particles. On further heating the metal–oxide–organic network begins to decompose at temperatures in the range 200–500°C to produce amorphous metal–oxide particles. Further heating in the range 500–650°C converts these amorphous particles into the crystalline ceramic material (Figure 5.4) which can be used in subsequent ceramic processes.

While the need to create ceramic particles is important, the drive for greater integration and co-processing has led to the direct integration of ceramic films onto device substrates using wet chemical routes [41]. The real advantage of sol-gel processing, therefore, lies in the ability to produce thin films of ceramics [41] directly on a variety of substrate materials. Using sol-gel processing large-scale uniform thickness films to be deposited with ease. Such films can be used as coatings, e.g. wear or optical, or as a processing stage in the creation of MEMS. The coatings can be deposited using a variety of techniques including spin, spray and dip coating. Once the wet films have been deposited, processing proceeds through drying/gelation and pyrolysis to create an amorphous ceramic network bonded to the substrate. Subsequent layers are deposited to build up the film thickness. Finally, a crystallization thermal treatment is used to convert the amorphous ceramic network into the crystalline ceramic material (Figure 5.5). If a suitable substrate is selected it is possible to crystallize the ceramic through heterogeneous nucleation at the substrate interface and so grown columnar grains through the thickness of the film. This can lead to enhanced properties in certain functional ceramics such as PZT [41]. Where heterogeneous nucleation does not occur, homogeneous nucleation occurs and an equiaxed grain structure is obtained. Due to the stresses generated during the drying–pyrolysis–crystallization process sol-gel films are typically limited to a few micrometres in thickness.

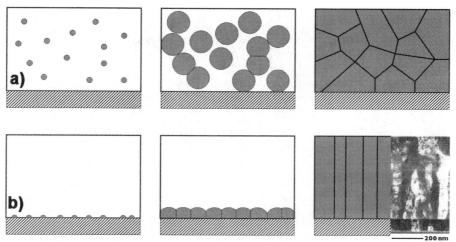

Figure 5.5 Schematic of thin-film sol-gel growth mechanisms leading to: (a) randomly orientated equiaxed structure; (b) preferentially oriented columnar structure.

To further increase the capabilities of sol-gel processing, a modification to the route has been proposed. This composite sol-gel approach [42] makes use of the low temperature processing capacity of sol-gel processing and incorporates the thick film forming capability of powder processing to yield a technology that is capable to producing films ranging in thickness form a few micrometres to a few tens of micrometres. In such films, powders produced from a variety of different, including wet chemical synthesis, routes can readily be used. These powders are mixed with the sol to create an ink or paste which is deposited using techniques such as spin coating, spraying, screen printing and dip coating. Again films are built up by successive deposition of numerous layers to chive the final thickness.

As with the sol-gel route the sol-gel phase nucleates and grows to produce a crystalline phase. However, unlike the pure sol-gel system heterogeneous nucleation occurs on the ceramic powder particles leading to lower crystallisation temperatures and a randomly orientated crystal structure (Figure 5.6).

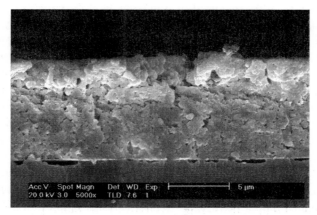

Figure 5.6 SEM micrograph of a cross-section of a composite sol-gel film.

Conclusions

Wet chemical synthesis of multinary oxide ceramics exhibits a number of advantages over conventional solid-state chemical routes for producing oxide ceramic materials. These advantages include:

- Chemical homogeneity and high purity of the final product ensured through high purity starting reagents.
- Flexibility in altering the composition and introducing small quantities of dopants to modify the properties of the parent material.
- A large variety of deposition processes allowing different morphologies of ceramic powder and film to be created.
- Low processing temperature and high sinterability of final powders ensures high quality final ceramic components.

The thermal two-step and sol-gel routes described here, highlight two of the wet chemical routes that can be used to produce complex oxide functional ceramics. Both routes involve first producing a solution of metal salts or metal–organic compounds which is then converted to the oxide ceramic using a thermal treatment. During this thermal treatment the metal salts or metal organic compounds decompose to produce the oxide ceramic. The main difference between the two routes is that in sol-gel synthesis the metal organic compounds react together to produce a 3D network of metallic atoms linked by oxygen atoms, while the metal salts remain as discrete atomic species. Thermal decomposition of these compounds can be accomplished at temperatures as low as 550°C and, due to the small diffusion distances, highly homogeneous ceramic materials can be produced at temperatures as low as 650°C.

References

[1] Günther E. and Maciejewski U. 1994. *Proceedings of 8th CIMTEC*, 29 June–4 July, 1994, Florence, Italy.

[2] Günther E., Linder D. and Maciejewski U. 1996. *Proceedings of CHISA'96*, 25–29, August, 1996, Prague, Czech Republic.

[3] Hennige V.D., Weddigen A., Günther E. and Ritzhaupt-Kleissl H.J. 1999. Production of piezoceramic powders by the thermal two-stage process. *Journal of Materials Science*, **34,** 3461–3465.

[4] Eyraud P., Eyraud L., Gonnard P. and Troccaz M. 1987. Piezoceramics of the lead metaniobate type prepared by a wet method. In *High Tech Ceramics*, Vincencini, P.ed, Elsevier, Amsterdam, 1487–1492.

[5] Nishio K., Sei T. and Tsuchiya T. 1996. Preparation and electrical properties of ITO thin films by dip-coating process. *Journal of Materials Science*, **31,** 1761–1766.

[6] Kayukawa T., Shigetani H. and Senna M. 1995. Preparation of high-density ITO ceramics by an *in situ* precipitation method. *Journal of Materials Science Letters*, **14,** 252–255.

[7] Lemoine L., Leriche A., Thierry T. and Tronc P. 1994. *Proceedings of the 8th CIMTEC*, 29 June–4 July, 1994, Florence, Italy.

[8] Lu, C.H. and Chyi N. 1996. Fabrication of fine lead metaniobate powder using hydrothermal processes. *Materials Letters*, **29,** 101–105.

[9]   Walton R.I. 2002. Subcritical solvothermal synthesis of condensed inorganic materials. *Chemical Society Reviews*, **31**, 230–238.

[10] Rabenau A. 1985. Die Rolle der Hydrothermalsynthese in der präparativen Chemie. *Angewandte Chemie*, **97**, 1017–1032.

[11] Eckert J.O. 1996. Kinetics and mechanisms of hydrothermal synthesis of barium titanate. *Journal of the American Ceramic Society*, **79**, 2929–2939.

[12] Oren E.E. and Tas A.C., *Metallurgical and Materials Transactions*, **30**(12), 1089–1093.

[13] Dias A., Buono V.T.L., Ciminelli V.S.T. and Moreira R.L. 1999. Hydrothermal synthesis and sintering of electroceramics. *Journal of the European Ceramic Society*, **19**, 1027–1031.

[14] Lencka M.M. and Riman R.E. 1993. Thermodynamic modeling of hydrothermal synthesis of ceramic powders. *Chemistry of Materials*, **5**(1), 61–70.

[15] Lencka M.M., Riman R.E. and Anderko A. 1995. Hydrothermal precipitation of lead zirconate titanate solid solutions: Thermodynamic modeling and experimental synthesis. *Journal of the American Ceramic Society*, **78**, 2609–2618.

[16] Gopalakrishnan J., Uma S., Kattsthuri Rangan K. and Bhuvanesh N.S.P. 1994. Soft-chemical routes to synthesis of solid oxide materials. In *Proceedings of the Indian Academy of Sciences*, 106(3), 609–619.

[17] Yoshimura M. 1989. Soft solution processing: Environmentally benign direct fabrication of shaped ceramics (Nano-crystals, whiskers, films, and/or patterns) without firing. *Journal of Material Research*, **13**(4), 796–802.

[18] Messing G.L. and Minehan W.T. 1989. *Sol-Gel Science and Technology*, Aegerter M.A., Jafelicci Jr M., Souza D.F. and Zanotto E.D.,(Eds.), *World Scientific*, Hackensack, New Jersey, 402.

[19] Yamaguchi O. and Mukaida Y. 1990. Formation and characterization of alkoxy-derived lead metaniobate. *Journal of Material Science Letters*, **9**, 556–558.

[20] Corriu R.J.P. 2000. *Angewandte Chemie*, **112**, 1432–1455.

[21] Bradley D.C., Mehrotra R.C. and Gaur D.P. 1978. *Metal Alkoxides*, Academic Press New York.

[22] Livage J., Henry M. and Sanchez C. 1988. *Progress in Solid State Chemistry* **18**, 259–341.

[23] Corriu R., Leclercq D., Vioux A., Pauthe M. and Phalippou J. 1988. *Ultrastructure Processing of Advanced Ceramics*. **113**, Wiley, Chichester, United Kingdom.

[24] Brinker C. J. and Scherer G.W. 1990. *The Physics and Chemistry of Sol-gel Processing*. Academic Press, San Diego, California, USA.

[25] Hench L.L. and West J.K. 1990. *Chemical Reviews 90*, 33–72.

[26] Chandler D.C., Roger C. and Hampden-Smith M.J. *Chemical Reviews 93*, 1205–41.

[27] Ghosal S., Emami-Naeini A., Harn Y.P., Draskovich B.S. and Pollinger J.P. 1999. *Journal of the American Ceramic Society*, **83**(3), 513–520.

[28] Stuijts A.L. 1970. *Science of Ceramics*, **5**, 335–362.

[29] Koppens L. 1976. *Science of Ceramics*, **8**, 101–109.

[30] Johnson D.W. and Gallagher P.K. 1978. *Ceramic Processing before Firing*. Wiley, New York, USA, 125–139.

[31] Mehrotra R. C. 1993. *7th International Workshop on Glasses and Ceramics from Gels*, 19–23 July, 1993, Paris, France.

[32] Hubert-Pfalzgraf L. G., Boulmaaz S., Daniele S. and Papiernik R. *7th International Workshop on Glasses and Ceramics from Gels*, 19–23 July, 1993, Paris, France.

[33] Sholokhovich M. L. and Dugin V. E. 1983. *Russian Journal of Inorganic Chemistry*, **28**, 890–892.

[34] Jaffe H., Jaffe B. and Cook W. R. 1971. *Piezoelectric Ceramics*. Academic Press, London, England, United Kingdom.

[35] Vossen J.L. 1971. RF sputtered transparent conductors the system In_2O_3-SnO_2. *RCA Review*, **32**, 289–295

[36] Ghosh A. K., Fishman C. and Feng T. 1978. SnO_2/Si solar cells—heterostructure or Schottky-barrier or MIS-type device. *Journal of Applied Physics*, **49**, 3490.

[37] Singh J. P., Joo J., Vasanthamohan N. and Poeppel R. B. 1993. Role of Ag additions in the microstructural development, strain tolerance, and critical current density of Ag-sheathed BSCCO superconducting tapes. *Journal of Materials Research*, **8**, 2458–2464.

[38] Sayer M., Yi G. and Sedler M. 1995. Comparative sol gel processing of PZT thin films. *Integrated Ferroelectrics*, **7**, 246–258.

[39] Linardos S., Zhang Q. and Alcock J.R. 2007. An investigation of the parameters effecting the agglomerate size of a PZT ceramic powder prepared with a sol–gel technique. *Journal of the European Ceramic Society*, **27**, 231–235.

[40] Whatmore R.W., Zhang Q., Huang Z. and Dorey R.A. 2003. Ferroelectric thin and thick films for microsystems. *Materials Science in Semiconductor Processing*, **5**, 65–76.

[41] Calame F. and Muralt P. 2007. *Applied Physics Letters*, **90**, 062907 (29 July, 2007).

[42] Dorey R.A., Stringfellow S.B. and Whatmore R.W. 2002. Effect of sintering aid and repeated sol infiltrations on the dielectric and piezoelectric properties of a PZT composite thick film. *Journal of European Ceramic Society*, **22**, 2921–2926.

6

Synthesis of nanoscaled powders for applications in microsystems technology

D. V. Szabó, M. Bruns,
R. Ochs and S. Schlabach

In this chapter a short overview on the current state of gas phase processes for the synthesis on nanoparticles is presented. The Karlsruhe microwave plasma process, a gas phase process in a nonthermal plasma at low pressures is described in detail, as this process allows us to obtain small nanoparticles with narrow particle size distributions. The physical basics of this process are explained. Examples of nanoparticles, synthesized with the microwave plasma process, and possible application of such nanoparticles in microsystems technology, especially in microoptics and sensor technology are given.

6.1 Introduction

Nanoparticles are particles with diameters below the micron dimension: in general their size is below 0.1 μm (100 nm). A more stringent definition considers nanoparticles as particles with properties depending directly on their size. Examples include optical, electrical, or magnetic properties. Therefore, in many cases the latter definition restricts nanoparticles to particles with sizes below 10–20 nm. Additionally, with decreasing particle size the ratio of surface/volume increases, so that surface properties become crucial. In this context it is important to realize that 5 nm particles consist of only a few thousand atoms or unit cells and possess approximately 40% of their atoms at the surface. In contrast, 0.1 μm particles contain some 10^7 atoms or unit cells, and only 1% of their atoms at the surface. For applications in microsystems technology, e.g. in microoptics, homogeneous and extremely narrow particle or agglomerate size distributions with mean sizes below 1/10 of the applied wavelength are of great importance. Therefore one needs synthesis processes which are able to produce nonagglomerated nanoparticles significantly below 20 nm with narrow particle size distribution.

Numerous methods are available for the synthesis of nanoparticles with various requirements. They can roughly be classified into:

- Top-down approaches such as mechanical processes [1], starting from conventional, coarse-grained materials. Here, the particles are broken into ultrafine grains by introducing high energy attrition. Unfortunately, these methods lead to hard agglomerates thus means they cannot be separated.

Therefore, mechanical processes are not well suited to producing nanopowders for applications in microsystems technology.

- Bottom-up approaches include wet chemical methods [2], and various gas phase methods [3–27]. Both synthesis strategies start from atomic or molecular precursors to create larger building blocks. In general, wet chemical routes lead to agglomerated particles whereas a reduced agglomeration is attained when using gas phase reactions.

6.2 Gas phase methods for the synthesis of nanoparticles

In the following sections gas phase methods will be described in more detail, as they are able to produce small nanoparticles with reduced agglomeration for the needs in microsystems technology.

The most common and historically most important gas phase process is the inert gas condensation (IGC) [3–5]. For metal nanoparticle fabrication a metal source is evaporated thermally in a vacuum chamber, filled with inert gas of typically 1–20 mbar pressure. The metal nanoparticles nucleate in the thermal zone just above the evaporative source, due to interactions between the hot vapour species and the much colder inert gas atoms. After condensation, coagulation and agglomeration the particles are transported by thermophoresis to a cold finger or substrate. For collection, the particles are scrapped off. Ceramic nanopowders are usually produced by a two-stage process: evaporation of a metal source, or preferably a metal suboxide of high vapour pressure, followed by slow oxidation to develop the desired ceramic powder nanoparticles. Physical vapour synthesis (PVS) and chemical vapour synthesis (CVS) are variants of the IGC process. In PVS an electron beam [6] or a laser beam [7] can be used to evaporate the metal. For CVS [8, 9] the evaporating heating source is replaced by a hot-wall tubular reactor which decomposes a precursor/carrier gas mixture to form a continuous stream of clusters or nanoparticles. Typically metal chlorides, metal carbonyls or metal-organic compounds are used as precursors.

The process with the highest technological and commercial importance is the flame synthesis [10, 11], an aerosol process. It is commercially used for the large-scale synthesis of submicron carbon black [12], titania pigment (TiO_2), and silica (SiO_2) [13]. In the basic set-up a flame is fueled with hydrogen (H_2), or methane (CH_4) with additions of oxygen (O_2), and argon (Ar). For the synthesis of TiO_2 titanium tetrachloride ($TiCl_4$) is introduced as vapour into the flame, for the synthesis of SiO_2, either silane (SiH_4), or silicon tetrachloride ($SiCl_4$) are introduced as vapours into the flame to form the nanoparticles. A broad variation of setups and synthesis parameters results in particles with different morphology, size, and crystal structure. A very comprehensive overview on flame synthesis has been given by Kammler and co-authors [11].

Powders synthesized by these methods are suitable in microoptics for tuning the refractive index of polymers [1, 2] (see Figure 9.7 of Chapter 9 in this book). However, there are some limitations: as the agglomerates are quite large, the transmission of the modified polymer is already reduced at low particle concentrations. Nanoparticles made in flames are also frequently used as gas sensing films. Examples include the nanoparticles with primary particle sizes of around 10–20 nm used for thick film sensor deposition by drop coating methods [3, 4] or the direct deposition of flame made nanoparticles with primary particle size of around 10 nm forming 10–40 μm thick films on interdigitated Pt-electrodes [18].

Another approach for the gas phase synthesis of nanoparticles is the broad field of non-thermal plasma-methods [19–27], using either microwaves [19–21], radio frequency (RF) [22] or dielectric barrier discharge [23] for plasma generation. A common feature of all nonthermal plasma methods is, that precursors are evaporated, introduced into the reaction zone (plasma) where a chemical reaction takes place in a few milliseconds. With nonthermal plasma methods oxides [19, 20], nitrides [21, 24], carbides [25], pure Si [22], sulfides and selenides [26] and core/shell nanoparticles [27] have successfully been produced.

Amongst the gas phase methods, nanoparticle synthesis in microwave plasmas exhibits the lowest tendency towards agglomeration. This is because the temperature of electrons is several times higher (some 10,000 K) than the temperature of ions or uncharged species (generally around room temperature): the microwave plasma is quite "cold". Additionally the nanoparticles leave the plasma zone as isolated particles probably repelling each other due to equal charging [28, 29]. Therefore, this route fulfills the basic requirements for the synthesis of nanoscaled powders for applications in microsystems technology, especially in microoptics and in sensor technology. For this reason it is worthwhile to have a more detailed look into the features of the microwave plasma process, which has been developed at the Karlsruhe Research Centre (Forschungszentrum Karlsruhe, (FZK)).

6.3 The Karlsruhe microwave plasma process

The Karlsruhe microwave plasma process (KMPP) is a nonthermal, low-pressure process. It is highly applicable for the synthesis of nanoparticles with particle size <10 nm and very narrow particle size distribution [20]. Beside the low intrinsic temperature in the plasma, a central feature of this process is the short residence time of the reactants in the plasma of only a few milliseconds. Due to the combination of short residence time in the reaction zone, low temperature, and equally charged particles, growth and formation of hard agglomerates is reduced. Another advantage of this process is that the synthesis of ceramic/ceramic, and inorganic/organic hybrid core/shell nanoparticles with sizes below 10 nm is possible. In such a case each nanoparticle is covered either with a ceramic or an organic layer, adding functionality to each particle [30, 31].

To clarify the advantages of this process, it is necessary to understand the physical basics in a microwave plasma. The temperature in a microwave plasma is significantly lower than in an AC or DC plasma because the energy E, transferred to a charged particle of a mass m in an oscillating electrical field is inversely proportional to its mass and the squared frequency (Equation (6.1)):

$$E \propto \frac{1}{m \cdot f^2} \tag{6.1}$$

As the mass of the electrons is small compared to that of ions, at high frequencies a substantially larger amount of energy is transferred to the electrons, as compared to the energy transferred to the ions. Besides ions and free electrons, a microwave plasma consists of neutral gas species, dissociated gas and precursor molecules. Therefore, collisions between charged and uncharged particles influence the energy transfer to the particles. In this case, the collision frequency z has to be considered (Equation (6.2)):

$$E \propto \frac{1}{m} \frac{z}{f^2 + z^2}$$
(6.2)

This does not alter the mass relationship of the energy transfer; however, it introduces an additional dependency of the collision frequency in the plasma (the collision frequency z is proportional to the gas pressure). Thus, the energy of the free electrons, the "temperature", is significantly higher than the "temperature" of the ions [32]. Additionally, the "temperature" of the neutral gas molecules, obtaining their energy from collisions with the charged ones, is even lower. This leads to a significantly lower "overall temperature" of a gas passing a microwave plasma compared to DC or RF plasma, where temperatures in the range 5000–15,000 K are obtained. According to formula (2) the gas pressure has a significant influence on the synthesis as the collisions between energy-rich charged particles and uncharged ones control the energy transfer. Analysing Equation (6.2) one learns that for $z < f$ the energy transfer increases, whereas for $z > f$ the energy transfer decreases with increasing gas pressure in the KMPP. Figure 6.1 shows schematically the relationship between collision frequency and energy transfer E on a charged particle of the mass m for two different industrial microwave frequencies.

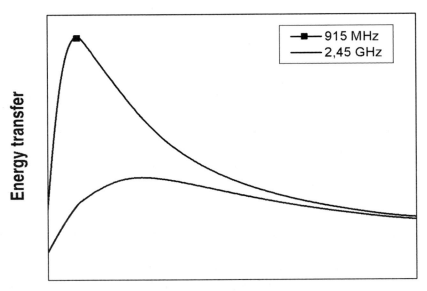

Collision frequency

Figure 6.1 Relationship between collision frequency and energy transfer, according to Equation (6.2), using the two common industrial microwave frequencies of 915 MHz and 2.45 GHz. Lower temperatures are achieved with the higher frequencies.

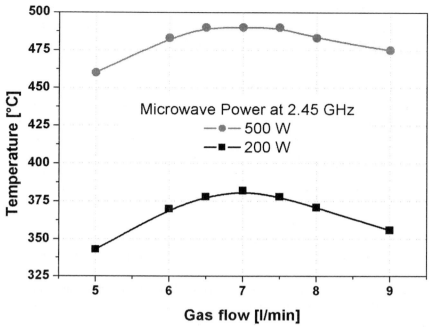

Figure 6.2 Experimentally determined temperatures as a function of gas flow in a microwave plasma reactor using 2.45 GHz at two different powers. Gas flow (related to pressure in system) is directly related to collision frequency. Run of the curves are in good agreement with the theoretical curve shown in Figure 6.1.

The temperature is determined directly after the plasma zone and can be adjusted in a range of approximately 150–900°C depending on the microwave frequency, the gas pressure and the input of microwave energy. Thus, frequency, gas pressure, energy input, and precursor concentration are the main parameters which must be adjusted in order to maintain optimal conditions for the necessary chemical reactions leading to the desired products. Figure 6.2 shows experimentally determined temperatures depending on the gas flow (equivalent to pressure or collision frequency), using two different constant microwave powers at 2.45 GHz. The data were determined without chemical reaction. With increasing microwave power an increasing temperature is measured. First increasing and then decreasing temperature with increasing gas pressure (collision frequency) is observed. This is in good agreement with the relationship shown in Figure 6.1.

The realization of an experimental set up is usually done with two standard industrial microwave frequencies: 0.915 GHz with a wavelength of approximately 30 cm and 2.45 GHz with a wavelength of approximately 10 cm, corresponding to the household microwave oven. Due to the smaller energy transfer at higher frequencies (see Equations (6.1) and (6.2) and Figure 6.1) lower temperatures can be realized using 2.45 GHz. This is of particular importance for hybrid inorganic/organic nanoparticles as well as for reduced hard agglomeration.

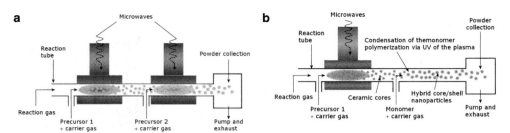

Figure 6.3 Schematic depiction of the experimental set-up for the microwave plasma synthesis:
(a) ceramic core/shell nanoparticles, ceramic core and ceramic shell are synthesized in the plasma;
(b) Hybrid core/shell nanoparticles, the ceramic core is synthesized in plasma, whereas organic
compounds condense outside plasma on ceramic nanoparticles leaving the plasma.

The equipment consists of different, commercially available microwave components as the
microwave generator, wave guides, directional coupler, isolator, and tri-stub-tuner. The main
part of the system is the specially designed cavity, using a rotating TE_{11} mode [33]. This cavity
has the advantage of easy plasma ignition and good microwave efficiency during the synthesis
process.

A schematic drawing of the KMPP process for the synthesis of ceramic core/shell as
well as hybrid core/shell nanoparticles is given in Figure 6.3 (a) and (b), respectively. The
basic element is a reaction tube made of quartz glass crossing the cavity. The plasma is
ignited at this intersection. Volatile and water-free precursors (e.g. chlorides, carbonyls,
metal alkoxides, or metal alkyls) are evaporated outside the reaction tube and mixed with
an inert carrier gas. The components are introduced as gases into the system just in front of
the plasma zone. The chemical reaction in the gas phase and the formation of nanoparticles
occurs here. For the synthesis of oxides one uses Ar as a carrier gas and a gas mixture of
80 vol% Ar and 20 vol% O_2 as reaction gas. The reaction gas flow is usually set to values in
the range 5–10 l/min, leading to pressures below 20 mbar.

By using consecutive reaction zones, core/shell nanoparticles and multilayer nanoparticles can be produced in consecutive synthesis steps. Core/shell and multilayer nanoparticles
may be either ceramic nanoparticles, consisting of a ceramic core and a ceramic shell, or
hybrid nanoparticles, consisting of a ceramic core and an organic shell. Figure 6.3 (a) shows
schematically the setup for the synthesis of ceramic core/shell nanoparticles. Cores and shell
are synthesized in the plasma. The cores are formed by homogeneous nucleation, whereas
the coating is deposited via heterogeneous nucleation on the cores. In contrast, the organic
shell of hybrid nanoparticles condenses (also via heterogeneous nucleation) and polymerizes
outside of the plasma zone on the cores synthesized in the plasma. This is shown schematically in Figure 6.3 (b).

The nanoparticles are usually collected by thermophoresis. As they leave the reaction
zone as isolated particles, they form only soft (van der Waals) agglomerates during
thermophoresis. The formation of hard agglomerates is not possible, due to the quite low
temperatures. In conclusion, the problem of agglomeration still persists, but in a reduced
form. Particles collected by this way may be incorporated into polymers or resins to form

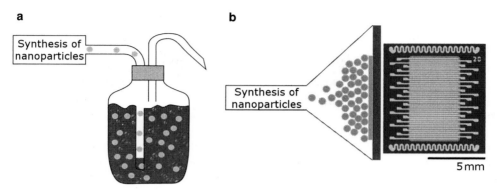

Figure 6.4 Sketches of (a) precipitation of nanoparticles in a liquid. Liquid may be a polymer resin, or a diethylene glycol. Rectangle named "Synthesis of nanoparticles" may be a setup equivalent as shown in Fig. 6.3 (a or b); (b) Deposition of nanoparticles on substrates to build thin nanoparticle films. Rectangle named "Synthesis of nanoparticles" may be a setup equivalent as shown in Fig. 6.3 (a or b). Right-hand side shows a pre-structured sensor field. Nanoparticles are deposited on top of such a sensor structure.

microtechnological parts for modifying optical or mechanical properties. With a modified set-up direct precipitation of nanoparticles in liquids is possible. This is shown schematically in Figure 6.4 (a). Suspensions can be immediately produced by this method. This collection method strongly reduces the tendency of nanoparticles to agglomerate in liquids, polymers and resins. Therefore it is well suited for applications in microoptics. The third collection possibility allows the deposition of thin and thick nanogranular layers on substrates, shown schematically in Figure 6.4 (b). This is appropriate for the deposition of nanogranular layers on substrates for instance in sensor applications.

6.4 Examples of nanoparticles and their application possibilities in microsystems technology

The following sections will consider examples of different nanoparticles produced by the KMPP with their application potential in microsystems technology. First, pure powders particles, and particles collected in liquids, as well as their application potential in microoptics will be presented. Finally, sensor application of SnO_2 nanoparticles will be demonstrated. The examples presented here are taken from ongoing research projects at the FZK, Institute of Materials Research III. Hence, the results are sometimes only single viewpoints, as not all the experiments to characterize a sample or a material have yet been carried out. But, the results show trends which promise the success of applying nanoparticles made by the KMPP in microsystems technology.

High resolution transmission electron microscopy (HRTEM), scanning transmission electron microscopy (STEM), and X-ray diffraction (XRD) are mainly applied to characterize the structure and morphology of "as produced" nanoparticles and nanocomposites. These are the most important tools for the characterization of nanoparticles. Scanning electron microscopy (SEM) is used to characterize the morphology of nanogranular layers.

6.4.1 Application potential of nanoparticles in microoptics

To clarify the problem of commercial nanopowders (large particles, broad particle size distribution), Figure 6.5 shows a typical transmission electron micrograph of commercially produced nanopowders. In this case SnO_2 (Sigma-Aldrich, product number 549657) is used. Bulk SnO_2 has a refractive index of 1.997 [34]; therefore SnO_2 nanoparticles are well suited to applications in microoptics. The micrograph shows large particles with sizes up to 200 nm, and a broad particle size distribution. Nanoparticles of such quality are not really suitable for high-end applications in microoptics.

ZrO_2, zirconium dioxide, also has an inherent high refractive index (see Table 9.3 in Chapter 9 of this book). The reaction of Zr-t-butoxide, $Zr(OC_2H_5)_4$, in a microwave plasma at temperatures around 300°C and pressures in the range 5–10 mbar leads to a very fluffy, white powder. The size of the nanoparticles is generally in the range of 2–5 nm. Figure 6.6(a) shows a typical high resolution electron micrograph of ZrO_2 nanoparticles, produced with the KMPP. In contrast to the commercial nanoparticles, shown exemplarily with SnO_2 in Figure 6.5, the ZrO_2 nanoparticles produced with the KMPP show primary particle sizes of around 5 nm and a very narrow particle size distribution. These features are typical for all nanoparticles produced with the KMPP. Lattice fringes with a spacing of 0.29 nm (imaging the [111] lattice planes) can clearly be seen in some particles. Electron diffraction as well as XRD (Figure 6.6(b)) indicates the cubic/tetragonal phase, although the nanoparticles are not stabilized by dopants. Due to the broad lines, the cubic and tetragonal phases cannot be distinguished. Primary particle sizes of ≤ 5 nm are determined from the diffraction images by evaluating the full width half maxima (FWHM) of the diffraction peaks with the Scherrer formula.

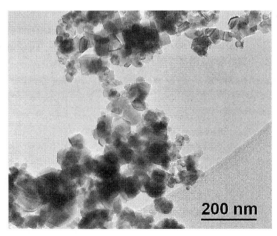

200 nm

Figure 6.5 Typical transmission electron micrograph of commercially produced nanopowder, using SnO_2 (Sigma-Aldrich, product number 549657) as an example. This micrograph clearly shows large particle sizes and broad particle size distribution.

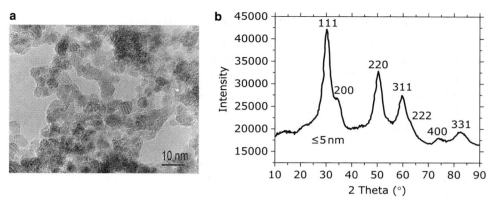

Figure 6.6 (a) Typical high resolution TEM micrograph of nanoscaled ZrO_2 particles, produced by the KMPP. Particles are $\leq$ 5 nm in diameter and exhibit a very narrow particle size distribution. They are crystalline: fringes visible in some particles have a distance of 0.29 nm and are [111] lattice planes of the ZrO_2; (b) Typical XRD diagram of ZrO_2 nanoparticles synthesized with the KMPP. The FWHM are broad, due to the small primary crystal size of the particles. Diagram clearly shows that ZrO_2 nanoparticles are crystalline. ZrO_2 is in the cubic/tetragonal phase. The monoclinic phase, the standard room temperature modification in coarse grained ZrO_2, is not detectable. Due to line broadening, phases cannot be distinguished. The evaluation of FWHM by the Scherrer formula leads to crystallites $\leq$ 5 nm.

As can be seen in the high resolution image the nanoparticles are not isolated. The primary particle size is small enough to avoid the Raleigh scattering: the particles are smaller than 1/10 of the applied wavelength (λ/10 criterion), but the agglomerates are significantly larger. Thus, pure powders are not really applicable in microoptics, e.g. for optical waveguides with tunable refractive indices.

Nevertheless, significant effects can be observed in refractive indices and in transmission. As reported in Chapter 9 of this book, successful attempts have been made to apply pure, commercial nanopowders with primary particle sizes around 30 nm in microoptics [14, 15]. Using hybrid core/shell nanoparticles made by the KMPP seems more promising as the polymer coating is supposed to improve the incorporation and dissociation of the nanoparticles into a polymer resin. For these experiments Ta_2O_5 with primary particle sizes below 5 nm was selected as the ceramic core, as Ta_2O_5 has an inherent refractive index of 2.11 in thin films [35]. The monomer for the coating was methylmethacrylic acid (MMA). The refractive index of polymethylmethacrylate (PMMA) resin can be increased, even with small additions ($\leq$ 1 wt%, corresponding to $\leq$ 0.15 vol%) of Ta_2O_5/PMMA hybrid nanoparticles [36]. Details of the processing chain to introduce nanoparticles into polymer resins can be found elsewhere [15]. Figure 6.7 (a) shows the increase of refractive index in nanoparticle/ PMMA composites, measured at 633 nm, for hybrid Ta_2O_5/PMMA nanoparticles compared to commercial Al_2O_3-C (primary d_{50} = 13 nm), ZrO_2 (primary d_{50} = 30 nm) and TiO_2-P25 nanopowders. Data for commercial powders are taken from Chapter 9 (Figure 9.7) of this book and from references [14, 15]. The actual increase in the refractive index is slightly higher than the increase measured with commercial ZrO_2 or TiO_2 at comparable weight

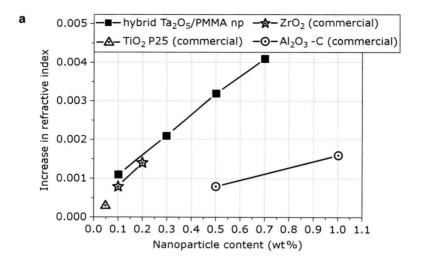

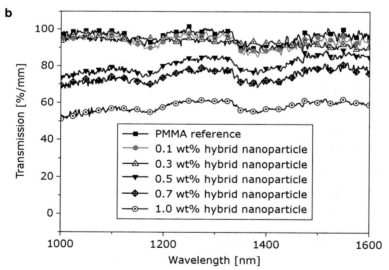

Figure 6.7 (a) Increase of refractive index (@ 633 nm) for different nanoparticle/ PMMA composites. Hybrid Ta_2O_5/PMMA nanoparticles were synthesized with the KMPP; data for refractive index of hybrid nanoparticles are taken from [36], commercial nanoparticles are described in detail in [14, 15] and in Chapter 9 of this book. The hybrid nanoparticles exhibit the best effect concerning refractive index; (b) Transmission in near infrared as a function of wavelength for hybrid nanoparticle/ PMMA composites containing different amounts of hybrid nanoparticles [36]. Up to a particle loading of 0.3 wt% transmission is not degraded compared to pure PMMA. Addition of 1 wt% of hybrid nanoparticles reduces the transmission significantly, but less than using commercial nanoparticles [14].

content. The increase in refractive index is significantly higher than that observed with commercial Al_2O_3-C at higher weight fraction. This is evident, because the refractive index of Al_2O_3-C is lower than the refractive index of Ta_2O_5. The transmission (Figure 6.7(b)) of the Ta_2O_5-composite compared to the PMMA remains nearly unchanged up to a particle concentration of 0.3 wt% [36]. Hence, for a low hybrid nanoparticle content the process of particle incorporation works relatively well. With increasing hybrid nanoparticle content the transmission is reduced. Compared to nanocomposites made of PMMA and commercial ZrO_2 nanoparticles [14] the transmission of the Ta_2O_5-nanocomposite is significantly higher at comparable particle concentrations. This is probably an effect of the small hybrid nanoparticles. Additionally, it has to be considered that the transmission depends strongly on the processing [15].

The significant reduction in transmission is due to the presence of agglomerates larger than $\lambda/10$. These agglomerates are a result of the typically high agglomeration tendency of nanoparticles and also the non optimal processing of nanoparticle incorporation. To visualize such agglomerates by TEM 50 nm thin samples are prepared by ultramicrotomy. Figure 6.8 shows a STEM dark field micrograph of a nanocomposite containing 1 wt% of Ta_2O_5/PMMA hybrid nanoparticles in the PMMA resin. In this micrograph the hybrid nanoparticles appear bright in the surrounding PMMA resin which is disrupted by the

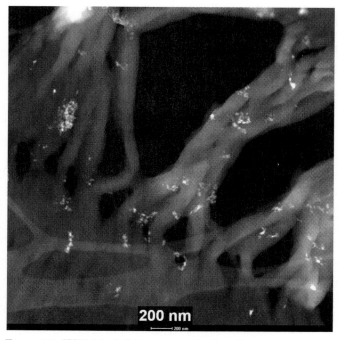

200 nm

Figure 6.8 STEM dark field micrograph of an ultramicrotome section of a hybrid nanoparticle/PMMA composite, containing 1.0 wt% of nanoparticles, used to determine of optical properties. Presence of agglomerates larger than 200 nm explains the significant reduction in transmission shown in Fig. 6.7 (b).

a

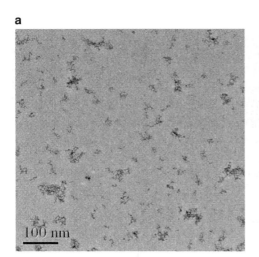

b

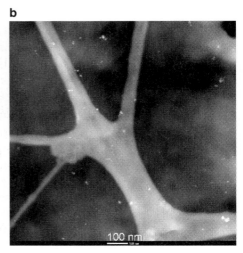

Figure 6.9 (a) TEM image of ZrO_2 nanoparticles, directly precipitated in diethylene glycol after synthesis process, as shown schematically in Fig. 6.4 (a). Dimension of the agglomerates is reduced; (b) Representative STEM dark field image of a nanocomposite made of directly deposited γ-Fe_2O_3 in EPR 162®. Nanoparticles are the bright dots in the image. The size of the agglomerates is considerably reduced.

sectioning technique. Very small agglomerates, but also some agglomerates with dimensions of ≥ 200 nm, can be found. The presence of larger agglomerates explains the decrease in transmission.

As already shown schematically in Figure 6.4 (a), one way to reduce the agglomeration tendency of nanoparticles is the instantaneous precipitation of the particles from the gas stream into liquids or resins. In this case the nanoparticles spread in the liquid before they have the opportunity to significantly agglomerate. As an example for the reduced agglomeration of nanoparticles Figure 6.9 (a) shows a TEM micrograph of ZrO_2 nanoparticles, collected in diethylene glycol. In this overview image all the agglomerates are smaller than 100 nm. Figure 6.9 (b) presents a STEM dark field micrograph of a 50 nm thin ultramicrotome section of a cured nanocomposite made of Fe_2O_3 nanoparticles, collected in EPR162®, an epoxy resin. It can be clearly seen that the nanoparticles are regularly distributed in the resin. These figures demonstrate the potential of nanoparticle collection in liquids for applications in microoptics.

6.4.2 Application of nanoparticles in sensor technology

Tin dioxide, SnO_2, is a wideband gap n-type semiconducting metal oxide, which is commonly used as gas-sensitive material. Sensor layers are usually produced by sputtering techniques [37–39], leading to grain sizes in the range 30–300 nm. Due to their high surface/volume ratio, nanoparticles with sizes < 10 nm are expected to be promising for an improvement in sensor performance. The signal strength of metal-oxide gas sensors has successfully been improved further by decreasing the crystallite size of the metal oxide, as shown by Kennedy *et al.* [40]. Nanoparticles have already be applied via spin-coating [41],

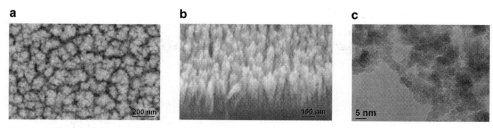

Figure 6.10 (a) Typical SEM image of the surface (top view) of deposited SnO_2 nanoparticles on a sensor structure; (b) SEM image of a broken edge of a deposited SnO_2 nanoparticle layer under an angle of 45°. This image clearly shows the club-shaped growth of the layer, and the layer thickness of around 200 nm; (c) Typical HRTEM image of SnO_2 nanoparticles produced by the KMPP and used for the formation of nanoparticle layers. In some particles, lattice fringes with a distance of 0.33 nm, corresponding to [110] lattice planes, are visible. Difference in quality, compared to commercial nanoparticles (shown in Fig. 6.5) is clearly visible.

drop-coating [16, 42] or screen-printing [42, 43] to form the sensing layer. But, deposition processes using colloidal solutions of nanoparticles pose the problem of particle growth, as one has to eliminate the binding phase by a temperature treatment. A further problem may be due to residuals of the binding phase.

These problems can be avoided by direct deposition of nanoparticles on a substrate [44], as shown schematically in Figure 6.4 (b). Figure 6.10 (a) shows a typical SEM top view image of an approximately 200 nm thin SnO_2 nanoparticle porous layer, deposited via mask technique on a Si-substrate already equipped with an electrode array consisting of 38 single detector fields. Figure 6.10 (b) shows a SEM image of a typical morphology of such thin layers viewed under an angle of 45°. A homogeneous, uniform particle deposition and porous structure of the layers is clearly visible in both images. The layer is around 200 nm thick. These thin layers exhibit club-shaped growth. The clubs are around 90 nm wide. Such layers typically grow in a time scale of 2–5 min, depending on the experimental conditions. Figure 6.10 (c) shows a typical TEM micrograph of the SnO_2 nanoparticles. Again it becomes clear that, in contrast to the commercial SnO_2 nanoparticles shown in Figure 6.5, small particles and a narrow particle size distribution are the predominant features of the particles produced by the KMPP. The SnO_2 particles are crystalline; they exhibit tetragonal cassiterite structure. The size of the nanoparticles can be estimated to be 3–4 nm. This is in good agreement with the results of electron as well as XRD. Particles of that size have approximately 50–70% of their atoms at the surface. Thus they seem ideal for applications in sensor technology.

However, the problem of particle growth during sensor application at operating temperatures in the range of 300–350°C still exists. Annealing experiments show that, uncoated SnO_2 nanoparticles grow slightly to approximately 5–6 nm after 3 months. This minor problem can be overcome by depositing core/shell SnO_2/SiO_2 nanoparticles instead of bare particles, using the SiO_2 shell as a growth barrier [44].

Prototype sensors made of SnO_2 nanoparticles and SnO_2/SiO_2 core/shell nanoparticle thin layers have been successfully fabricated [45]. In the first case (bare particles), the size of

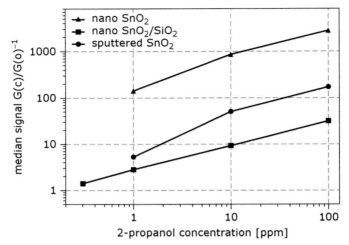

Figure 6.11 Sensitivity of sensors made from bare SnO_2 and core/shell SnO_2/SiO_2 nanoparticles in comparison to sputtered sensor KAMINA as function of the gas concentration using iso-propanol as test gas, after [45].

the SnO_2 nanoparticles is around 3–4 nm. In the latter case (core/shell particles) the core has a similar diameter of approximately 3 nm; the coating is 0.5 nm thin. First sensor tests are very promising in terms of sensitivity as well as long-term stability. Figure 6.11 compares the sensitivity of three types of sensors for iso-propanol: a standard sputtered gas sensor microarray of Karlsruhe micro nose (KAMINA) [46] is used as a reference. It is compared to a sensor made of bare SnO_2 nanoparticles and a sensor made of SnO_2/SiO_2 core/shell nanoparticles. The sensitivity of the bare SnO_2 nanoparticle sensor is one order of magnitude higher than for the sputtered reference sensor. The sensitivity is excellent and is reproducible even at concentrations around 1 ppm [45]. The sensor made of SnO_2/SiO_2 core/shell nanoparticles shows less sensitivity compared to the sputtered reference sensor. The reduced sensitivity is supposed to be due to the SiO_2 coating. Optimization of the core particle size and the coating thickness are in progress.

In contrast to the reduced sensitivity the long-term stability of the prototype sensor made of core/shell particles is very promising. This is shown in Figure 6.12. Even after 8.5 months continuous operation the sensor signal is still 40% of the initial signal. For the sputtered reference KAMINA sensor a reduction to 50% of the initial sensitivity is quite common already after 1.5–2 months continuous operation. The response increased by a factor of 2–3, but is still very good compared to a sputtered reference sensor; the recovery time also increased slightly.

As already mentioned at the beginning of this section, these results are snapshots of ongoing research on layers of nanoparticles deposited on substrates. The picture is not yet complete, but single images show the promising application potential of nanoparticles produced using the KMPP in the field of gas sensors.

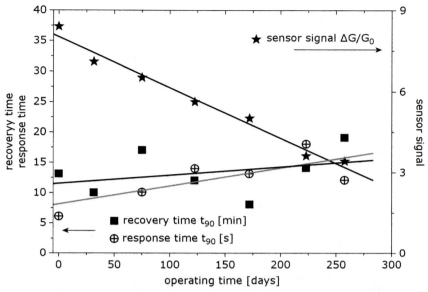

Figure. 6.12 Long-term performance of a sensor made of core/shell SnO_2/SiO_2 nanoparticles, measured at 10 ppm iso-propanol. After 8.5 months continuous operating time the sensor signal is reduced to approximately 40% of initial signal. Response time and recovery time increased slightly, respectively. Degradation is significantly slower than for sputtered KAMINA sensors.

Conclusions

A general overview on nanoparticle synthesis methods has been given in this chapter, with special emphasis on the application of nanoparticles in microsystems technology. A detailed description of the KMPP was given, as this process is able to produce extremely small nanoparticles with narrow particle size distribution as well as functionalized core/shell nanoparticles, all owing a large application potential in microsystems technology.

Examples of ongoing research in the field of applying mainly different core/shell nanoparticles of the KMPP in microsystems technology were presented. These examples focused on applications in microoptics and as gas sensors. The results are very promising in both application areas. Nevertheless, research has to be continued in both fields to obtain complete images. Fundamental questions, as e.g. the influence of particle sizes (quantum phenomena) on refractive indices and on sensing properties as well as systematic studies of the parameters and optimization of the processes are still open.

References

[1] Suryanarayana C. 2001. Mechanical alloying and milling. *Progress in Materials Science*, **46**(1–2), 1–184.
[2] Brinker C.J. and Scherer G.W. 1990. *Sol–Gel Science: The Physics and Chemistry of Sol–Gel Processing*. Academic Press, Boston, Massachusetts, USA.

[3]   Hahn H., Eastman J.A. and Siegel R.W. 1988. Processing of nanophase ceramics. In *Ceramic Powder Science II* Ceramic Transactions 1B, Messing G.L., Hausner H. and Fuller Jr. E.R. (Eds) *American Ceramic Society*, 1115–1123, Westerville, Ohio, USA.

[4]   Skandan G. 1995. Processing of nanostructured zirconia ceramics. *Nanostructured Materials*, **5**(2), 111–126.

[5]   Gleiter H. 1989. Nanocrystalline materials. *Progress in Material Sciences*, **33**(4), 223–315.

[6]   Eastman J.A., Thompson L.J. and Marshall D.J. 1993. Synthesis of nanophase materials by electron beam evaporation. *Nanostructured Materials*, **2**(4), 377–382.

[7]   Ullmann M., Friedlander S.K. and Schmidt-Ott A. 2002. Nanoparticle formation by laser ablation. *Journal of Nanoparticle Research*, **4**(6), 499–509.

[8]   Chang W., Skandan G., Hahn H., Danforth S.C. and Kear B.H. 1994. Chemical vapour condensation of nanostructured ceramic powders. *Nanostructured Materials*, **4**(3), 345–351.

[9]   Hahn H. 1997. Gas phase synthesis of nanocrystalline materials. *Nanostructured Materials*, **9**(1–8), 3–12.

[10] Pratsinis S. 1998. Flame aerosol synthesis of ceramic powders. *Progress in Energy and Combustion and Science*, **24**(3), 197–219.

[11] Kammler H., Mädler L. and Pratsinis S. 2001. Flame synthesis of nanoparticles. *Chemical Engineering and Technology*, **24**(6), 583–596.

[12] What is Carbon Black®. 200?. Corporate publication. Evonik-Degussa AG, Frankfurt, Germany.

[13] Aerosil®. 200?. Corporate publication. Evonik-Degussa AG, Frankfurt, Germany

[14] Böhm J., Haußelt J., Henzi P., Liftin K. and Hanemann T. 2004. Tuning the refractive index of polymers for polymer waveguides using nanoscaled ceramics or organic dyes. *Advanced Engineering Materiasl*, **6**(1–2), 5257.

[15] Ritzhaupt-Kleissl E., Böhm J., Haußelt J. and Hanemann T. 2005. Process chain for tailoring the refractive index of thermoplastic optical materials using ceramic nanoparticles. *Advanced Engineering Materials*, **7**(6), 540–545.

[16] Sahm T., Mädler L., Gurlo A., Barsan N., Pratsinis S.E. and Weimar U. 2004. Flame spray synthesis of tin dioxide nanoparticles for gas sensing. *Sensors and Actuators B*, **98**(2–3), 148–153.

[17] Mädler L., Sahm T., Gurlo A., Grunwaldt J.D., Barsan N., Weimar U. and Pratsinis S.E. 2006. Sensing low concentrations of CO using flame-spray-made Pt/SnO$_2$ nanoparticles. *Journal of Nanoparticle Research*, 8(6), 783–796.

[18] Mädler L., Roessler A., Pratsinis S.E., Sahm T., Gurlo A., Barsan N. and Weimar U. 2006. Direct formation of highly porous gas-sensing films by *in situ* thermophoretic deposition of flame-made Pt/SnO$_2$ nanoparticles. *Sensors and Actuators B*, **114**(1), 283–295.

[19] Vollath D. and Sickafus K.E. 1993. Synthesis of ceramic oxide powders in a microwave plasma device. *Journal of Materials Research*, **8**(11), 2978–2984.

[20] Vollath D. and Szabó D.V. 2006. The microwave plasma process – a versatile process to synthesize nanoparticulate materials. *Journal of Nanoparticle Research*, **8**(3–4), 417–428.

[21] Anthony R., Thimsen E., Johnson J., Campbell S. and Kortshagen U. 2006. Fast high-density low-pressure plasma synthesis of GaN nanocrystals. In *GaN, AlN, InN and Related Materials*, Kuball M., Myers T.H., Redwing J.M. and Mukai T. (Eds) *Materials Research Society*, Warrendale, Pennsylvania, USA. FF11-05.01 FF11-05.04.

[22] Mangolini L., Thimsen E. and Kortshagen U. 2005. High-yield plasma synthesis of luminescent silicon nanocrystals. *NanoLetters*, **5**(4), 655–659.

[23] Vons V., Creyghton Y. and Schmitt-Ott A. 2006. Nanoparticle production using atmospheric pressure cold plasma. *Journal of Nanoparticle Research*, **8**(5), 721–728.

[24] Vollath D. and Sickafus K.E. 1993. Synthesis of nanosized ceramic nitride powders by microwave supported plasma reactions. *Nanostructured Materials*, **2**(5), 451–456.

[25] Schlabach S., Szabó D.V., Shi Z., Wang D. and Vollath D. 2004. Synthesis of nanoparticulate SiC in a low temperature microwave plasma process. *VDI-Berichte*, **1839**, 167–170.

[26] Szabó D.V. and Vollath D. 1999. Morphological characterization of nanocrystals with layered structure. *Nanostructured Materials*, **12**(1–4) 1999, 597–600.

[27] Vollath D. and Szabó D.V. 1994. Nanocoated particles: a special type of ceramic powder. *Nanostructured Materials* **4**(8), 927–938.

[28] Kortshagen U. and Bhandarkar U. 1999. Modeling of particulate coagulation in low pressure plasmas. *Physical ReviewE*, **60**(1), 887–898.

[29] Matsoukas T. and Russell M. 1995. Particle charging in low-pressure plasmas. *Journal of Applied Physics*, **77**(9), 4285–4292.

[30] Vollath D., Szabó D.V. and Schlabach S. 2004. Oxide/polymer nanocomposites as new luminescent material. *Journal of Nanoparticle Research*, **6**(2), 181–191.

[31] Vollath D. and Szabó D.V. 2004. Synthesis and Properties of Nanocomposites. *Advanced Engineering Materials*, **6**(3), 117–127.

[32] MacDonald A.D. 1966. *Microwave Breakdown in Gases*. Wiley, New York, New York, USA.

[33] German Patent 195 28 343 C2. 1995. Möbius A. And Mühleisen M. Vorrichtung zur reflexionsarmen absorption von mikrowellen.

[34] Lide D.R. 1993. *CRC Handbook of Chemistry and Physics*, CRC Press, Boca Raton, Florida, USA, **73**.

[35] Babeva T., Atanassova E. and Koprinarova J. 2005. Optical characteristics of thin rf sputtered Ta_2O_5 layers. *Physica Status Solidi A*, **202**(2), 330–336.

[36] Szabó D.V., Ochs R., Schlabach S., Ritzhaupt-Kleissl E. and Hanemann T. 2008. New core/shell Ta_2O_5-PMMA nanocomposites for applications as polymer waveguides. In *Nanophase and Nanocomposite Materials V*, (Symposium Proceedings 1056E), Komarneni S, Kaneko K., Parker J.C. and O'Brien P. (Eds.), Materials Research Society, Warrendale, Pennsylvania, USA, HH10-09.01–HH10-09.06.

[37] Chouwdhuri A., Gupta V. and Sreenivas K. 2003. Fast response H_2S gas sensing characteristics with ultra-thin CuO islands on sputtered SnO_2. *Sensors and Actuators B*, **93**(1–3), 572–579.

[38] Lee D.S., Rue G.H., Huh J.S., Choi S.D. and Lee D.D. 2001. Sensing characteristics of epitaxially-grown tin oxide gas sensor on sapphire substrate. *Sensors and Actuators B*, **77**(1–2), 90–94.

[39] Althainz P., Goschnick J., Ehrmann S. and Ache H.J. 1995. Multisensor microsystem for contaminants in air. *Sensors and Actuators B*, **33**(1–3), 72–76.

[40] Kennedy M.K., Kruis F.E., Fissan H., Mehta B.R., Stappert S. and Dumpich G. 2003. Tailored nanoparticle films from monosized tin oxide nanocrystals: Particle synthesis, film formation, and size-dependent gas-sensing properties. *Journal of Applied Physics*, **93**(1), 551–560.

[41] Gong J., Chen Q., Sei W. and Seal S. 2004. Micromachined nanocrystalline SnO_2 chemical gas sensors for electronic nose. *Sensors and Actuators B*, **102**(1), 117–125.

[42] Sahm T., Mädler L., Gurlo A., Barsan N., Pratsinis S.E. and Weimar U. 2005. Flame spray synthesis of tin oxide nanoparticles for gas sensing. In *Semiconductor Materials for Sensing*, (Symposium Proceedings 828) Seal S., Baraton M.I., Parrish C. and Murayama N. (Eds), Materials Research Society, Warrendale, Pennsylvania, USA, A1.3.1–A1.3.6.

[43] Carotta M.C., Giberti A., Guidini V., Magalù C., Vendemiati B. and Martinelli G. 2005. Gas sensors based on semiconductor oxides: basic aspects onto materials and working principles. In *Semiconductor Materials for Sensing*, (Symposium Proceedings 828), Seal S., Baraton M.I., Parrish C. and Murayama N. (Eds), Materials Research Society, Warrendale, Pennsylvania, USA, A4.6.1–A4.6.12.

[44] Schumacher B., Szabó D.V., Schlabach S., Ochs R., Müller H. and Bruns M. 2006. Nanoparticle SnO_2 films as gas sensitive membranes. In *Nanoparticles and Nanostructures in Sensors and Catalysis*, (Symposium Proceedings 900E), Zhong C.J., A. Kotov N., Daniell W. and Zamborini F.P. (Eds) Materials Research Society, Warrendale, Pennsylvania, USA, O08-06.01–O08-06.06.

[45] B. Schumacher B., Ochs R., Tröße H., Schlabach S., Bruns M., Szabó D.V. and Haußelt J. 2007. Nanostructured SnO_2 layers for gas sensing applications by in-situ deposition of nanoparticles produced by the Karlsruhe Microwave Plasma Process. *Plasma Processes and Polymers*, **4**(S1), S865–S870.

[46] Bruns M., Frietsch M., Nold E., Trouillet V., Baumann H., White R. and Wright A. 2003. Surface analytical characterization of SiO_2 gradient membrane coatings on gas sensor microarrays. *Journal of Vacuum Science and Technology A*, **21**(4), 1109–1114.

7

Near net shaping of ceramics in microsystems technology

J. R. Binder, R. A. Dorey, M. Müller, H. Geßwein, S. A. Rocks and D. Wang

There is an increasing interest in fabrication technologies for ceramic microcomponents with reduced or negligible shrinkage during processing. This is motivated by the difficulties and high costs which arise when narrow tolerances cannot be met and a finishing operation has to be performed. In this chapter several processing technologies for achieving minimized sintering shrinkage are described including methods which are suited for 2D structured functional ceramics and such which are predestinated for the production of 3D microparts in structural applications. For the latter one the principle of reaction bonding has been applied, as it allows the fabrication of oxide as well as nonoxide ceramic components with remarkable mechanical strength. It is also shown that shrinkage-free 2D microcomponents can be produced by infiltration methods using sol-gel techniques.

7.1 Introduction

In a variety of technical applications, tailored ceramics, which are optimized for specific demands, are increasingly gaining an important place in the market. They represent key materials for the realization of many new technologies. These modern high-performance materials possess their particular properties through the use of synthetic starting materials as well as subsequent processing on controlled conditions.

Normally, the production of ceramic components is based on the densification of porous green compacts by a high-temperature process. During this sintering process a linear shrinkage of typically 15–20% occurs. In order to obtain the required size of components after sintering, the green compacts have to be oversized accordingly. Although compensation of sintering shrinkage, and thus net shape sintering, is state of the art in large-volume production, for narrow dimensional tolerances, an additional step of postsintering machining is necessary, which may account for more than 30% of the production costs [1, 2]. In the case of ceramic microcomponents, with small details in the range of few micrometres, such finishing processes are extremely difficult if not completely impossible. Therefore, alternative approaches which reduce the sintering shrinkage are advantageous for the use in microsystem engineering.

Figure 7.1 Fracture surface of C/SiC manufactured by liquid silicon infiltration of a C/C preformed at 1450°C; dark grey circles: cross-section of carbon fibres, light grey: reaction-bonded SiC matrix.

One possibility for manufacturing of these so-called near net shape ceramics is the infiltration of porous preforms with gaseous compounds (chemical vapour infiltration (CVI)), liquid or dissolved ceramic precursors with subsequent pyrolysis (liquid polymer infiltration (LPI)), or molten metal. In the field of ceramic matrix composites (CMCs) this is the preferred processing route as the reinforcing fibres inhibit any dimensional changes in the direction of fibre orientation [3]. In Figure 7.1 the microstructure of a liquid silicon infiltrated (LSI) carbon fibre reinforced silicon carbide (LSI-C/SiC) structure is shown as an example.

Instead of mere infiltration another approach to reduced porosity is to utilize chemical reactions between the bulk material and a second gaseous or liquid phase which results in a volume increase of the solid. The most common gaseous reactants are oxygen and nitrogen while molten aluminum and silicon are often applied as liquid phases for reactive melt infiltration [2]. Their special importance is not only based on the dimensional stability during processing, but also on the significantly reduced processing temperatures (and often costs) compared to conventional processing [4].

Two innovative approaches are described in detail below. On the one hand, a 2D net shaping process for the fabrication of functional ceramics in terms of microstructured thick films and on the other hand a reaction bonding processes for the synthesis of structural ceramics as well as the fabrication of corresponding 3D microparts.

7.2 2D net shaping

Two routes can be envisaged to create 2D ceramic microfeatures. The first consists of first depositing a continuous film and then machining it to create the desired shape through a subtractive patterning route. The second route would be to deposit material where it is required in an additive process. A refinement of this second route would be to use a process whereby the material that is deposited does not undergo a dimensional change on continued processing. Such a process would be described as a 2D net shaping process.

2D net shaping can be differentiated from other additive processing routes through the requirement for the dimensional changes during processing to be minimized so that the final sintered dimensions are approximately equal to those of the green body. Many 2D shaping techniques such as screen printing, ink-jet printing, embossing or casting yield ceramic structures that undergo significant dimensional changes as the ceramic particles densify. In addition, only moulding or casting routes can be used to create fine-scale ceramic features as the pixel size (ink-jet drop size or screen printer weave size) is sufficiently small. The pixel size of embossing or moulding routes will be dictated by the ability to create the fine moulds using machining or photolithography. 2D-shaped MEMS devices can be created by forming the mould on the surface of the device wafer so that it forms a part of the mould. When the mould is filled and then removed the 2D microstructures will remain on the surface of the substrate. In order to achieve a net shape process using the micromoulding approach it is necessary to achieve densification without lateral dimensional changes.

The two requirements for a 2D net shape process are that the mould should be easily shaped at the appropriate scale and removed, and that the ceramic system should not undergo significant dimensional changes.

Conventional ceramic processing routes that rely on consolidating the ceramic particles will always exhibit a high degree of shrinkage as the densification process progresses. Alternative processing routes such as sol-gel also exhibit a significant dimensional change associated with the drying, pyrolysis (removal of organic components) and sintering. Physical vapour deposition (PVD) routes, such as pulsed laser and sputtering, can be used to deposit ceramics onto substrates in a near net shape using a moulding-type approach termed "lift off". However, the thickness of the resultant structures is limited by the deposition rate, expense, small area coverage, generation of stresses within the system and maximum achievable thickness.

An alterative route has to be used in order to achieve 2D ceramic structures with appreciable thicknesses. As with many macroscopic net shape processing routes, the 2D micromoulding process relies on the formation of a rigid skeletal network which is then filled with a secondary material to achieve densification without shrinkage. This approach, known as composite sol-gel [5], involves the creation of a composite ink consisting of a ceramic powder and ceramic producing sol. By matching the composition of the sol to that of the powder it is possible to produce a homogeneous microfeature. The composite ink operates in two ways: first, the ceramic powder particles fill the mould cavity to define the shape of the 2D microfeatures; secondly the ceramic producing sol is infiltrated into the porous skeletal network and converted to an oxide ceramic through heating. In conventional sol-gel processing such conversion usually results in significant shrinkage. While this also occurs in this instance, the presence of the skeletal network prevents any significant shape change from occurring as the temperature used to convert the sol-gel material into an oxide ceramic is too low to result in densification of the ceramic powder particles. Due to the shrinkage of the sol-gel material during pyrolysis, repeated sol infiltration treatments are usually required to ensure that a high degree of densification is achieved.

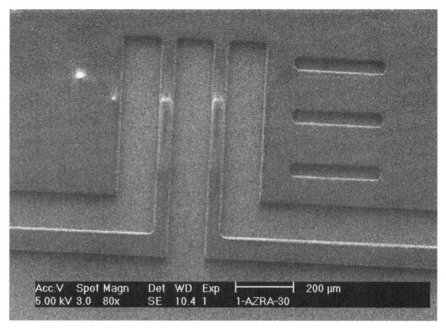

Figure 7.2 SEM micrograph of a polymeric mould on a silicon substrate.

The mould material selected for this technique does not need to have a very high thermal stability due to the low processing temperatures employed. While a metallic mould could be used, the high thermal expansion coefficient and oxidation potential of most metals makes their use problematic. Conventional wisdom holds that it would not be possible to coprocess a ceramic and a polymer material. However, in this instance a polymeric mould (Figure 7.2), in the form of a photoresist material can successfully be used with a ceramic material [6–8] due to the low processing temperature. While it is not possible to fully process the ceramic and maintain the polymeric mould, the drying stages can successfully be accomplished with the final pyrolysis/crystallization stage used to both create the fully crystalline ceramic structure and also remove the mould material. In the micromoulding process the ceramic ink and sol can be dried at temperatures below 200°C without damaging the polymeric moulds. This drying removes the carrier fluids and solvents and partially decomposes the sol-gel materials. At temperatures above 200°C the photoresist material is unstable and undergoes structural and dimensional changes which will affect the moulding process and the subsequent features. As a general rule, photoresists that are easier to remove at the end of the moulding process are generally also attacked by the solvents found in the sol requiring either a modification in the sol chemistry or the use of a more resilient photoresist. Conversely, the photoresists that are more stable to chemical attack are also more difficult to remove at the end of the moulding process.

To account for the shrinkage of the wet ink on drying, the structure is gradually built up to the required thickness using repeated filling/layering with intermediate drying stages (below the critical temperature). Once the required thickness is achieved the structure is

Figure 7.3 SEM micrograph of a fracture cross-section of a micromoulded structure that has been infiltrated with sol-gel to increase green density and bind structure together.

subjected to a heat treatment in the range 600–700°C. During this process the polymeric mould material is burned off and any ceramic material on its surface is removed. The final microstructure (Figure 7.3) of the micromoulded features is homogeneous and has a high density despite the low ceramic processing temperature. This is achieved due to the high reactivity of the sol-gel phase which increases the green density and binds the ceramic powder together and to the substrate.

7.3 3D net shaping

While functional ceramics are often applied as thin and thick films, the use of structural ceramics in microengineering is focused on the production of 3D parts. Both infiltration and reaction bonding processes are suitable for achieving shape and dimensional accuracy. However, it should be noted that one shortcoming of reaction-bonded materials is that, in general, the mechanical properties in general cannot compete with the corresponding densely sintered materials. The drawbacks and opportunities of different reaction-bonded materials are discussed in detail in the following sections, separated into nonoxide and oxide ceramics. The adjustment of zero-shrinkage and the special features of diverse shaping methods are also considered.

7.3.1 Nonoxide ceramics

The most prominent and best investigated examples of the nonoxide fraction are reaction-bonded silicon carbide (RBSC) and reaction-bonded silicon nitride (RBSN). RBSC which is also known as siliconized silicon carbide (SiSiC) may be one of the first industrially applied nonoxide ceramics, described in detail by Popper [9]. It represents a fully dense engineering ceramic formed by bonding together SiC powder with further SiC produced *in situ* by reaction between Si and C [10]. The infiltration of a mixed green compact of primary α-SiC and C powder by liquid Si, is driven by capillary forces and excellent wetting of C and SiC by the Si melt. In the course of the reaction between C and Si melt secondary β-SiC is formed which bonds the compact together (Equation 7.1).

$$C(s) + Si(l) \rightarrow SiC(s) \qquad (7.1)$$

The original porosity of the compact is reduced by the formation of secondary SiC (the increase of molar volume from C to SiC amounts from 60% to 130% depending on whether graphite or glassy carbon is taken as raw material), and by the residual unreacted silicon filling the pore channels and voids. The lack of open porosity is beneficial for oxidation resistance of RBSC, but the presence of free Si limits the application temperature to values below the melting point of silicon (1410°C). Whilst for liquid-phase or solid-phase sintering of SiC temperatures about 2000°C are required, it is possible to produce the reaction-bonded material at 1450–1500°C.

In contrast to the afore mentioned SiC processing routes which aims to produce a dense microstructure, attractive approaches exist to exploit a high porosity with favorable pore structure and pore size distribution. Potential applications are: catalyst supports, machinable ceramics or light structural ceramics. Wood is typically used as carbon source as it keeps its highly anisotropic cellular structure after carbonization and serves as template for SiC formation (biomimetic materials). The Si counterpart can again be delivered via liquid or vapour phase. Ota *et al.* [11] have described an artificial fossilization process where charcoal is impregnated with tetraethyl orthosilicate (TEOS) to form SiO_2 which will react with the C skeleton in a solid–solid reaction to SiC at 1400°C (Equation (7.2)):

$$3C(s) + SiO_2(s) \rightarrow SiC(s) + 2CO(g) \qquad (7.2)$$

Other processing techniques include the solid-gas reaction with silicon monoxide at 1550–1600°C (Equation (7.3)), or with Si vapour (Equation (7.4)), whereas in contrast to the liquid melt infiltration no residual silicon limits the application temperature [12, 13]:

$$2C(s) + SiO(g) \rightarrow SiC(s) + CO(g) \qquad (7.3)$$
$$C(s) + Si(g) \rightarrow SiC(s) \qquad (7.4)$$

The base reaction for the formation of silicon nitride can be described as follows (Equation (7.5)):

$$3Si(s) + 2N_2(g) \rightarrow Si_3N_4(s) \qquad (7.5)$$

Due to the strong exothermic character of this reaction there is a risk of melting and coalescence of Si particles; this would clearly reduce the available surface for the reaction and would also allow the shape of the powder compact to collapse. Lowering the temperature, on the other side, retards the reaction rate and results in incomplete conversion or extremely extended and uneconomic processing times. A trade-off is generally found in a step wise increase of the nitriding temperature starting with about 1200°C.

The influencing factors and kinetics of the silicon nitride formation have been intensively studied by several authors [14–17]. It was found that impurities of the raw material, as well as the composition of the nitriding atmosphere (flowing or static, hydrogen addition, oxygen content), had a strong influence on the conversion rate and phase composition

(α/β-Si$_3$N$_4$ ratio). It is generally accepted that lower temperatures (<1400°C) favor the formation of α-Si$_3$N$_4$ and that Si vapour is involved in that reaction. At higher temperatures, which also result in the presence of liquid silicon, the β-modification is increasingly formed.

The conversion of Si to Si$_3$N$_4$ is accompanied by a volume increase of about 22%; however, this volume increase does not lead to significant volume changes of the powder compact. The original shape of the moulded Si green body is conserved by a bridging Si$_3$N$_4$ network and only the voids between the particles are reduced; sintering in the normal sense does not take place [18].

Owing to typically achievable green densities of 55–65% for moulded parts, even after complete conversion of the Si starting material into the nitride the increase of molar volume of about 22% is not high enough to result in a dense final material; residual porosity in the range 25–30% is the average. Although strength values of 200–300 MPa can be achieved with such porosity, this is not sufficient for many structural applications. As a consequence, intensive research has been carried out to develop RBSN with less porosity by means of postsintering [19]. The resulting sintered RBSN has properties between RBSN and conventional liquid phase sintered silicon nitride.

For the manufacturing of sintered RBSN a mixture of Si and sintering additives (e.g. Y$_2$O$_3$ + Al$_2$O$_3$) instead of pure silicon is prepared and employed for the moulding process. Although the role of the sintering aid is to supply a liquid phase for Si$_3$N$_4$, in combination with SiO$_2$ which is always present at the surface of Si particles, to allow liquid phase sintering and α- to β-Si$_3$N$_4$ transformation, it cannot be considered as inert material during the nitridation process [20, 21]. Due to the formation of SiO$_2$ containing phases (e.g. Y$_5$Si$_3$O$_{12}$N) even at temperatures below 1300°C, the silica coverage of the Si particles may be removed and the nitridation kinetics are modified.

On the one hand a certain amount of sintering shrinkage has to be accepted during densification, hence drawbacks arising from the disappearance of dimensional stability will occur; however the total amount of shrinkage is clearly reduced (about a factor of two), owing to the conversion of Si into Si$_3$N$_4$ the "green density" after nitridation is correspondingly higher. On the other hand 93–99% of theoretical density can be realized by postsintering RBSN and therefore the mechanical properties are not far from those for hot pressed or hot isostatically pressed silicon nitride [22]. Hence sintered RBSN may be a good compromise between net shape manufacturing and outstanding mechanical properties. Figure 7.4 shows a typical fracture surface of sintered RBSN; elongated β-Si$_3$N$_4$ grains are a characteristic feature of this material and bring about relatively high fracture toughness.

7.3.2 Oxide ceramics

The reaction-bonded aluminum oxide (RBAO) process developed by Claussen and his co-workers [23–25] is a condensed-phase/gas-phase reaction forming technique based on the oxidation reaction of a compacted mixture of attrition-milled aluminium/Al$_2$O$_3$ or aluminium/Al$_2$O$_3$/ZrO$_2$ powders. The heat treatment in an oxidizing atmosphere results in sintered Al$_2$O$_3$-based ceramics. The oxidation reaction described by

$$4 \text{ Al (s)} + 3 \text{ O}_2 \text{ (g)} \rightarrow 2 \text{ Al}_2\text{O}_3 \text{ (s)} \tag{7.6}$$

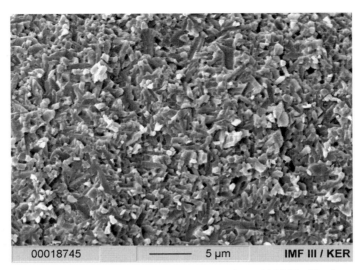

Figure 7.4 SEM micrograph of a fracture surface of SRBSN, Y_2O_3/Al_2O_3 as sintering aid, 1750°C, 2 h, N_2 atmosphere (0.15 MPa).

is extremely exothermic (H = 1669.4 kJ/mol) and results in a net volume expansion of 28%. Usually RBAO green bodies contain 30–80 vol% of aluminium metal powder. Due to the plastic deformation of Al particles during compaction high green body densities and strengths can be achieved which enables the production of complex shaped parts by green body machining. The main drawback of this process is that after complete oxidation and densification a sintering shrinkage of about 15% may remain when the final densities exceed 95% of the theoretical density. The sintering shrinkage can be further reduced by the addition of metals which exhibit larger volume expansions during oxidation such as Zr (49%), Ti (76%), Cr (102%) or Nb (174%) [26]. A modification of the RBAO process is the production of reaction-bonded mullite (RBM) ceramics where aluminium and silicon metal or silicon carbide are used as reactive precursor materials. These near net shape mullite-based ceramics exhibit mechanical strengths up to 600 MPa [27].

Net shape reaction-bonded oxide ceramics can be produced with a combined approach where silicon containing polymers, so-called siloxanes, are used as low-loss binder materials together with intermetallic compounds in the systems Al–Zr and Si–Zr. Intermetallic compounds such as $ZrSi_2$, Zr_2Si and $ZrAl_3$ offer the advantage of a much higher volume expansion during the oxidation step compared to pure metals like aluminum or silicon. This process is capable of providing reaction-bonded oxide ceramics with high density and zero-shrinkage [28, 29]. The starting powders are ball- or attrition-milled in an organic milling liquid to reduce the particle size. The suspensions are then converted into granules by spray-drying. Green compacts have to be shaped by isostatic pressing to avoid green density gradients. The relatively high amount of siloxanes in the precursor compacts makes easy green machining possible. So even complex shapes can be formed in the green state and can be reproduced exactly by the reaction bonding process [30, 31].

During heat treatment in an oxidizing atmosphere the siloxane is pyrolized to SiO_2 according to the simplified reaction scheme:

$$[Si(CH_3O_{1.5})]_n (s) + O_2(g) \rightarrow SiO_2(s) + ... \tag{7.7}$$

At temperatures above 450°C the intermetallics are oxidized according to:

$$ZrSi_2(s) + 3O_2(g) \rightarrow ZrO_2(s) + 2\,SiO_2(s) \tag{7.8}$$
$$Zr_2Si(s) + 3O_2(g) \rightarrow 2ZrO_2(s) + SiO_2(s) \tag{7.9}$$
$$ZrAl_3(s) + 13/4O_2(g) \rightarrow ZrO_2(s) + 3/2Al_2O_3(s) \tag{7.10}$$

These oxidation reactions are associated with net volume expansions of 122%, 84% and 42%, respectively. The newly formed Al_2O_3 or ZrO_2 phases react with SiO_2 to form mullite ($3Al_2O_3 \cdot 2SiO_2$) and zircon ($ZrSiO_4$), respectively. The oxidation reactions and the mullite-zircon formation are associated with a volume expansion which effectively compensates the sintering shrinkage. An overview of the substages, binder pyrolysis, oxidation reaction, phase formation and sintering, that take place during thermal processing is shown in Figure 7.5.

Through adjusting the initial Zr/Al/Si ratio composites belonging to different phase fields in the ternary oxide system Al_2O_3–SiO_2–ZrO_2 can be produced. The mechanical strengths of these reaction-bonded oxide ceramics depend on the phase composition of the sintered materials and vary between approximately 300 MPa for zircon-based ceramics [32], 630 MPa for Al_2O_3–mullite–ZrO_2 or mullite–ZrO_2 ceramics [29] and up to 780 MPa for Al_2O_3–ZrO_2 ceramics [33]. The microstructure of the zircon-based ceramic shown

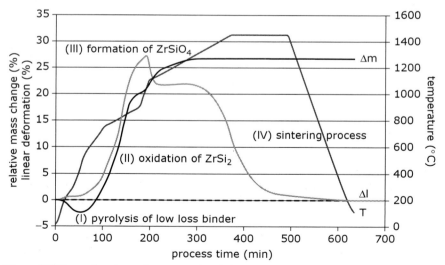

Figure 7.5 Thermal analysis of a reaction-bonded zircon-based ceramic. Reaction sintering stages: pyrolysis, oxidation, phase formation and sintering, can be deduced from linear deformation curve (Δl) and from relative mass change curve (Δm). After reaction sintering sample returns to its original length.

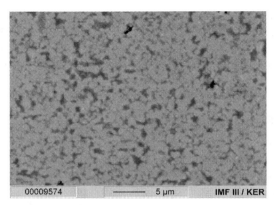

Figure 7.6 SEM micrograph showing microstructure of a reaction-bonded zircon-based ceramic (material contrast); gray: zircon matrix, black: glassy silica phase, light gray: zirconia.

in Figure 7.6 consists of zirconium silicate grains (gray matrix), an amorphous SiO_2 like phase (black) and some finely dispersed ZrO_2 (light gray phase). The microstructures of the final sintered ceramics in the system Al_2O_3–mullite–ZrO_2 are shown in Figure 7.7. The zirconia phases in the SEM micrographs appear brighter due to atomic number contrast in comparison with Al- and Si-containing alumina and mullite phases which appear gray. The zirconia crystals exhibit equiaxed grains with sizes < 1 μm and some of the ZrO_2 grains are distributed in the alumina and mullite grains. The mullite particles in Figure 7.7 (c) show an elongated morphology with relatively large sizes of about 2–3 μm.

The inevitable heterogeneous gas–solid reactions are one of the main disadvantages or rather limitations on the fabrication of reaction-bonded ceramics. These reactions are usually extremely exothermic and thus a large amount of heat can be released. As a result local overheating of the compacted green bodies may be observed. This uncontrolled reaction behaviour leads to typical processing problems like sample cracking and bloating. Therefore, the detailed knowledge of the oxidation kinetics of the reactive solids is an essential prerequisite for understanding and control of processing [34–36]. Depending on the sizes of the reacting green bodies and the applied temperature programme the rate limiting process of the oxidation of the sample can

a   b   c

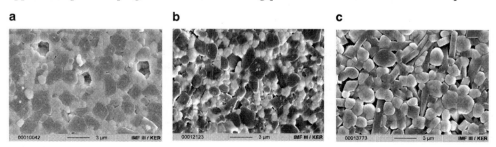

Figure 7.7 SEM micrographs showing microstructure of reaction-bonded ceramics in system Al_2O_3–mullite–ZrO_2; (a) Al_2O_3–ZrO_2, (b) Al_2O_3– mullite–ZrO_2; and (c) mullite–ZrO_2.

change from oxygen-limited, where the diffusion of oxygen through the porous matrix is the rate determining step, to reaction-limited, where the oxidation of the powder particles governs the oxidation behaviour. Due to the small dimensions of microparts and thus the small diffusion paths of the reactive gases, these net shaped reaction-bonded ceramics are particularly suited to applications in microsystem technology.

7.3.3 Shaping of 3D microparts

In order to benefit from these reaction-bonded materials in microengineering suitable shaping processes for the production of microcomponents have to be available. Here powder-technological fabrication processes such as ceramic injection moulding [37, 38], the embossing method [39], or electrophoretic deposition [39] which have already been successful in the production of ceramic microparts can be used. Furthermore material removal techniques which take in account the special characteristics of the developed reaction-bonded oxide ceramics [30, 31] have also been considered.

The basic conditions to realize zero-shrinkage are generally independent of the moulding or shaping method and can be calculated according to the following equation:

$$\Delta \widetilde{V} = (1 + \Delta \widetilde{m}) \, \frac{\rho_{green \, compact}}{\rho_{ceramic}} - 1 \qquad (7.11)$$

where $\Delta \widetilde{V}$ is the relative volume change during the thermal process; $\Delta \widetilde{m}$ is the relative weight change during the thermal process; and ρ is the green and sinter density of the bodies in g/cm^3.

When the sintering shrinkage is supposed to be exactly compensated during thermal processing ($\Delta \widetilde{V} = 0$) it is obvious that the green density, the sinter density, and the relative mass change have to be adjusted. The adjustment of the green density plays a decisive role. Consequently, the shaping method can be considered the crucial factor for achieving shape and dimensional fidelity.

Powder compaction and sintering is a widely-used process in the ceramic technology. Powders or granulates are densified to compacts by uniaxial or isostatic pressing. The green density of the densified bodies depends on both the chemical composition of the starting materials and the squeezing pressure. For a defined composition of the starting materials, there is a level of compaction that will achieve the necessary green density for zero-shrinkage, which can be determined through a so-called compressibility curve. Based on these principles microstructured green compacts can be produced by a combined pressing and embossing method. Figure 7.8 shows the comparison between a structured green compact produced with a coin as a metallic stamping die and the corresponding shrinkage-free sintered ceramic part. However, this simultaneous pressing and embossing method is only suitable for the use of microstructured dies with a low aspect ratio.

Microparts with a low aspect ratio can also be realized by electrophoretic deposition. This shaping technique is generally suited to achieve high green densities even by using nanopowders. But it is also necessary for the production of shrinkage-free reaction-bonded ceramics to choose appropriated slurries and process parameters carefully to attain the desired green density.

Figure 7.8 Green compact (left) and reaction-sintered ceramic (bottom right) structured by a combined pressing and embossing method using a coin as metallic stamping die (in middle); the three parts have nearly equal diameters.

Material removal techniques like micromilling are interesting methods for the fabrication of free form 3D microparts. Because of the hard and brittle nature of ceramics, mechanical machining can be best performed in the green state. If the above mentioned compacts have to be shaped anyway by green machining these green bodies should exhibit the adjusted density. In order to achieve not only optimized densities but also a high stability and a good machinability of the green compacts a suitable binder and pressing additives have to be used. Besides the material composition, the quality of the milled microparts as well as the feasible dimensions depends on the milling tools used as well as the milling parameters [40]. Ritzhaupt-Kleissl *et al.* [30] describe the opportunities and limits for net shape manufacturing of reaction-bonded ceramic microparts. It could be shown that details down to about 40 μm can be realized by an optimized micromilling strategy. The attainable precision was demonstrated on a micromilled gear wheel with an outer diameter of $2943 \pm 4\,\mu m$. After the reaction sintering process the sintered gear wheel possesses a diameter of $2941 \pm 7\,\mu m$. A further example of dimensional accuracy of these reaction-bonded ceramics is illustrated in Figure 7.9. This nozzle plate was produced by micromilling, except the narrow nozzle outlet which was drilled in green state. After reaction sintering, the size accuracy of few micrometres is in the range of the metering precision of the used light-optical microscope. Figure 7.10 shows three reaction-bonded ceramic microparts with different milled structures such as microcyclone mixers. The test specimen in the middle exhibit triangular columns which have lateral dimensions of about 400 μm and are 1000 μm high. These examples show the technical potential to realize microparts with small dimensions, fine details, and high aspect ratios by this technique.

Ceramic injection moulding (CIM) is an economic process for a wide range of applications with diversified requirements. In contrast to the aforesaid methods the green density of a hot moulding feedstock can only be varied by a change in composition. Provided that the

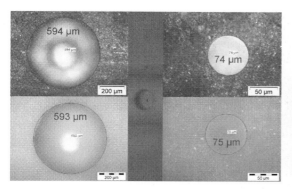

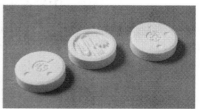

Figure 7.9 Detail of a reaction-sintered ceramic nozzle plate (in middle), machined at the FZK (T. Gietzelt and L. Eichhorn). Light-optical micrographs show the size accuracy of the milled circular notch of nozzle plate before (top left) and after sintering (bottom left) as well as dimensional accuracy of drilled nozzle outlet, before (top right) and after sintering (bottom right).

Figure 7.10 Reaction-sintered microparts, produced by green machining at FZK (T. Gietzelt and L. Eichhorn); in middle: different test structures such as a meander, a gear wheel, and triangular columns; on left and on the right; two microcyclone mixers.

green density of the feedstock corresponds to its theoretical density (i.e. without bubbles), the volume change can be predicted according to the following equation:

$$\Delta \widetilde{V} = \frac{\widetilde{V}_{powder} \cdot \sum_i \widetilde{m}_i (1 + \Delta \widetilde{m}_i)}{\rho_{ceramic} \cdot \sum_i \frac{\widetilde{m}_i}{\rho_i}} - 1 \qquad \text{Eqn (7.12)}$$

where $\widetilde{V}_{powder}$ is the volume fraction of the powders in the feedstock; $\widetilde{m}_i$ is the weight fraction of the component i in the mixed powder; ρ_i: is the density of the component i in the mixed powder; and $\Delta \widetilde{m}_i$ is the relative weight change of the component i during the thermal process.

Consequently the maximum achievable volume fraction of the powder in the feedstock defines whether a ceramic micropart with the desired composition can be produced by CIM.

Conclusions

Near net shape manufacturing is an interesting alternative to conventional ceramic processes. The inevitable sintering shrinkage during thermal processing can be reduced or even completely avoided by both, infiltration techniques and reaction bonding processes. In the case of infiltration, a porous ceramic preform is filled with gaseous or liquid precursors. The prospects for this method have been successfully demonstrated by the fabrication of functional ceramic microstructured thick films. During reaction bonding processes the porosity of green compacts is reduced by chemical reactions between bulk material and a second gaseous or liquid phase which result in an increase in the volume of the corresponding bodies. Due to the initial low mechanical properties reaction bonding materials cannot compete against the corresponding densely sintered materials. However, this gap could be nearly closed by the choice of suitable starting materials and an optimized reaction sintering process. It has already been shown that accurate 3D microparts can be produced by reaction bonding processes using adjusted green compacts. The realization of this net-shape processing techniques for the manufacturing of ceramic devices at industrial scale is always dependent on the prices of the starting materials used as well as the reaction kinetics, or rather the processing time which once again influences the production costs. Because of the small dimensions near net shape techniques are especially suited for the fabrication of microparts.

References

[1] Sinhoff V., Schmidt C. and Bausch S. 2001. Machining components made of advanced ceramics: prospects and trends. *cfi*/Ber. DKG **78(6)**, E12–E18.

[2] Greil P. 1999. Near net shape manufacturing of ceramics. *Materials Chemistry and Physics*, **61**, 64–68.

[3] Naslain R. 2004. Design, preparation and properties of non-oxide CMCs for application in engines and nuclear reactors: an overview. *Composites Science and Technology*, **64(2)**, 155–170.

[4] Chiang Y.M., Messner R. P., Terwilliger C. D., Behrendt D. R. 1991. Reaction-formed silicon carbide. *Materials Science and Engineering*, **A144**(1–2), 63–74.

[5] Dorey R. A., Stringfellow S. B. and Whatmore R. W. 2002. Effect of sintering aid and repeated sot infiltrations on the dielectric and piezoelectric properties of a PZT composite thick film. *Journal of the European Ceramic Society*, **22**, 2921–2926.

[6] Navarro A., Rocks S. A. and Dorey R. A. 2007. Micromoulding of lead zirconate titanate (PZT) structures for MEMS. *Journal of Electroceramics*, DOI 10.1007/s10832-007-9170-y.

[7] Tyholdt F., Dorey R. A. and Ræder H. 2007. Novel patterning of composite thick film PZT, *Journal of Electroceramics*, DOI 10.1007/s10832-007-9318-9.

[8] Frood A. J. M., Beeby S. P., Tudor M. J. and White N. M. 2007. Photoresist patterned thick-film piezoelectric elements on silicon, *Journal of Electroceramics*, DOI 10.1007/s10832-007-9049-y.

[9] Popper P. 1960. The preparation of dense self-bonded silicon carbide. In *Special Ceramics*, Popper P. (Ed.), The British Ceramic Research Association, Heywood, London, United Kingdom, 209–219.

[10] Ness J. N. and Page T. F. 1986. Microstructural evolution in reaction-bonded silicon carbide. *Journal of Material Science*, **21**(4), 1377–1397.

[11] Ota T., Takahashi M., Hibi T., Ozawa M., Suzuki S. and Hikichi Y. 1995. Biomimetic process for producing SiC wood. *Journal of the American Ceramic Society*, **78**(12), 3409–3411.

[12] Vogli E., Mukerji J., Hoffman C., Kladny R., Sieber H. and Greil P. 2001. Conversion of oak to cellular silicon carbide ceramic by gas-phase reaction with silicon monoxide. *Journal of the American Ceramic Society*, **84**(6), 1236–40.

[13] Vogli E., Sieber H. and Greil P. 2002. Biomorphic SiC-ceramic prepared by Si-vapour phase infiltration of wood. *Journal of the European Ceramic Society*, **22**(14–15), 2663–2668.

[14] Moulson A. J. 1979. Review: Reaction-bonded silicon nitride: its formation and properties. *Journal of Materials Science*, **14**(5), 1017–1051.

[15] Jennings H. M. 1983. On reactions between silicon and nitrogen, *Journal of Materials Science*, **18**(4), 951–967.

[16] R. G. Pigeon R. G. and Varma A. 1993. Some factors influencing the formation of reaction-bonded silicon nitride, *Journal of Materials Science*, **28**(7), 1919–1936.

[17] Pigeon R. G. and Varma A. 1993. Quantitative kinetic analysis of silicon nitridation, *Journal of Materials Science*, **28**(11), 2999–3013.

[18] Riley F. L. 2000. Silicon nitride and related materials, *Journal of the American Ceramic Society*, **83**(2), 245–265.

[19] Ziegler G., Heinrich J. and Wötting G. 1987. Review: Relationships between processing, micro structure and properties of dense and reaction-bonded silicon nitride, *Journal of Materials Science*, **22**(9), 3041–3086.

[20] Kleebe H. J. and Ziegler G. 1989. Influence of crystalline secondary phases on the densification behaviour of reaction-bonded silicon nitride during post-sintering under increased nitrogen pressure, *Journal of the American Ceramic Society*, **72**(12), 2314–17.

[21] Zhu X., Zhou Y. and Hirao K. 2004. Post-densification behaviour of reaction-bonded silicon nitride (RBSN), *Journal of Materials Science*, **39**(18), 5785–5797.

[22] Hampshire S. 1994. Nitride ceramics. In *Material Science and Technology*, **11**. Cahn R. W., Haasen P. and Kramer E. J., (Eds.). VCH Verlagsgesellschaft, Weinheim, Germany, 119–171.

[23] Claussen N., Le T. and Wu S. 1989. Low-shrinkage reaction-bonded alumina. *Journal of European Ceramics Society*, **5**(1), 29–35.

[24] Claussen N., Travitzky N. A. and Wu S. 1990. Tailoring of reaction-bonded Al_2O_3 (RBAO) ceramics. *Ceramic Engineering and Science Proceedings*, **11**(7–8), 806–820.

[25] Wu S. and Claussen N. 1992. Reaction-bonding of ZrO_2-containing Al_2O_3. *Solid State Phenomena*, **25–26**, 293–300.

[26] Hlavacek V. and Puszynski J. A. 1996. Chemical engineering aspects of advanced ceramic materials, *Industrial and Engineering Chemistry Research*, **35**(2), 349–377.

[27] She J. H., Schneider H., Inoue T., Suzuki M., Sodeoka S. and Ueno K. 2001. Fabrication of low-shrinkage reaction-bonded alumina-mullite composites. *Materials Chemistry and Physics*, **68**(1–3), 105–109.

[28] Hennige V. D., Haußelt J., Ritzhaupt-Kleissl H. J. and Windmann T. 1999. Shrinkage-free $ZrSiO_4$-ceramics: characterisation and applications, *Journal of the European Ceramics Society*, **19**(16), 2901–2908.

[29] Geßwein H., Binder J. R., Ritzhaupt-Kleissl H. J., Haußelt J. 2006. Fabrication of net shape reaction bonded oxide ceramics *Journal of the European Ceramics Society*, **26**(4–5), 697–702.

[30] Ritzhaupt-Kleissl H. J., Binder J. R., Gietzelt T. and Kotschenreuther J. 2006. Net shape reaction bonded ceramic micro parts by mechanical microstructuring, *Advanced Engineering Materials*, **8**(10), 983–988.

[31] Binder J. R., Schlechtriemen N., Jegust S., Pfrengle A., Geßwein H. and Ritzhaupt-Kleissl H. J. 2007. Green machining of net-shape reaction-bonded ceramics: materials and applications. *cfi*/Ber. DKG **84**(6), E57–E60.

[32] Ritzhaupt-Kleissl H. J., Binder J. R., Klose E. and Haußelt J. 2002. Net-shape ceramic microcosm ponents by reaction bonding. *cfi*/Ber. DKG **79**(10), E9–E12.

[33] Geßwein H. 2005. Entwicklung hochfester Net shape Oxidkeramiken im System Al_2O_3-SiO_2-ZrO_2. Doctoral Thesis, University of Freiburg, Baden-Württemberg, Germany.

[34] Gaus S. P., Harmer M. P., Chan H. M. and Caram H. S. 1999. Controlled firing of reaction-bonded aluminium oxide (RBAO) ceramic. *Journal of the American Ceramics Society*, **82**(4), 897–908.

[35] Aaron J. M., Chan H. M., Harmer M. P., Abpamano M., Caram H. S. 2005. A phenomenology cal description of the rate of the aluminum/oxygen reaction in the reaction-bonding of alumina. *Journal of the European Ceramics Society*, **25**(15), 3413–3425.

[36] Geßwein H., Binder J. R., Ritzhaupt-Kleissl H. J. and Haußelt J. 2006. Reaction–diffusion model for the reaction bonding of lumina–zirconia composites using the intermetallic compound ZrAl$_3$. *Thermochimica Acta*, **451**(1–2), 139–148.

[37] Bauer W. and Knitter R. 2002. Development of a rapid prototyping process chain for the production of ceramic microcomponents. *Journal of Materials Science*, **37**(15), 3127–3140.

[38] Bauer W., Haußelt J., Merz L., Müller M., Örlygsson G. and Rath S. 2005. Micro ceramic injection moulding. In *Microengineering of metals and ceramics - Part I: Design, tooling, and injection molding*. Löhe D. and Haußelt J.(Eds.), Wiley-VCH Verlag, Weinheim, Germany. 325–356.

[39] Ritzhaupt-Kleissl H. J., von Both H., Dauscher M. and Knitter R. 2005. Further ceramic replication techniques. In *Microengineering of metals and ceramics - Part II: Special replication techniques, automation and properties*. Löhe D. and Haußelt J., (Eds.), Wiley-VCH Verlag, Weinheim. Germany, 421–447.

[40] Pfrengle A., Binder J. R., Ritzhaupt-Kleissl H. J., Gietzelt Th. and Haußelt J. 2007. Green machining of net shape ceramics. In proceedings of the 10th International Conference of the European Ceramic Society, Berlin, 17–21 June 2007. Heinrich, J. G., Anezris C. G. (Eds), Göller Verlag, Baden–Baden, Germany, 487–492.

8

(Ba/Sr)TiO$_3$: A perovskite-type ferroelectric for tunable dielectric applications at radio frequencies

F. Paul, A. Giere and J. R. Binder

This chapter briefly reviews the materials properties and potential applications of tunable Ba$_{1-x}$Sr$_x$TiO$_3$ (BST). Predominately bulk ceramics and thick films are considered. Basic knowledge about the underlying physical mechanisms in tunable dielectrics, especially ferroelectrics is given in the light of their application in radio frequency (RF) components and systems. Materials, dielectric and tunable properties of BaTiO$_3$ and BST, especially at GHz frequencies, are discussed. Tunable RF components based on BST are presented. Some methods for the dielectric characterization of such devices are also mentioned, and the potential for commercial exploitation of BST-based devices is considered.

8.1 Tunable dielectrics for radio frequency components and systems

In the past dielectric materials were essential for applications at radio frequencies. Moreover, they still are and most probably will remain essential in the future. In addition, the tunability of dielectrics is one of the key functionalities of future radio frequency (RF) components and systems. This helps to meet the requirements of emerging trends in technology and markets. Various existing and upcoming applications are focused on mobile communication within the frequency range 0.5–10 GHz. An indicator for a dynamic increase in this field is the continuously growing number of mobile radio services like GSM900/1800, UMTS, Bluetooth, WLAN, WiMAX and CWUSB. These additional functionalities have to be integrated in a cost efficient manner and densely packed into mobile handsets. Such integration can be supported by using a single tunable filter in the RF front-end to tune the frequency response on the demands of the actual used service [1].

A simple and basic component within such a tunable RF circuit is a tunable capacitor, also known as a varactor. Its tunability can be defined by the continuously tunable capacitance in relation to its untuned capacitance value. This varactor can be realized for applications within the RF range by using components based on several different principles or technologies. Using semiconductor diodes as the varactor is the most common technique

[2], causes some difficulties with the usable frequency range, the achievable device quality factor, the power consumption and the power handling capability. In MEMS [3] cantilevers have often been used in recent years to realize continuously tunable capacitors. However, MEMS technology still has to face challenges concerning power handling capability, component size, tuning voltage, reliability and production costs [4]. In addition varactors for RF applications can be realized by using tunable dielectrics based on perovskite-type ferroelectrics or nematic liquid crystals. While the nematic liquid crystals are suitable for applications beyond 20 GHz due to their dielectric properties [5], ferroelectric-based ceramics with a tunable permittivity are suitable for various types of tunable RF components in the frequency band of mobile communications. The ferroelectric material has the advantage, that most of the required ceramic processing techniques are already established in microsystem technology. However, significant challenges remain in the fields of processing, interconnection and device design, if ferroelectric ceramics are to be integrated with semiconductors and other materials.

To understand what tunable behaviour in ceramics (and especially in perovskites) means and to give an idea of which material properties have to be taken into account if perovskite ceramics are introduced into RF-components and devices, one has to look at the polarization processes and dielectric properties.

Dielectric materials are electrically insulating materials which can modify the dielectric function of a vacuum by polarization. In general, this behaviour can be described by the dielectric displacement $\vec{D}$:

$$\vec{D} = \varepsilon_0 \vec{E} + \vec{P} \tag{8.1}$$

which depends on the vacuum distribution $\varepsilon_0 \vec{E}$ of the dielectric displacement and the electric polarization $\vec{P}$. This electric polarization can be further expressed by the susceptibility χ_e of the material to the electric field strength:

$$\vec{P} = \varepsilon_0 \chi_e \vec{E} \tag{8.2}$$

In a linear dielectric material, the susceptibility has a constant value. In RF applications many different materials are used [6]. Linear dielectrics with a high χ_e value are commonly used in capacitors to increase the charge density in the dielectric. But also dielectrics with a quite low susceptibility are also important at RF, for instance as substrates where signal lines are built on top. In this case the low susceptibility helps to increase the speed of signal propagation.

Ferroelectrics are a subclass of the vast class of dielectric materials. A ferroelectric material shows a spontaneous electric polarization which can be reoriented by an external electric field between crystallographically defined states [7]. A typical polarization curve is shown in Figure 8.1(a) where the dashed line displays a ferroelectric material without hysteresis and the opened loop shows the typical ferroelectric hysteretic behaviour of the spontaneous polarization.

The susceptibility of such a ferroelectric material can be expressed by:

$$\chi_e = \frac{1}{\varepsilon_0} \cdot \frac{\delta|\vec{P}|}{\delta|\vec{E}|} \tag{8.3}$$

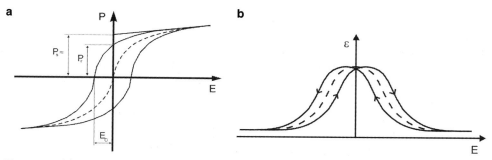

Figure 8.1 (a) Polarization loop (P–E loop) of nonlinear dielectric, ferroelectric material; and (b) tunable permittivity.

Equation (8.3) shows that the susceptibility is dependent on the external electric field. All materials exhibiting a nonlinear behaviour described by Equation (8.3) are known as nonlinear dielectrics.

For the realization of tunable RF components this nonlinear behaviour of the dielectric:

$$\varepsilon_r\left(\vec{E}\right) = 1 + \chi_e\left(\vec{E}\right) \tag{8.4}$$

material is used to tune the relative permittivity , by applying an external electric field.

Figure 8.1(b) shows the resulting E-field dependent permittivity derived from a polarization loop as shown in Figure 8.1(a). The tunability τ_ε of the dielectric material, which is based on this field-dependency of the permittivity can be defined by:

$$\tau_\varepsilon\left(\vec{E}\right) = \frac{\varepsilon_r(0) - \varepsilon_r\left(\vec{E}\right)}{\varepsilon_r(0)} \tag{8.5}$$

To realize tunable components based on ferroelectrics the dielectric properties of the material have to meet the basic requirements on dielectrics at RF in addition to the mandatory tunable behaviour. Where the demand on the tunability of the material is dependent on the application, the dielectric loss factor (tanδ) has to be as low as possible for all tuning states to avoid power loss in the RF signal. Typical values for minimum values of tanδ are below 0.01–0.02 in the frequency range of 0.5–10 GHz. At radio frequencies the nonlinear effects of the materials used and thus the components have a significant impact on tuning speed and power handling capability. Whereas the typically fast tuning speed for permittivity in ferroelectrics (which is in the nano- to picosecond range) is necessary for fast tuning on the one hand, the nonlinear behaviour of permittivity results in a harmonic generation by the RF power dependent signal [8] on the other hand.

To choose a ferroelectric material for an RF application, one has to distinguish between the different crystallographic phases of the material. Ferroelectrics with a huge remnant polarization P_r at the operating temperature, like those used for ferroelectric random access memories (FE-RAM) [9], are not suitable for RF applications, because continuous tuning is essential. Furthermore, the temperature characteristic of ferroelectrics must be considered, because in RF applications the variation in permittivity versus temperature should be as low as possible. A typical temperature characteristic for a ferroelectric single crystal material

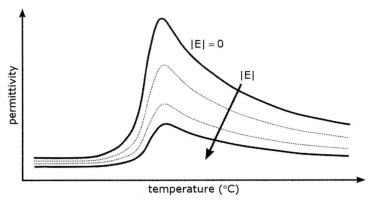

Figure 8.2 Dependency of permittivity on temperature and applied electric tuning field in a ferroelectric material.

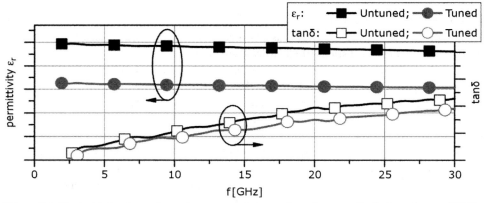

Figure 8.3 Typical behaviour of permittivity and dielectric loss-factor tanδ of a ferroelectric (BST) thick-film in the GHz range with and without tuning.

is shown schematically in Figure 8.2. This temperature dependency can be reduced by a proper processing of ferroelectric thin- and thick-films. In such films the porosity, stress at interfaces and the grain boundaries affect the crystal structure and its interaction with the electrical fields and therefore the temperature dependency of the permittivity [7, 10].

The permittivity and tunability in thin- and thick-films show little dependency on frequency at frequencies above 2 GHz over a broad frequency range when measured by using a coplanar waveguide (CPW). However, a slight decrease in both, permittivity and tunability with increasing frequency can be observed. In contrast to permittivity, dielectric loss increases with increasing frequency (see Figure 8.3).

8.2 Materials, dielectric and tunable properties of BT and BST

Crystal structure of ferroelectric BaTiO$_3$ (BT): The spontaneous polarization in ferroelectrics is connected with a noncentrosymmetric crystal structure [11]. BaTiO$_3$ (BT) was

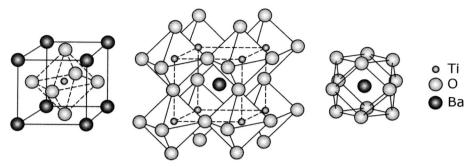

Figure 8.4 Cubic perovskite structure of BaTiO$_3$. Each Ti$_4^+$-cation is placed in the middle of an octahedron and is surrounded by six O^{2-}-anions, while each Ba^{2+} is coordinated within a cubic octahedron, consisting of 12 O^{2-}-anions.

the first ferroelectric ceramic to be discovered [12], therefore we will focus on this ferroelectric prototype first. Pure BT shows, as all other ferroelectrics do, a noncentrosymmetric crystal structure, due to a distinguished lattice distortion.

The ideal perovskite structure of cubic (centrosymmetric) BaTiO$_3$ is shown in Figure 8.4. It is called "perovskite" due to the fact that it is derived from the isomorph mineral Perovskite (CaTiO$_3$). Within this structure the Ba^{2+}-cations occupy the A-sites and the Ti^{4+}-cations occupy the B-sites. Each Ti^{4+}-cation is coordinated octahedrally by six O^{2-}-anions, placing the Ti^{4+} in the middle, while each Ba^{2+} is coordinated within a cubic octahedron, consisting of 12 O^{2-}-anions. This cubic structure in polycrystalline ceramics is stable up to a temperature of about 120°C at normal pressure. At temperatures below 120°C (the so-called Curie-point, T_c) the centrosymmetric, cubic structure is distorted along the [001]-direction and becomes noncentrosymmetric, tetragonal [13]. The cubic form is paraelectric, whereas the tetragonal structure is ferroelectric. There are conflicting reports about the exact transition temperature [12–15]. This is due to the high sensitivity of this transition point and the dielectric properties due to impurities in the sample, heating and cooling rates during measurement and, most importantly, the microstructure [12, 16, 17].

In contrast to the diffusion controlled phase transition at 1432°C [18] the cubic to tetragonal phase transition at 120°C is of displacive nature [12, 17].

At the Curie-point the lattice is distorted tetragonally along the c-axis ([001]-direction). This causes a deformation of the oxygen-octahedra, which shifts the minimum energy position of the Ti^{4+} along one axis (usually called the c-axis) towards one oxygen-anion as schematically illustrated in Figure 8.5.

Upon cooling below room temperature two further phase transitions occur at 0°C and −90°C as illustrated in Figure 8.6. The corresponding lattice parameters are listed in Table 8.1.

This shift together with a shift of the barium-cations opposite to it causes a charge separation, thus creating dipoles. The equilibrium position of the Ti^{4+}-cation may be influenced by applying an external electrical field. This effect is used when pooling ferroelectric materials

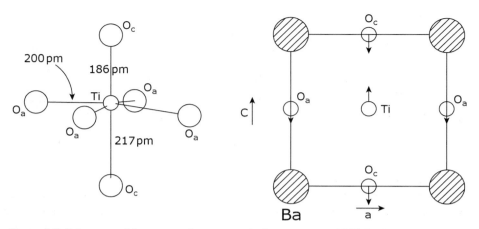

Figure 8.5 Schematic of distortion of oxygen octahedra in tetragonal BT. O_a denotes oxygen anions, on equivalent a- or b-axis, O_c denotes oxygen anions shifted along tetragonal c-axis ([001]-direction) (after Jona and Shirane [19]).

for the use as piezoelectric actuators, transducers or FE-RAMs. This spontaneous polarization leads to the inherent characteristic of ferroelectric P-E behaviour, the ferroelectric P-E hysteresis loop as shown in Figure 8.1(a). The overall direction of the permanent dipoles is directly connected to the direction of the lattice distortion. Therefore their direction changes, when the respective temperatures of the tetragonal-to-orthorhombic and orthorhombic-to-rhombohedral transition points at 0°C and −90°C are reached. Both low temperature modifications of BT are also ferroelectric (see Figure 8.6).

These phase transitions have a significant influence on the dielectric properties of BT and all other systems derived from it (e.g. BST). In particular, permittivity is directly coupled to the crystal structure and therefore is heavily dependent on the temperature as shown in

Table 8.1 Lattice parameters of BaTiO$_3$ with transition temperatures after Kay and Vousden [15]).

Phase	Transition-temperature	Lattice parameters (pm and °)			
		a	b	c	Angle
cubic	130°C	400.9	400.9	400.9	90°
tetragonal	130°C	400.3	400.3	402.2	90°
	0°C	399.2	399.2	403.5	90°
orthorhombic	0°C	401.2	398.9	401.2	89° 51,6'
	−90°C	401.3	397.6	401.3	89° 51'
rhombohedral	−90°C	399.8	399.8	399.8	89° 52,5'

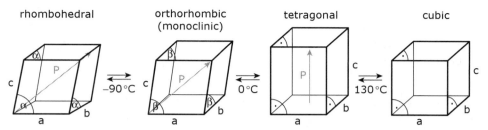

Figure 8.6 Sketch of four phases occuring in BT. P denotes direction of spontaneous polarization (rhombohedral: [111], orthorhombic: [110], tetragonal: [001]) (after Moulson and Herbert [20]).

Figure 8.7. This is very obvious in bulk ceramics. At all three phase transition temperatures there is a sudden change of permittivity with a global maximum at the temperature of the cubic–tetragonal phase transition. However, the value of permittivity is strongly dependent on grain size as small-grained ceramics and films show very shallow maxima and significantly lower permittivity and a lowered Curie-point [21–23].

At temperatures above the Curie-point, the perovskite structure becomes isotropic and permittivity is decreased to values well below those values at temperatures below the Curie-point. In this temperature range the value of permittivity follows an (empirical) Curie–Weiss behaviour [14, 19, 25, 26].

Ferroelectric behaviour and ferroelectric domains in BT: A uniform and parallel aligned ferroelectric distortion together with a uniformly directed spontaneous polarization, which one might expect at the Curie-point, would lead to a maximum in separated charges and strong electrical fields within the crystal, thus leading to an unfavourable increase in free energy and depolarization effects. This leads to the formation of alternately polarized regions with the direction of the polarization opposed by 180° [27]. These regions with

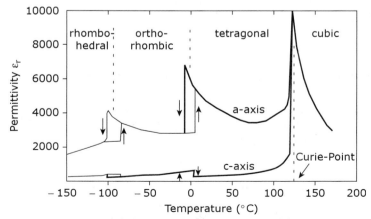

Figure 8.7 Permittivity vs. temperature of single crystal BT. Anisotropic behaviour of permittivity along the different crystallographic directions is obvious (after Merz [24]).

uniform distortion and polarization are the so-called ferroelectric domains [20, 25, 27, 28] and diminish depolarizing fields (which may be of the order of kV/mm [27]).

However, in bulk ceramics the grains are wedged by the surrounding matrix. It is unfavourable for the crystal to be distorted in only one direction, due to the built up of mechanical stress. Therefore the crystal forms additional regions, which are distorted and thus polarized in directions perpendicular to the initial 180° domains. These regions are called 90° domains [27].

The regions between two domains are called domain walls. There are two types of domain walls: ferroelectric and ferroelastic. Purely ferroelectric domain walls separate domains, which only show different directions in their polarization vectors. Regions that show only a difference in their strain tensor are called ferroelastic [27]. BT has ferroelectric and mixed ferroelectric–ferroelastic domain walls [27]. The type of domain walls which occur in a ceramic material depends on different parameters e.g. on the difference in symmetry between the ferroelectric and paraelectric phases [29].

Domain wall configurations are not stable and hence, the configurations change with time [30], under external electrical fields, under changing temperature [31–33], with decreasing grain size [16, 34–36] and upon the addition of dopants [37]. The dielectric properties of all ferroelectrics, especially at low frequencies (up to the MHz range), are influenced significantly by the characteristics of their domain structure at dynamic operation. At higher frequencies, i.e. in the GHz range, these properties seem to become less dominant. However, the dynamic behaviour of domain walls is not completely understood [27], due to the large variety of influencing parameters.

The BST system: The efficiency of the package of the constituent ions within the cubic perovskite lattice, which may be described by the general formula ABO_3 may approximately be quantified by the Goldschmidt tolerance factor, t [38, 39]:

$$t = \frac{r_A + r_O}{\sqrt{r_B + r_O}} \tag{8.6}$$

with r_i ($i = A, B, O$) denoting the ionic radii of stiff spheres on the A-, B- or oxygen site. In pure $BaTiO_3$ t equals 1.062 and in pure $SrTiO_3$ t equals 1.002, which is very close to 1, due to a more favourable ratio of the ionic radii of the ions occupying the A-site and B-site. This indicates that a substitution of Ba^{2+} by Sr^{2+} on the A-site leads to a more favourable packing within the perovskite structure.

Therefore and due to the fact, that the chemistry of Sr is very similar to the chemistry of Ba, Ba-cations can easily be substituted by Sr-cations within the perovskite lattice. This alters the general formula ABO_3 of the perovskite structure to $A^1_{1-x}A^2_xBO_3$. $SrTiO_3$ is miscible with $BaTiO_3$ in all proportions [30, 40, 41].

By exchanging Ba with Sr the properties of $BaTiO_3$ can gradually be shifted to those of $SrTiO_3$. For example, the Curie-point T_C [42, 43], the maximum permittivity value at T_C [12] (see Figure 8.8(b)) and the lattice parameters nearly linearly change from the values of $BaTiO_3$ to those of $SrTiO_3$ [30] (see Figure 8.8(a)). In addition the formation of the hexagonal, high temperature phase is suppressed by the addition of only 0.5 mass% $SrTiO_3$ [41], indicating the induced formation of a well-packed lattice.

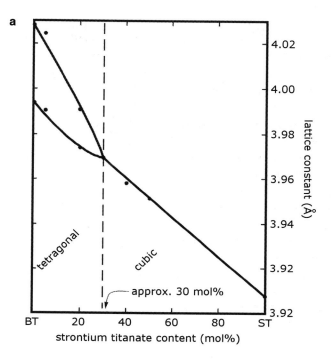

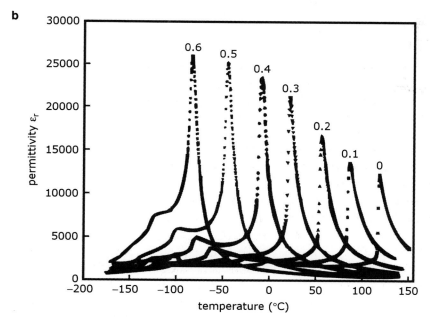

Figure 8.8 (a) Change in lattice constants with changing SrTiO$_3$ content at room temperature in BT (after [30]); (b) Influence of substitution of Ba by Sr according to the formula Ba$_{1-x}$Sr$_x$TiO$_3$ (after Jeon [42]).

However, not only dielectric or crystallographic values are influenced, but also sintering properties like grain growth are influenced by an increasing content of the refractory $SrTiO_3$ [42] ($MP(SrTiO_3) = 2050$–$2100°C$ [41]).

High frequency properties of bulk BT and BST: The influence of domain-wall movement or ferroelastic domain wall vibrations in BT and BST bulk ceramics or single crystals on their low frequency properties has been investigated thoroughly as described. However, the influence of the dynamic behaviour of the domain walls on the high frequency properties of BT and BST ceramics and films has been examined by a much smaller number of investigators [21, 44–49]. This is probably due to the difficulties of exactly measuring permittivity and dielectric loss over a broad frequency range, especially at frequencies higher than several GHz. Moreover, the pronounced impact of density and microstructure of the examined specimen at RF can be seen in Figure 8.9, where the temperature dependence of permittivity of $Ba_{0.6}Sr_{0.4}TiO_3$ (BST) is significantly different in dense bulk ceramic samples and porous ceramic thick-films, especially in the vicinity of the Curie point.

So far most of these examinations showed that BT ceramics and single crystals show a pronounced Debye-like relaxation step at frequencies higher than 10^8 Hz. Arlt *et al.* [49] expect that this relaxation is independent from temperature and becomes less pronounced at frequencies above 10^8 Hz. In contrast Kazaoui *et al.* [47, 50] found a slight temperature dependence of the relaxation-frequency in BT und BST at temperatures below the Curie point. At temperatures slightly above the Curie point they found a pronounced minimum of the relaxation frequency, which, however, increased again with rising temperature. However, doubts remain about the general appearance and frequency range of this relaxation step [51].

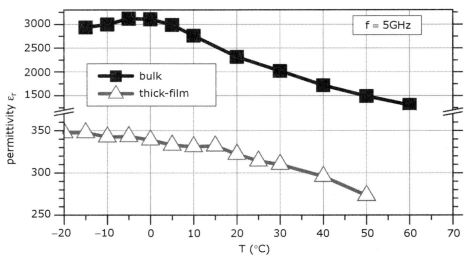

Figure 8.9 Difference in permittivity vs. temperature T (at 5 GHz) for a bulk ceramic $Ba_{0.6}Sr_{0.4}TiO_3$ and for a (porous) thick-film of $Ba_{0.6}Sr_{0.4}TiO_3$.

Arlt *et al.* [49] expect this relaxation step to stem from ferroeleastic domain-wall-vibrations. The domain-walls are not able to follow the changing electric field at higher frequencies. In contrast to Arlt's model Kazaoui *et al.* [50] expect the relaxation step to originate from an intrinsic and structure independent mechanism. Poplavko *et al.* [52] find no relaxation of permittivity in paraelectric BST25 thin-films.

The height of the relaxation step of permittivity is influenced by the grain size [21, 49, 53], isovalent substitution of Ba by Ca [47] or Sr [48] and by doping with Fe [54]. With decreasing grain size the relaxation frequency is shifted to higher frequencies and with increasing Sr-substitution the minimum of the relaxation frequency at temperatures above the Curie point is shifted to lower frequencies.

To date, the only accessible results about dielectric investigations on BST-single crystals have been published by Bethe [55].

Thin-films made of BT, especially at low frequencies, have been studied by many researchers. Therefore it is impossible to give an adequate overview within the scope of this chapter. Tagantsev *et al.* [56] have given a good overview on ferroelectric films and bulk ceramics for tunable application. However, information about tunable BST thick-films at RF is scarce in the literature. To the best of our knowledge, Weil *et al.* [57] were the first to publish results on this topic.

Stojanovic *et al.* [58] have investigated slightly doped BT thick-films. Undoped BST thick-films have been investigated by some authors in the low frequency range [59, 60] and in the RF range [57, 60–63]. The investigations of Weil [62] and Zimmermann [61] are the most profound, however, focusing mainly on methods of measuring the RF properties of thick-films. In summary one can say that undoped BST thick-films still show dielectric losses, which cause one to have reservations about their suitability for high frequency applications.

There have been no investigations of doped BST thick-films at RF except those of Paul [10, 78]. These results show that suitable doping lowers losses tremendously, thus making such doped BST thick-films suitable for tunable RF applications.

8.3 Tunable RF components based on BST

Tunable capacitance: The elementary component which is based on tunable dielectrics is a varactor. It is commonly realized in a metal–insulator–metal (MIM) topology. The capacitance value $C(U)$ of an ideal MIM capacitor can be computed by the basic formula:

$$C(U) = \varepsilon_0 \varepsilon_r (U) \frac{A}{d} \tag{8.7}$$

where the permittivity tunability is directly related to the tunability of the varactor. The electric tuning field for permittivity tuning can be computed from the applied tuning voltage U by the simple relation $E = U/d$ due to the homogeneous field distribution in the dielectric layer. A top view on such MIM varactor is shown in Figure 8.10(a) with the lower Pt electrode from the right, the upper Pt electrode from the left and the (transparent) ferroelectric BST thin-film in pink/red to form the MIM varactor in the overlapping area in the centre.

a
b

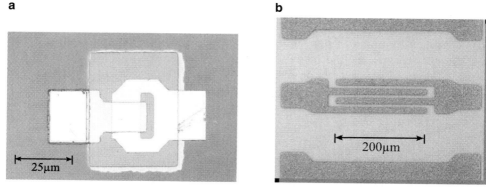

Figure 8.10 BST varactor; (a) vertical (MIM) technology using BST thin-films; (b) in planar technology by a coplanar IDC on a BST thick-film.

For processing a MIM varactor the bottom electrode must withstand the conditions during processing of the BST film on top of it. This means that it must withstand the deposition temperature or sintering of the ceramic film. Either a low deposition or sintering temperature of the ceramic film is required or a high melting temperature of the electrode is necessary. Using a standard thin-film deposition technique like e.g. RF sputtering, its process temperature has to be lower than the melting or decomposition temperature of the electrode. The microtechnology structuring of the metal and dielectric layers is possible by wet-/dry-etching and lift-off techniques.

Besides the MIM topology, planar structured capacitors are also suitable for the realization of varactors for RF applications. As an example an interdigital-capacitor (IDC) is shown in Figure 8.10(b). This type of varactor is even suitable for those BST films, which require a processing temperature, which is in the order or higher than the melting or decomposition temperature of the electrode material (usually gold). Such films usually are BST thick-films, processed by screen-printing technology and sintered at temperatures between 1100°C and 1300°C [10]. The advantage of these thick-films is that they provide an increased power handling capability, sufficiently small component size, high applicable tuning voltages and highly competitive production costs compared to thin-films.

Whereas the electric field in the dielectric is homogeneously distributed, if a MIM topology is used, the electric fields between the fingers of an IDC are inhomogeneous. Therefore conformal mapping (CM) methods must be used [64, 65] to compute the capacitance values for the untuned state, if the permittivity in the tunable film is constant. The calculation of the tuned varactor value of such a planar structure and its tunability is proposed in [66, 67] using an adapted EM-solver.

Regardless of the chosen topology for an application, the kind of metal chosen for the electrodes determines the achievable quality factor Q of the RF component. As shown in Equation (8.8), the varactor's Q-factor results from an addition of the dielectric ($\tan\delta_{dielectric}$) and the metal induced loss factor ($\tan\delta_{metal}$) [68]:

$$Q \approx \frac{1}{\tan\delta_{dielectric} + \tan\delta_{metal}} \qquad (8.8)$$

Where the dielectric loss factor is a property of the used BST film, the metal loss factor can significantly affect the Q-factor and limits the overall device performance in the RF range. The finite conductivity and thickness of the metal electrodes, in contact with the dielectric film, results in a serial resistance which is related to the metal loss factor. At RF the skin effect also has to be considered, which reduces the effective conductivity when the electrodes thickness $d_{electrode}$ is less than $d_{electrode} < 3\delta_{skin}$. The frequency-dependent skin depth, δ_{skin}, can be calculated by using Equation (8.9):

$$\delta_{skin} = \sqrt{\frac{1}{\pi f \mu_0 \mu_r \sigma}} \tag{8.9}$$

Within the RF range of mobile communication from 0.5 GHz up to 10 GHz the required thickness must be of the order of 22 μm (0.5 GHz) to 4.9 μm (10 GHz) using Pt electrodes to fulfil this requirement. It can be reduced by using metals like Au or Al with a higher conductivity. This calculation illustrates that for a usual metallization where (due to technological reasons) the electrode thickness is in the order of 1–4 μm, so the effective conductivity of the electrode is reduced. This increases the overall losses, which are caused by the electrode.

For the practical realization of high performance varactors the challenge is within the used microsystem technology. For instance, the metallic losses can be reduced by the combination of different metals to increase the effective conductivity of the electrodes. Therefore the Pt electrode which is often used as interface electrode for sputtered BST films can be enhanced by electroplating of Au, Cu or Al on the top where the electrodes are accessible.

A further optimization and integration of BST varactors by new material combinations has been shown [69, 70]. Kuylenstierna et al. [69] realized an integration of the BST varactor in a multilayer BCB substrate, Giere et al. [70] have improved the varactor performance by sealing the surface with Parylene C (see Figure 8.11) and eliminating limitations from the air-gap between the planar electrodes.

a   **b**

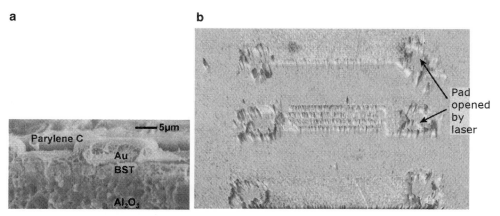

Figure 8.11 Parylene C coated ferroelectric planar IDC: (a) SEM picture of IDC's cross-section; (b) top-view by confocal microscope where colours are related to height. Image shows pads opened by laser to access IDC through Parylene C layer.

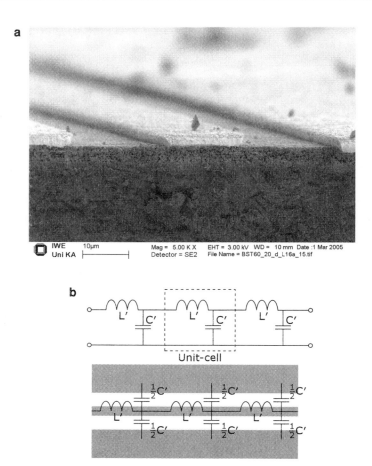

Figure 8.12 (a) Cross-section of a coplanar waveguide on a BST thick-film and; (b) equivalent circuit of this transmission line and its unit-cell formed by C' and L' (capacitance and inductance per unit-length) for lossless case.

Phaseshifters: In addition to a varactor as a discrete component, distributed elements are essential to build up modern RF communication systems. A typically distributed component is a transmission line for RF signals. In integrated RF circuits, CPWs are often used, realized by planar metallization (Figure 8.12(a)). The equivalent circuit of the CPW is shown in Figure 8.12(b).

Realizing the CPW on top of tunable dielectrics such as a BST thick-film, the capacitance of the unit cells becomes tunable which enables us to vary the signal propagation characteristics such as the phase velocity γ and the impedance Z of the line, as shown in Equation (8.10).

$$\gamma(U) = j\omega\sqrt{L'C'(U)} \qquad Z(U) = \sqrt{L'/C'(U)} \qquad (8.10)$$

This functionality can be further used to build up tunable phase-shifters which are commonly used in many RF circuits and systems. The tunability of the line impedance is used in tunable matching-networks for the transformation of characteristic impedances of connected circuits.

In addition to this distributed RF phase-shifter component, an equivalent device can be build up from discrete tunable elements positioned as in the equivalent circuit shown in Figure 8.12(b). Even though the integration of the discrete tunable components increases the complexity of the circuits and therefore the demands on the manufacturing process, the performance of the device can be increased due to a reduction in the parasitic losses like those deriving from the metal.

Characterization of tunable dielectrics: The discussed coplanar varactor concepts and the distributed phase shifters can be used for the characterization of the tunable dielectrics, which have been used as the substrate for these components. To be able to extract the complex relative permittivity of the dielectric, an analytic model for the component has to be available and measurement results with at least two suitable component characteristics like C and Q of a MIM varactor and the simplified equation (Equation (8.7)) are necessary to extract the complex relative permittivity. Further information on the characterization techniques applicable to tunable dielectrics at radio frequencies has been given elsewhere [56, 70, 71].

Further tunable RF-circuits based on tunable components: By integrating and combining these basic and tunable RF components complex RF circuits like tunable filters [72, 73], tunable voltage-controlled oscillators [74] and tunable matching networks [75–77] [Tom2007] can be built up.

However, there are significant challenges in microsystem technology, materials integration and materials science, if basic tunable components, as briefly described in this chapter, should be integrated into more advanced components, devices or even systems with an increased level of functionality. Nevertheless, an increased level of research on this material and its integration very likely will bring about the breakthrough of tunable, BST-based components from research to commercialization in the future.

References

[1] Prismark Partners LLC. 2004. The Prismark Wireless Technology Report, May 2004. Cold Spring Harbour, New York, USA.

[2] Fischer G., Eckl W. and Kaminski G. 2002. RF-MEMS and SiC/GaN as enabling technologies for a reconfigurable multi-band/multi-standard radio. In *Bell Labs Technical Journal*, 7, 169–189.

[3] Marsan E., Gauthier J., Chaker M. and Wu K. 2005. Tunable microwave device: status and perspective. In *Proceedings of the 3rd International IEEE-NEWCAS Conference*, 279–282.

[4] Chang C. 2003. MEMS for telecommunications: devices and reliability. In *Proceedings of the Custom Integrated Circuits Conference the IEEE 2003*, 199–206.

[5] Mueller S., Penirschke A., Damm C., Scheele P., Wittek M., Weil C. and Jakoby, R. 2005. Broad-band microwave characterization of liquid crystals using a temperature-controlled coaxial transmission line. *IEEE Transactions on Microwave Theory and Techniques*, **53**(6), 1937–1945.

[6] Wersing W. 1996. Microwave ceramics for resonators and filters, *Current Opinion in Solid State and Material Science*, **1**, 715–731.

[7] Waser R. 2003. In *Nanoelectronics and Information Technology*, Wiley–VCH, Verlag, Weinheim, Germany.

[8] Booth J. C., Ono R. H., Takeuchi I. And Chang K. 2002. Microwave frequency tuning and harmonic generation in ferroelectric thin film transmission lines. *Applied Physics Letters*, **81**(4), 718–720. **2006**

[9] Scott James F. 2000. In *Ferroelectric Memories*. Springer Verlag, Heidelberg, Germany. **3**

[10] Paul F. 2006. Dotierte $Ba_{0,6}Sr_{0,4}TiO_3$-Dickschichten als steuerbare Dielektrika -Pulversynthese und dielektrische Eigenschaften, PhD thesis Albert-Ludwigs-Universität Freiburg im Breisgau, Germany.

[11] Kittel C. 2005. *Introduction to Solid State Physics*, 8th E., Wiley, Chichester, United Kingdom.

[12] Jaffe B., Cook W. R. and Jaffe H. 1971. *Piezoelectric Ceramics*. Academic Press, London, United Kingdom.

[13] Xu Y. 1991. *Ferroelectric Materials and Their Applications*. Elsevier, Amsterdam, Holland.

[14] Lines M.E. and Glass A. M. 1977. *Principles and Applications of Ferroelectrics and Related Materials*. Oxford University Press, Oxford, United Kingdom.

[15] Kay H.F. and Vousden P. 1949. Symmetry changes in barium titanate at low temperatures and their relation to its ferroelectric properties. *Philosophical Magazine (Series 7)*, **40**(309), 1019–1040.

[16] Arlt G., Hennings D. and de With, G. 1985. Dielectric properties of fine-grained barium titanate ceramics, *Journal of Applied Physics*, **58**(4), 1619–1625.

[17] Kingery W., Bowen H. and Uhlmann D. 1976. *Introduction to Ceramics*, Wiley, Chichester, United Kingdom.

[18] Kirby K. and Wechsler B. 1991. Phase Relations in the Barium Titanate-Titanium Oxide System. *Journal of the American Ceramic Society*, **74**(8), 1841–1847.

[19] Jona F. and Shirane G. 1962. *Ferroelectric Crystals*, Pergamon Press. **1**

[20] Moulson A. and Herbert J. 1990. *Electroceramics*. Chapman & Hall.

[21] McNeal M.P., Jang S.J. and Newnham R.E. 1998. The effect of grain and particle size on the microwave properties of barium titanate ($BaTiO_3$). *Journal of Applied Physics*, **83**(6), 3288–3297.

[22] Hoffmann S. 1998. Modifizierte Erdalkalititanat-Dünnschichten für integrierte Bauelemente: Morphologie, dielektrische Eigenschaften und Ladungstransportmechanismen, PhD thesis, VDI Verlag, Düsseldorf, Germany.

[23] Zhang L., Zhong W. and Wang C. 1998. Dielectric relaxation in barium strontium titanate. *Solid State Communications*, **107**(12), 769–773.

[24] Merz W. 1949. The electrical and optical behaviour of $BaTiO_3$ single-domain crystals. *Physical Review*, **76**(8), 1221–1225.

[25] Sonin A. and Strukov B. 1974. *Einführung in die Ferroelektrizität*, Vieweg, Braunschweig, Germany.

[26] Rupprecht G. and Bell R. 1964. Dielectric constant in paraelectric perovskites. *Physical Review*, **135**(3A), A748–A752.

[27] Damjanovic D. 1998. Ferroelectric, dielectric and piezoelectric properties of ferroelectric thin films and ceramics. *Reports on Progress in Physics*, **61**(9), 1267–1324.

[28] Swartz S. 1990. Topics in Electronic Ceramics. *IEEE Transactions on Electrical Insulation*, **25**(5), 935–987.

[29] Fousek J. and Janovec V. 1969. The orientation of domain walls in twinned ferroelectric crystals. *Journal of Applied Physics*, **40**(1), 135–142.

[30] McQuarrie M. and Buessem W. 1955. The aging effect in barium titanate. *Ceramic Bulletin*, **34**, 402–406.

[31] Plessner K. W. 1956. Ageing of the dielectric properties of barium titanate ceramics. *Proceedings of the Physical Society (London)*, **69**(12), 1261–1268.

[32] Bradt R. and Ansell G. 1969. Aging in tetragonal ferroelectric barium titanate. *Journal of the American Ceramic Society*, **52**(4), 192–199.

[33] Schulze W. and Ogino K. 1988. Review of literature on aging of dielectrics. *Ferroelectrics*, **87**(1), 361–377.

[34] Bell A. and Moulson A. 1985. The effect of grain size on the dielectric properties of barium titanate ceramic. *British Ceramic Proceedings*, **36**, 57–66.

[35] Kinoshita K. and Yamaji A. 1976. Grain-size effects on dielectric properties in barium titanate ceramics. *Journal of applied Physics*, **47**(1), 371–373.

[36] Buscaglia V., Buscaglia M.T., Viviani M., Mitoseriu L., Nanni P., Trefiletti V., Piaggio P., Gregora I., Ostapchuk T., Pokorný J. and Petzel, J. Grain size and grain boundary-related effects on the properties of nanocrystalline barium titanate ceramics. *Journal of the European Ceramic Society*, **26**, 2889–2898.

[37] Robels U., Zadon C. and Arlt G. 1992. Linearization of dielectric nonlinearity by internal bias fields. *Ferroelectrics*, **133**, 163–168.

[38] Tsur Y., Dunbar T.D. and Randall, C.A. 2001. Crystal and defect chemistry of rare earth cations in BaTiO$_3$. *Journal of Electroceramics*, **7**, 25–3.

[39] Goldschmidt V.M. and Dybwad, J., (Eds.), 1926. *Die Gesetze der Kristallochemie Geochemische Verteilungsgesetze der Elemente (I-VIII)*, *Norske Videnskaps-Akademi*, Oslo, Norway, **VII**, 1–116.

[40] Kwestroo W. and Paping H. 1959. The Systems BaO–SrO–TiO$_2$, BaO–CaO-TiO$_2$, and SrO–CaO–TiO$_2$. *Journal of the American Ceramic Society*, **42**, 292–299.

[41] Basmajian J. and DeVries R. 1957. Phase Equilibria in the System BaTiO$_3$ -SrTiO$_3$. *Journal of the American Ceramic Society*, **40**, 373–376.

[42] Jeon J. 2004. Effect of SrTiO$_3$ concentration and sintering temperature on microstructure and dielectric constant of Ba$_x$Sr$_{1-x}$TiO$_3$. *Journal of the European Ceramic Society*, **24**, 1045–1048.

[43] Rushman D., Strivens M. 1946. Permittivity of polycrystalls of the perovskite type. *Transactions of the Faraday Society*, **42A**(2), 231–238.

[44] Turik A. and Shevchenko N. 1979. Dielectric spectrum of BaTiO$_3$ single crystals. *Physica Status Solidi (b)*, **95**(2), 585–592.

[45] Poplavko Y.M., Tsykalov V.G. and Molchanov V.I. 1969. Microwave dielectrics dispersion of the ferroelectric and paraelectric phases of barium titanate. *Soviet Physics - Solid State*, **10**(2), 2708–2710.

[46] Kazaoui S. and Ravez J. 1991. Céramiques de type BaTiO$_3$: Influence des substitions Zr-Ti et Hf-Ti sur la dispersion diélectrique en hyperfréquences. *Physica Status Solidi (a)*, **123**(1), 165–170.

[47] Kazaoui S. and Ravez J. 1992. Hyperfrequency dielectric relaxation in ferroelectric ceramics with composition (Ba$_{1-x}$Ca$_x$)TiO$_3$. *Physica Status Solidi (a)*, **130**, 227–237.

[48] Kazaoui S. and Ravez J. 1991. Dielectric dipolar - type relaxation in ferroelectric ceramics with composition (Ba$_{1-y}$Sr$_y$)TiO$_3$. *Physica Status Solidi (a)*, **125**(2), 715–722.

[49] Arlt G., Böttger U. and Witte S. 1994. Dielectric dispersion of ferroelectric ceramics and single crystals at microwave frequencies. *Annalen der Physik*, **506**(7), 578–588.

[50] Kazaoui S., Ravez J. and Elissalde C. 1992. High frequency dielectric relaxation in BaTiO$_3$ derived materials. *Ferroelectrics*, **135**(1), 85–99.

[51] Elissalde C. and Ravez J. 2001. Ferroelectric ceramics: defects and dielectric relaxations. *Journal of Materials Chemistry*, **11**, 1957–1967.

[52] Poplavko Y. and Cho N. 1999. Clamping effect on the microwave properties of ferroelectric thin films. *Semiconductor Science and Technology*, **14**(11), 961–966.

[53] Zhang L., Zhong W.L. Wang C.L., Zhang P.L. and Wang Y.G. 1998. Dielectric properties of $Ba_{0.7}Sr_{0.3}TiO_3$ ceramics with different grain size. *Physica status solidi (a)*, **168**(2), 543–548.

[54] Maglione M., Böhmer R., Loidl A. and Höchli U. 1989. Polar relaxation mode in pure and iron-doped barium titanate. *Physical Review B*, **40**(16), 11441–11444.

[55] Bethe K. and Welz F. 1971. Preparation and properties of (Ba, Sr) TiO_3 single crystals. *Materials Research Bulletin*, **6**, 209–218.

[56] Tagantsev A., Sherman V., Astafiev K., Venkatesh J. and Setter N. 2003. Ferroelectric materials for microwave tunable applications. *Journal of Electroceramics*, **11**(1), 5–66.

[57] Weil C., Wang P., Downar H., Wenger J. and Jakoby R. 2000. Ferroelectric thick film ceramics for tunable microwave coplanar phase shifters. *Frequenz*, **11–12**, 250–256.

[58] Stojanovic B., Foschini C.R. and Pavloviv V. 2002. Barium titanate screen-printed thick films. *Ceramics International*, **28**(3), 292–298.

[59] Zimmermann F., Voigts M. and Weil C. 2001. Investigation of barium strontium titanate thick films for tunable phase shifters. *Journal of the European Ceramic Society*, **21**, 2019–2023.

[60] Ditum C.M. and Button T.W. 2003. Screen printed barium strontium titanate films for microwave applications. *Journal of the European Ceramic Society*, **23**(14), 2693–2697.

[61] Zimmermann F. 2003. Steuerbare Mikrowellendielektrika aus ferroelektrischen Dickschichten, PhD thesis, Universität Karlsruhe, Baden Württemberg, Germany.

[62] Weil C. 2003. Passiv steuerbare Mikrowellenphasenschieber auf der Basis nichtlinearer Dielektrika, PhD thesis Fachbereich Elektrotechnik und Informationstechnik der Technischen Universität Darmstadt.

[63] Su B., Holmes J.E., Meggs C. 2003. Dielectric and microwave properties of barium strontium titanate (BST) thick films on alumina substrates.n *Journal of the European Ceramic Society*, **23**(14), 2699–2703.

[64] Heinrich W. 1993. Quasi-TEM description of MMIC coplanar lines including conductor-loss effects. *IEEE Transactions on Microwave Theory and Techniques*, **41**(1), 45–52.

[65] Gevorgian S., Martinsson T., Linner P. and Kollberg E. 1996. CAD models for multilayered substrate interdigital capacitors. *IEEE Transactions on Microwave Theory and Techniques*, **44**(6), 896–904.

[66] Giere A., Scheele P., Damm C. and Jakoby, R. 2005. Optimization of uniplanar multilayer structures using nonlinear tunable dielectrics. In *Proceedings of the 35th European Microwave Conference*, 4–6 October 2005, Paris, France, 561–564.

[67] Giere A., Scheele P., Zheng Y. and Jakoby R. 2007. Characterization of the field-dependent permittivity of nonlinear ferroelectric films using tunable coplanar lines. *IEEE Microwave and Wireless Components Letters*, **17**(6), 442–444.

[68] Maria J.P., Boyette B.A., Kingon A.I., Ragalia C., Stauf G. 2005. Low loss tungsten-based electrode technology for microwave frequency BST varactors. *Journal of Electroceramics*, **14**(1), 75–81.

[69] Kuylenstierna D., Vorobiev A. and Gevorgian S. 2004. Integration of parallel-plate ferroelectric varactors with BCB-on-silicon microstrip circuits. *IEEE MTT-S International Microwave Symposium Digest*, 6–11 June 2004, Fort Worth, Texas, USA, **3**, 1907–1910.

[70] Giere A., Zheng Y., Gieser H., Marquardt K., Wolf H., Scheele P. and Jakoby R. 2007. Coating of planar barium-strontium-titanate thick-film varactors to increase tunability. In *Proceedings of the 37th European Microwave Conference*, 9–12 October, 2007, Munich, Germany, 114–117.

[71] Chen L., Varadan V.K. 2004. Microwave electronics: measurement and materials characterization. Wiley, Chichester, United Kingdom.

[72] Tombak A., Maria J.P., Ayguavives F.T., Jin Zhang, Stauf G.T., Kingon A.I. and Mortazawi A. 2003. Voltage-controlled RF Filters employing thin-film barium-strontium-titanate tunable capacitors. *IEEE Transactions on Microwave Theory and Techniques*, **51**(2), 462–467.

[73] Papapolymerou J., Lugo C., Zhao Z., Wang X. and Hunt A. 2006. A miniature low-loss slow-wave tunable ferroelectric bandpass filter from 11–14 GHz. *Proceedings of the IEEE MTT-S International Microwave Symposium Digest*, 11–16 June 2006, San Francisco, California, USA 556–559.

[74] Norling M., Vorobiev A., Jacobsson H. and Gevorgian S. 2007. A low-noise K-Band VCO based on room-temperature ferroelectric varactors. *IEEE Transactions on Microwave Theory and Techniques*, **55**(2), 361–369.

[75] Scheele P., Goelden F., Giere A., Mueller S. and Jakoby R. 2005. Continuously tunable impedance matching network using ferroelectric varactors. *Microwave Symposium Digest, 2005 IEEE MTT-S International*, 12–17 June 2005, Long Beach, California, USA.

[76] Schmidt M., Lourandakis E., Leidl, A., Seitz S. and Weigel R., 2007. A comparison of tunableferroelectric Pi and T-matching networks. *Proceedings of the 37th European Microwave Conference*, 9–12 October 2007, Munich, Germany.

[77] Tombak A. 2007. A ferroelectric-capacitor-based tunable matching network for quad-band cellular power amplifiers. *IEEE Transactions on Microwave Theory and Techniques*, **55**(2), 370–375.

[78] Paul F., Giere A., Menesklou W., Binder J.R., Scheele P., Jakoby R., HauBelt J. 2008. Influence of Fe-F-codoping on the dielectric properties of Ba$_{0.6}$Sr$_{0.4}$TiO$_3$ thick films. In *International Journal of Materials Research*, **93**(10) 1119–1128.

9

Properties and application of polymer–ceramic-composites in microsystem technologies

T. Hanemann, J. Böhm, K. Honnef, R. Heldele and B. Schumacher

Besides the established material classes in microsystem technologies like silicon, polymers, metals and ceramics polymer-based composites combine the tailoring of new material properties and hence new applications with the simple processability and mass fabrication potential of polymers. Two main application fields of polymer–ceramic-composites are investigated in detail: first the use of these composites as feedstocks for the realization of dense ceramic parts applying the different variants of ceramic injection molding exploiting the good flow properties of polymer melts. Secondly the modification of polymer properties by the addition of nano- and microsized fillers or dopants.

In this chapter the changes in different polymer properties such as the coefficient of thermal expansion, Vickers hardness, flow behaviour, refractive index or relative permittivity with solid ceramic load will be discussed in detail.

9.1 Introduction

Transferring the 4M name (Multi-Material Micro Manufacturing) to materials established in microsystem technology, four basic groups of materials can be identified:

- silicon and other semiconductors;
- polymers;
- metals; and
- ceramics and glasses.

Devices made from silicon can be realized easily exploiting the established MEMS technologies. Mechanical microengineering techniques like micromilling or fly cutting, electric discharge machining, electroplating, casting, and, quite recently, powder-derived techniques enable the fabrication of metal microcomponents. Different variants of casting and moulding methods like spin coating, reaction moulding, hot embossing or injection moulding allow a low-cost replication of polymeric microcomponents. Comprehensive overviews of the state-of-the-art can be found in the proceedings of the four 4M Conferences [1–4]. In the last few years polymer-based composites containing passive or active fillers

120

as well as special dopants have become increasingly important in microsystem technologies within two main application fields: first, feedstocks used in PIM consist of a binder based on easy flowing organic moieties, different additives, and a high solid metal or ceramic load up to 60 vol%. After replication and thermal postprocessing, i.e. pyrolysis of the organic content, dense metal or ceramic parts can be obtained [5–8]. In general, the polymer is only used as liquefier of the solid powder enabling a moulding process and will be removed after replication. Secondly, in polymer-matrix composites the physical properties of a polymer are improved by the addition of micro- or nanosized inorganic or organic dopants enabling new functionalities and applications [9–11]. Examples for both composite types and related applications will be presented in the following sections.

9.2 Composites with high solid load for feedstock development

In the past two types of PIM have been established for the realization of microstructured ceramic parts: low pressure CIM for rapid prototyping or small-scale fabrication using wax-based binders and high pressure injection moulding (HPIM) on the basis of thermo-plastic feedstocks for large-scale or mass production [5–8, 12–14]. A new replication technique, composite reaction moulding, suitable for the realization of prototypes has been developed; dense microstructured alumina or zirconia parts have been produced in good quality [15, 16].

All moulding processes have the following process sequence in common:

(1)   ceramic (or metal) powder conditioning: milling, drying, surface modification, a.o.;
(2)   feedstock preparation: applied equipment (dissolver, mixer-kneader, extruder) depends on feedstock viscosity and process temperature;
(3)   moulding: applied replication technique depends on feedstock viscosity and process temperature;
(4)   debinding: method (thermal, solvent assisted) depends on binder composition; and
(5)   sintering.

Table 9.1 lists the different binder composition and compounding process characteristics for the three moulding techniques; Figure 9.1 shows the related compounding equipment

Table 9.1 Comparison of feedstock composition and processing.

	Composite reaction moulding	LPIM	HPIM
Main binder component	curable reactive polymer resin	wax	thermoplastics, wax
Compounding temperature (°C)	25	70–90	150–200
Feedstock viscosity (Pa s)	0.1–10	2–20	< 4000
Processable ceramic load (vol%)	< 50	< 70	< 70
Suitable for	Rapid prototyping	Rapid protyping Small-scale fabrication	Mass fabrication

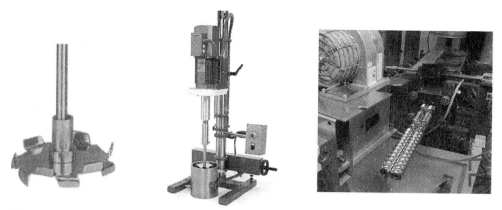

Figure 9.1 Compounding equipment for polymer-based composites sorted with increasing composite viscosity: laboratory dissolver-like stirrer, high power dissolver, extruder.

starting with a simple laboratory stirrer for low viscosity composites, suitable for rapid proto-typing using reactive resins or wax-based binders and ending up with extrusion techniques for thermoplastic binder-based feedstocks for HPIM suitable for mass fabrication in industry.

Polymer-based reactive resins consist of a polymer-like polymethylmethacrylate or unsaturated polyester, solved in its reactive monomer like methylmethacrylate or styrene. These resins show a low viscosity below 10 Pa s under ambient conditions. A further reduction in viscosity down to values around 0.1 Pa s can be achieved by the addition of the reactive monomer. After solidification by radical polymerization a thermoplastic polymer can be obtained. Hence reactive resins can be treated as model binders for HPIM exploiting, in contrast to thermoplastic feedstock development, the simple processing capabilities like low viscosity under ambient conditions. In addition, small volume processing, which is essential for testing new or expensive fillers and intense material screening experiments like the influence of dispersants on the feedstock flow behaviour, is possible. The obtained results can be transferred after feedstock optimization like dispersant type and concentration to thermoplastic feedstocks suitable for industrial use.

The processability of a composite or feedstock depends strongly on the solid load, on the average particle size, the particle size distribution, and especially on the specific surface area. With decreasing particle size towards the nanoscale the specific surface area measured via the Brunauer–Emmett–Teller (BET) method increases disproportionately. It was expected that the surface roughness of a sintered ceramic part could be reduced by using nanosized instead of microsized particles or mixtures of both in the feedstock base material. Table 9.2 lists the specific surface area and average particle size for a series of commercial micro- and nanosized alumina. Despite the fact that the d50-values of the TM DAR, the Nanotek and the type C alumina are quite similar, the specific surface areas differ by a factor up to 10.

Figure 9.2 shows the relative viscosity change of alumina-filled composites with an unsaturated polyester resin as binder at 60°C and a shear rate of 10 s^{-1}. The relative viscosity is defined as the apparent viscosity of the composite divided by the pure binder viscosity allowing a better comparison of different composite systems. The addition of the microsized

Table 9.2 Specifications of micro- and nanosized alumina (*vendor's data).

Type	Vendor	Specific surface area (m²/g)	Average particle size d$_{50}$ (µm)
CT3000SG	Almatis	6–8*	0.266
RCHP	Baikowski	7–9*	0.335
TM DAR	Taimei	12.8	0.165
Nanotek	Nanophase	34	0.155
C	Degussa	107	0.155

filler does not significantly affect the relative viscosity up to a load of 30 vol%, once this point has been reached a pronounced rise then follows. In contrast, the nanosized alumina causes a strong relative viscosity increase even at low loads below 10 vol%, especially the type C alumina with its tremendous surface area beyond 100 m²/g. Hence the processing of composites with nanosized inorganic fillers is quite difficult due to the large increase in viscosity with load [17–21]. Feedstocks containing only nanosized ceramics cannot be used for dense ceramic parts due to the small accessible solid load. Bimodal mixtures of micro- and nanosized ceramics enable the realization of dense ceramic parts by exploiting the enhanced sinter activity of nanosized particles [16].

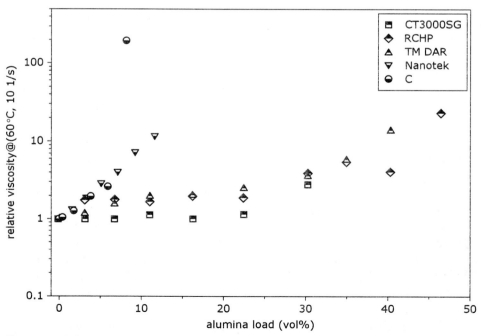

Figure 9.2 Relative viscosity change with alumina load at 60°C and a shear rate of 10 1/s.

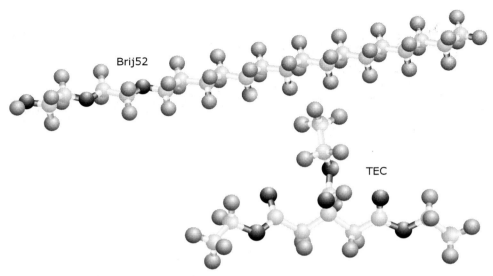

Figure 9.3 Dispersant molecules investigated for composite development (top: Brij52, bottom: triethylcitrate (TEC)).

Large solid loads and a homogeneous filler distribution can only be achieved if dispersants have been selected which connect the hydrophilic ceramics and the hydrophobic polymer matrix. Suitable organic molecules must possess an amphiphilic molecular structure like the monofunctional polyethylene-glycol-alkylethers (Brij®-type dispersants) which carry a nonpolar aliphatic tail and a polar polyethylene-glycole head (Figure 9.3, top). The citric acid derivatives (Figure 9.3, bottom) are a quite different group of suitable molecules with a larger number of potential coupling positions. The amphiphilic character of a molecule can be estimated empirically from the molecular structure calculating the hydrophilic–lipophilic balance (HLB) value introduced by Griffin [22, 23] (see Equation (9.1)). $M_{hydrophilic}$ is the sum of the mass of all hydrophilic groups in the molecule, M_{tot} is the total molecular mass. A HLB value of 0 corresponds to a completely hydrophobic molecule, a value of 20 would correspond to a molecule carrying only hydrophilic moieties. Molecules with a HLB value in the range 3–6 can be used as a water-in-oil emulsifier, those in the range 7–9 can be used as surfactants, and in the range 8–12 they can be used as an oil-in-water emulsifier; molecules with HLB values larger than 12 behave as a detergent [24]:

$$HLB = 20 \ \frac{M_{hydrophilic}}{M_{tot}} \tag{9.1}$$

The polyethylene-glycol-alkylethers possess a HLB value of about 5, the citrates between 6 and 8 depending on the numbers of ester groups in the molecule. It has been shown that the monofunctional polyethylene-glycol-alkylethers allow a viscosity reduction at constant load while the polyfunctional citrates complicate the compounding process due to network formation resulting in an increase in viscosity [21, 25]. Figure 9.4 shows sintered ZrO_2 bending bars on substrate fabricated by composite reaction moulding and polyester-based

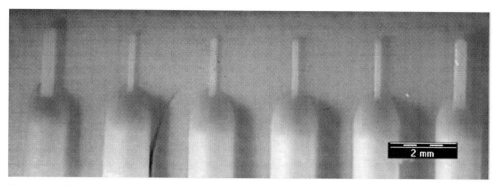

Figure 9.4 Zirconium dioxide bending bars fabricated by composite reaction moulding.

reactive resin binder applying a Brij®-type dispersant. The sinter density is around 98%, the Vickers hardness around 1700 HV0.4 [26]. Due to the relatively low solid load causing stress during the the the sintering a crack occurred in the ceramic part. As a result of the dispersing agent screening Brij®-type, citric acid derivatives and stearic acid-based dispersants have been introduced into thermoplastic feedstock development using mixtures of wax and polyethylene as main binder components suitable for micro-PIM of dense alumina or ZrO_2 parts [16, 25, 27, 28] enabling a mass production suitable for industrial purposes.

9.3 Functional composite development

9.3.1 Adjustment of the thermal expansion coefficient

With respect to the combination of polymer layers with ceramics an adjustment of the coefficient of thermal expansion (CTE) is necessary to prevent delamination during the thermal treatment due to the large mismatch in thermal expansion (polymers: 10^{-4} 1/K, silicon or ceramics 10^{-6} 1/K). In general, in binary composites consisting of two different components 1 and 2 with individual CTEs, α_1 and α_2, an approximate linear dependency of the CTE $\alpha_{composite}$ as function of the volume fractions can be found. Thermal expansion coefficient:

$$\alpha_{composite} = \frac{\alpha_1 V_1 + \alpha_2 V_2}{V_{composite}} \tag{9.2}$$

Following the two individual CTE-values for the polyester and the alumina (2×10^{-4} 1/K and 6×10^{-6} 1/K) an almost linear dependence of the CTE on the alumina load could be proved experimentally using dilatometry [16]. Hence a composite with adjustable thermal expansion can be realized e.g. as adhesive interface layer between a polymer and silicon or pure ceramics for improved stability of a laminate-type composite especially for alternating exposure to heat and cold.

9.3.2 Adjustment of Vickers hardness

In contrast to ceramics or metals, polymers show relatively low surface hardness values when measured by the Vickers test method [29]. The addition of ceramic fillers enables a significant increase in the thermomechanical properties like Young's modulus, tensile

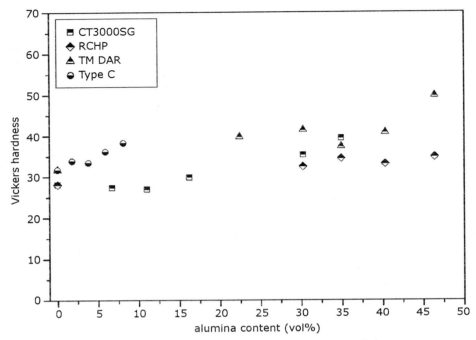

Figure 9.5 Vickers hardness adjustment as function of micro- or nanosized alumina content.

strength or specific stiffness [30]. The reliable estimation of the microhardness is somewhat difficult due to the presence of two different components with large differences in hardness and the small measuring area, hence only an average value coming from a large number of individual measured test results taken at different measuring positions can be used. Figure 9.5 shows the change in the Vickers hardness with solid load in the composite for the investigated alumina. The addition of the two alumina CT3000SG and RCHP with a BET-value below 10 m²/g causes a slight hardness increase. The doubling of the specific area (Table 9.2) for the TM DAR alumina has a strong impact on the polymers hardness resulting in a rise of around 60%. The nanosized alumina type C with a large specific surface area of 107 m²/g and the resulting immobilization of a huge number of polymer chains causes a pronounced hardness increase at low loads up to 8 vol%. Hence small amounts of nanosized alumina can be added in order to achieve an improved polymer surface hardness for special applications.

9.3.3 Refractive index modification

As described earlier the addition of small amounts of nanosized particles with large specific surface areas has a pronounced impact on viscosity or hardness. Nanosized ceramics with a large index of refraction can be used for the refractive index modification of highly transparent polymers, if all particles sizes (primary ones and agglomerates) are below 1/10 of the wavelength of interest, thus avoiding Rayleigh scattering [31]. In the case of waveguide

Table 9.3 Refractive index values of PMMA and different commercial nanosized ceramics.

Material	Refractive index @633 nm
PMMA	1.49
amorphous SiO$_2$ (Aerosil)	1.45
Al$_2$O$_3$	1.76
ZrO$_2$	2.20
TiO$_2$	2.49–2.90 (depending on crystal structure)

fabrication a low viscous polymer-based reactive resin, mostly methylmethacrylate, is used as core material, which can be polymerized by light immediately after filling in the waveguide channel [32, 33]. With respect to processing, i.e. waveguide channel filling, the addition of the nanosized ceramic filler must not affect the viscosity. But increasing amounts of nanosized fillers in polymers have a significant impact on the flow behaviour of the resulting composites. For instance, the addition of surface modified Aerosil R8200 (specific surface area: 142 m^2/g, primary particle size 15 nm, average measured particle size 15 µm [17]) changes the Newtonian flow to a pronounced pseudoplastic one and increases the absolute viscosity values over many decades [17]. The strong viscosity increase complicates the processing, especially the filling of the waveguide channel.

The influence of the nanosized ceramics on the refractive index depends on the difference between the individual values of the two composite components. Table 9.3 shows the index of refraction at 633 nm for PMMA and some commercial ceramics. The addition of Aerosil causes a depression in the refractive index, while the other nanoceramics show an increase in the refractive index increase (Figure 9.6). The latter ones tend to form large agglomerates,

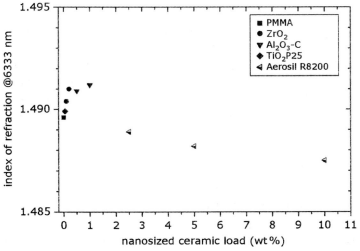

Figure 9.6 Refractive index change of an acrylate based composite with nanofiller load.

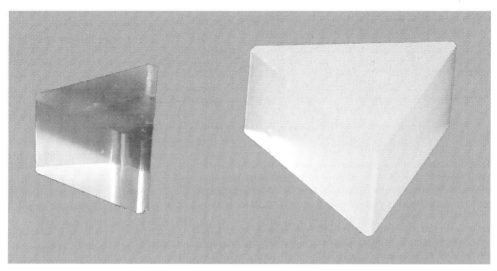

Figure 9.7 Prism devices containing 5 wt% Aerosil R8200 (left) and 1 wt% alumina type C.

therefore a pronounced scattering occurs which prevents acceptable transmission values even at very low loads in the visible range [34–37]. Figure 9.7 shows two different prism type parts fabricated by composite reaction moulding, the one on the right-hand side consists of PMMA, filled with 1 wt% nanosized alumina type C, the other is filled with 5 wt% nanosized Aerosil 8200. In the composite containing the alumina a pronounced scattering in the demonstrator part can be observed. This composite can be used as an optical element with almost perfect scattering characteristics [37]. Therefore due to scattering only small amounts of nanosized ceramics can be added, the resulting impact on the refractive index is small but sufficient for single mode waveguiding. Since the scattering problem due to agglomeration can now be solved by the *in situ* synthesis of isolated nanoparticles in the polymer matrix and subsequent *in situ* surface hydrophobization using dispersant or silanization agents [31, 38, 39]. Highly transparent acrylate-based composites e.g. with a titania load up to 75 wt% have been realized [38]. A further particle size reduction following the top-down approach ends at the molecular level, especially the use of electron-rich organic molecules enables a significant increase of the PMMA's refractive index from 1.49 up to 1.55, while retaining the excellent transmission properties [40]. An undesirable side effect, a reduction of the composite's glass transition temperature, can be observed (plasticizer effect). The polymerization of the doped PMMA-based reactive resin to PMMA and the subsequent pelletizing enable an injection moulding of modified PMMA with improved optical properties suitable for new applications.

9.3.4 Increase in relative permittivity

In the microelectronic industry the integration of different passive electronic elements (resistors, capacitors, inductors) into the printed circuit board (PCB) accelerates year by year due to the increasing functionalities of the devices and the related transmitted

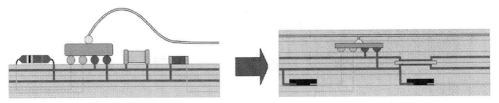

Figure 9.8 Actual (left) and aspired (right) PCB layout with improved functional element density.

data rate. One important target in PCB layout is the integration of surface-mounted devices (Figure 9.8, left) into the board (Figure 9.8, right) as embedded functional elements. As an example new high capacity materials are needed, which are compatible with the existing PCB process technology for the realization of embedded capacitors [41]. Besides a reduction in capacitor thickness, ceramic materials with large relative dielectric constant (permittivity) are necessary. Within the same material and the same phase a crystallite size reduction from the meso- down to the microscale yields a significant increase in permittivity [42, 43]. Below a grain size of 1 μm the permittivity falls significantly due to the stress-induced phase change [44]. The critical crystallite size depends individually on which ceramic is used. Figure 9.9 shows the change of the relative permittivity with solid load in a polyester matrix for a number of different nanosized ceramics (average particle sizes in the range 50–150 nm). The pure polymer shows a value around 3, a roughly linear increase with solid content can be observed. The poor relative permittivity values are due to the small particle sizes below the

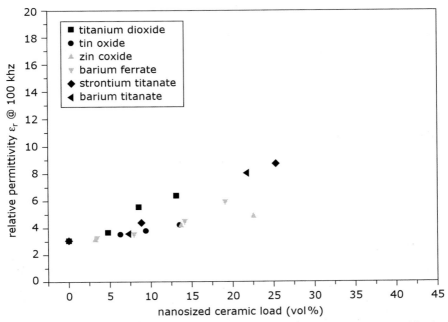

Figure 9.9 Refractive permittivity change of a polyester-based composite with nanofiller load.

critical particle size [45]. With respect to application in embedded capacitors the optimized crystallite size, which is between 100 nm and 1 µm, has to be found individually for each used ceramic e.g. by a controlled grain growth of nanosized material [46].

Conclusions

Polymer-based composites containing micro- or nanosized fillers have become increasingly important in microsystem technology, enabling either the fabrication of microstructured ceramic components via different variants of replication techniques or the realization of functional composites with particular physical properties. The use of nanosized particles seems to be particularly promising, but due to the large surface area the addition of nanosized fillers to polymers causes a pronounced increase in the viscosity of the composites even at small loads. Therefore, composites containing fillers with large specific areas with a solid load beyond 5 vol% are often difficult to process. As a consequence, depending on the filler used, the resulting property modification can be small. The realization of polymer-ceramic composites using micro- or nanosized fillers allows the development of materials with modified physical properties, which enable new applications. With respect to commercialization of polymer-filler composites with new functionalities the following questions or technical problems need solutions:

- commercial availability of special nanofillers;
- use of further additives, e.g. dispersants, possible for significant viscosity reduction for better processing, necessary;
- is a low cost device fabrication by e.g. replication methods possible;
- nanofiller costs (e.g. SWCNT: 350€/g [47]) hinders further commercialisation;
- composite formation process affects resulting composite properties due to particle agglomeration;
- composite properties reproducibility and comparability; and
- side effects on the flow behaviour, thermal stability etc.

Current research is trying to overcome these technical problems.

Acknowledgements

The authors gratefully acknowledge the financial support by the European Commission within the 4M Network of Excellence and the Deutsche Forschungsgemeinschaft DFG (SFB 499) as well as Morflex Inc. for the provision of dispersant samples and Evonik/Degussa for the supply of the nanosized ceramics.

References

[1]   Menz W. and Dimov S. 2005. *4M2005 – Proceedings of the First International Conference on Mutli-Nmaterial Micro Manufacture*, 29 June–1 July, 2005, Karlsruhe, Germany, Elsevier, Amsterdam.

[2]   Menz W., Fillon B. and Dimov S. 2006. *4M2006 - Proceedings of the Second International Conference on Multi-Material Micro Manufacture*, 20–22 September, 2006, Grenoble, France, Elsevier, Amsterdam.

[3] Dimov S., Menz W. and Toshev Y. 2007. *4M2007 - Proceedings of the Third International Conference on Multi-Material Micro Manufacture*, 03–05 October, 2007, Borovets, Bulgaria, Whittles Publishing, Scotland.

[4] Dimov S. Menz W. and 2008. *Proceedings of the Fourth International Conference on Multi-Material Manufacture*, 09–11 September, 2008, Cardiff, UK, Whittles Publishing, Scotland.

[5] German R. G. 1990. *Powder Injection Molding*. Metal Powder Industries Federation, Princeton, New Jersey, USA.

[6] Bauer W., Knitter R., Bartelt G., Emde A., Göhring D. and Hansjosten E. 2002. Replication techniques for ceramic microcomponents with high aspect ratios. *Microsystem Technologies*, **9**, 81–86.

[7] Zauner R. 2006. Micro powder injection molding. *Microelectronic Engineering*, **83**, 1442–1444.

[8] Rath S., Merz L., Plewa K., Holzer P., Gietzelt T. and Hausselt J. 2005. Isolated metal and ceramic micro parts in the sub-millimeter range made by PIM. *Advanced Engineering Materials*, **7**, 619–622.

[9] Talreja R. and Manson J.A. 2001. *Polymer Matrix Composites*, Elsevier, Amsterdam.

[10] Pinnavaia T.J. and Beall G.W. 2000. *Polymer-Clay-Nanocomposites*, Wiley, Chichester, United Kingdom.

[11] Ajayan P.M., Schadler L.S. and Braun P.V. 2003. *Nanocomposite Science and Technology*, Wiley-VCH, Weinheim, Germany.

[12] Hanemann T., Bauer W., Knitter R. and Woias P. 2006. Rapid prototyping and rapid tooling techniques for the manufacturing of silicon, polymer, metal and ceramic microdevices. In *MEMS/NEMS Handbook: Techniques and Applications, Vol. 3*, Leondes, C.T. (Ed), Springer, Berlin, Germany, 801–869.

[13] Piotter V., Merz L., Örlygsson G., Rath S., Ruprecht R. and Zeep B. 2005. Microinjection molding – principles and challenges. In *Advanced Micro and Nanosystems Vol. 3, Microengineering of Metals and Ceramics*, **289**, Löhe D. and Haußelt J. (Eds), Wiley-VCH, Weinheim, Germany.

[14] Bauer W., Haußelt J., Merz L., Müller M., Örlygsson G. and Rath S. Microceramic injection molding. In *Advanced Micro and Nanosystems Vol. 3, Microengineering of Metals and Ceramics*, Löhe D. and Haußelt J. (Eds), Wiley-VCH, Weinheim, Germany, 325–356.

[15] Hanemann T., Honnef K. and Haußelt J. 2007. Process chain development for the rapid prototyping of microstructured polymer, ceramic and metal parts: composite flow behaviour optimisation, replication via reaction molding and thermal postprocessing. *International Journal of Advanced Manufacturing Technology*, **33**, 167–175.

[16] Hanemann T., Heldele R., Honnef K., Rath S., Schumacher B. and Haußelt J. 2007. Properties and application of polymer-nanoparticle-composites. *CFI-ceramic forum international*, **84**, 49–54.

[17] Hanemann T., Heldele R. and Haußelt J. 2006. Particle size dependent viscosity of polymer-silica-composites. In *4M2006 –Proceedings of the Second International Conference on Multi-Material Micro Manufacture*, 20–22 September, 2006, Grenoble, France, Elsevier, Amsterdam, 191–194.

[18] Hanemann T. 2008. Influence of particle properties on the viscosity of polymer–alumina–composites. *Ceramics International*, **34**, 2099–2105.

[19] Hanemann T. 2006. Influence of disperants on the flow behaviour of unsaturated polyester–alumina–composites. *Composites A: Applied Science and Manufacturing*, **37**, 735–741.

[20] Hanemann T. 2006. Viscosity change of unsaturated polyester–alumina–composites using polyethylene glycol alkyl ether based dispersants. *Composites A: Applied Science and Manufacturing*, **37**, 2155–2163.

[21] Hanemann T., Heldele R. and Haußelt J. 2007. *4M2007–Proceedings of the Third International Conference on Multi-Material Micro Manufacture*, 03–05 October, 2007, Borovets, Bulgaria, Whittles Publishing, Scotland. 73–76.

[22] Griffin W. C. 1949. Classification of surface–active agents by HLB. *Journal of the Society of Cosmetic Chemists*, **1**, 311–326.

[23] Griffin W. C. 1954. Calculation of HLB values of non–ionic surfactants. *Journal of the Society of Cosmetic Chemists*, **5**, 259–262.

[24] http://en.wikipedia.org/wiki/Hydrophilic-lipophilic_balance (accessed November 2007).

[25] Heldele R., Hanemann T. and Haußelt J. 2007. In *Proceedings Eurofillers*, 26–30 August, 2007, Zalakaros, Hungary.

[26] Hanemann T. and Honnef K. 2009. Process chain development for the realization of zirconia microparts using composite reaciton molding. *Ceramics International*, **35**, 269–275.

[27] Rath S., Merz L., Plewa K., Holzer P., Gietzelt T. and Hausselt J. 2005. Isolated metal and ceramic micro parts in the sub-millimeter range made by PIM. *Advanced Engineering Materials*, **7**, 619.

[28] Merz L., Rath S., Piotter V., Ruprecht R. and Hausselt J. 2004. Powder injection molding of metallic and ceramic microparts. *Microsystem Technologies*, **10**, 202–204.

[29] Leitfaden zur Härteprüfung. 2007. http://www.hegewald-peschke.de/06_info/leitfaden.htm (accessed November 2007).

[30] Wen J. 2007. In *Physical Properties of Polymers Handbook*, Mark J. E., (Ed), 2nd Edn. Springer Science and Business Media, New York 2007, 487.

[31] Nakayama N. and Hayashi T. 2007. Preparation and characterization of TiO_2–ZrO_2 and thiol-acrylate resin nanocomposites with high refractive index via UV-induced crosslinking polymerization. *Composites A: Applied Science and Manufacturing*, **38**, 1996–2004.

[32] Dannberg P., Kley E.B., Knoche T. and Neyer A. 1993. Fabrication of monomode polymer waveguides by replication technique. In *Integrated Optics and Micro-Optics with Polymers*, Ehrfeld W., Wegner G., Karthe W., Bauer H.D., Moser H.O. (Eds), B.G: Teubner Verlagsgesellschaft, Leipzig, Germany, 211–218.

[33] Ma H., Jen A.K.Y. and Dalton L.R. 2002. Polymer-based optical waveguides: materials. processing, and devices. *Advanced Materials*, **14**, 1339–1365.

[34] Böhm J., Haußelt J., Henzi P., Litfin K. and Hanemann T. 2004. Tuning the refractive index of polymers for polymer waveguides using nanoscaled ceramics or organic dyes. *Advanced Engineering Materials*, **6**, 52–57.

[35] Hanemann T., Böhm J., Henzi P., Honnef K., Litfin K., Ritzhaupt-Kleissl E. and Haußelt J. 2004. From micro to nano: properties and potential applications of micro and nano filled polymer ceramic composites in microsystem technology. *IEE Proceedings Nanobiotechnology Special Issue*, **151**, 167–172.

[36] Ritzhaupt-Kleissl E., Böhm J., Haußelt J. and Hanemann T. 2005. Process chain for tailoring the refractive index of thermoplastic optical materials using ceramic nanoparticles. *Advanced Engineering Materials*, **7**, 540–545.

[37] Ritzhaupt-Kleissl E., Boehm J., Haußelt J. and Hanemann T. 2006. Thermoplastic polymer nanocomposites for applications in optical devices. *Materials Science and Engineering C*, **26**, 1067–1071.

[38] Wu S., Zhou G. and Gu M. 2007. Synthesis of high refractive index composites for photonic applications. *Optical Materials*, **29**, 1793–1797.

[39] Sumida K., Hiramatsu K., Sakamoto W. and Yogo T. 2007. Synthesis of transparent $BaTiO_3$ nanoparticle/polymer hybrid. *Journal of Nanoparticle Research*, **9**, 225–237.

[40] Hanemann T., Böhm J., Honnef K., Ritzhaupt-Kleissl E. and Haußelt J. 2007. Polymer-phenanthrene derivative-composites: rheological, optical and thermal properties. *Macromolecular Materials and Engineering*, **292**, 285–294.

[41] http://www.ats.net/en/index.php/technology/_key+topics/c-814-embedded+capacitors.html, (accessed March 2008).

[42] Moulson A.J. and Herbert J.M. 2003. *Electroceramics*, 2nd Edn. Wiley, Chichester, United Kingdom, 81.

[43] Zhang L., Zhong W. L., Wang C.L., Zhang P. L. and Wang Y.G. 1998. Dielectric Properties of $ba_{0.7}sr_{0.3}tio_3$ ceramics with different grain size. *physica status solidi (a)*, **168**, 543.

[44] Moulson A.J. and Herbert J.M. 2003. *Electroceramics*, 2nd Edn. Wiley, Chichester, United Kingdom, 315.

[45] Hanemann T., Böhm J., Ritzhaupt-Kleissl E., Schumacher B. and Haußelt J. 2007. Polymer-dopant-nanocomposites with improved physical properties. In *Proceedings Smart Systems Integration 2007*, Gessner T. (Ed), 27–28 March, 2007, Paris, VDE Verlag GmbH, Berlin, Germany. 485–487.

[46] Ying K.L. and Hsieh T.E. 2007. Polymer-dopant-nanocomposites with improved physical properties. *Materials Science and Engineering B*, **138**, 241–245.

[47] http://www.nanoamor.com, (accessed March 2008).

10

Micromoulds for injection moulding with ceramic feedstocks

J. Kotschenreuther and A. M. Dieckmann

Replication techniques such as injection moulding are one of the most promising technologies for the medium- to large-scale production of microparts. Typically for these replication processes, feedstock is injected to fill the negative forms, the so-called cavities. The mould inserts of these cavities can be produced either by conventional or removal processes. This chapter gives an overview of the state-of-the-art of special microproduction technologies and points out strategies to fabricate micromoulds and cavities which meet the specific requirements of mould production.

10.1 Introduction

Demand for reduced weight, reduced dimensions, higher surface quality and part accuracy, while at the same time reducing component costs for components or devices ranging from electromechanical instruments to medical devices has already been documented [1]. The worldwide market for complete microsystems is expected to grow from about 50 billion US$ in 2003 to more than 200 billion US$ in 2010 [2]. The success of these microtechnologies will depend both on the utilization of a broad range of polymeric, metallic and ceramic materials as well as the utilization of cost-effective process technology [1, 3, 4]. Micro-injection moulding, which has already achieved high precision standards for polymers, looks to be a very promising process technology which can meet these requirements. However, the processing of ceramic feedstocks of high abrasiveness in microPIM requires mould inserts with increased wear resistance compared to those used in polymeric injection moulding. Promising materials include tool steels or hard metals which are already being successfully used for macroscopic mould inserts [5].

So far microtechnological approaches to miniaturization have been based on processes which were taken over from the semiconductor technology. These include: photolithography, processes of depositing thin layers, etching techniques as well as the cost-efficient concepts of batch manufacturing.

For a higher degree-of-freedom regarding processed materials and geometries different machining technologies have to be used. Methods that have been derived from macromanufacturing processes like milling, turning, boring, EDM, ECM (electro-chemical machining) or LBM (laser beam machining) are used for a small and medium batch production of miniaturized systems and microcomponents. In order to reach higher outputs these processes have to be implemented into a new process chain comprising the structuring of micromoulds in wear resistant materials using the above mentioned technologies. This holds especially true for abrasive feedstock materials like ceramics which need to be subsequently formed using various replication technologies [6].

10.2 Tooling concepts

The basic principles of tooling technology for injection moulding have been standardized. The international standard ISO 12165 summarizes the typical mould base configurations as follows:

- standard mould;
- split mould;
- stripper plate mould;
- three-plate mould;
- stack mould; and
- hot runner mould.

Additionally, there are cold runner moulds. During the construction of the cold runner mould it has to be ensured that the colder material at the flow front of the material will be caught within so-called cold material catchers [7]. The entry of the material flow into the cavity determines the quality of the produced part. A variety of gating systems are in use [7]:

- pin gate;
- cone gate;
- diaphragm gate; and
- submarine gate.

Reworking the produced parts can be minimized with a properly selected gating system. If the gating is placed in nonfunctional areas of the produced parts tolerances can be met more easily [8]. An example of a three-plate mould for PIM ceramic bevel gears is shown in Figure 10.1. The three-plate mould has a two-stage opening. The first stage separates the sprue from the parts. After the second stage of tool opening the separated part can be demoulded from the cavity. The selection of a cone gate ensures that the remains of the separation of the parts from the sprue which might result in possible tolerance deviations are placed in a nonfunctional area.

Special attention has to be paid to venting the mould's cavities. The air within the cavity must be vented properly to ensure that the produced part is free of inclusions and that there are no diesel effects due to inclusions during the moulding process. There can be restrictions to tempering the mould due to the venting of the cavities [10]. If the cavities cannot be well vented evacuating the cavities and the runner system could be an option [7].

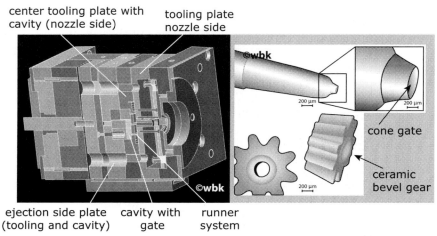

center tooling plate with cavity (nozzle side)

tooling plate nozzle side

ejection side plate (tooling and cavity)

cavity with gate

runner system

cone gate

ceramic bevel gear

Figure 10.1 Three-plate mould with gating system and ceramic bevel gears [9].

Temperature control of the mould is important for the layout and design of an injection mould. The stable temperature conditions are crucial to obtaining good quality parts. Depending upon the binder system within the used feedstock the injected material requires a minimum temperature during injection itself. However, the heat must be conducted away continuously from the part after injection [11]. A very uniform tempering of the mould can be reached if the tempering of the runner system is separated from the tempering of the part's cavity [7]. The main aim during tempering of the mould is to minimize distortion, residual stresses and total cycle time. Variation in cycle time results in variations in the cooling time and distortion of the parts [8, 12].

There are different kinds of ejectors available for ejecting the produced parts from the cavity. They include: ejector pins, ejector sleeves and shaped ejectors. The selection of the type of ejection system depends on the shape of the produced part. Deformation of the product should be minimized by reducing the contact pressure per unit area [8]. The surface roughness and the contact area of the ejected part act against the ejection force [12, 13]. Since the surface roughness of the produced part also depends on the surface roughness of the cavity special attention must be paid to the surface quality of the cavity.

10.3 Mould insert concepts

Micromoulds pose different demands on manufacturing methods. The most commonly used methods are LIGA, micromilling, EDM and LBM. LIGA is a process for making microstructures in a resist and with subsequent electroplating to reproduce these structures in an Ni-based metal, which can then be used itself, or be used as a replicate tool for molding processes.

The features that determine the manufacturing method are:

- mould material needed;
- quantities needed;
- smallest structure size (embossed or engraved);

Figure 10.2 REM picture of a microgear mould insert in steel produced by micromilling (source: collaborative research centre SFB 499).

- aspect ratio;
- manufacturing time;
- machining result (burr formation, surface quality, characteristics of surface layers);
- electrical conductivity;
- wall properties (steep, angled);
- available budget; and
- complexity of geometry (2.5 or 3D).

Each process has its specific advantages and setbacks. LIGA has the advantages of realizing details in the one-digit micrometre range, high aspect ratios and micro-accuracy, however, there are rather high costs, a limited palette of suitable materials and long processing times. Micromilling can only manufacture materials with a hardness up to approximately 65 HRC and is up to now limited to 30 µm tool diameter for the structuring of hardened steels and is also limited to aspect ratios of 1.5 for a 30 µm tool. Nevertheless, a 100 µm tool already features flute lengths of 1 mm [14]. Figure 10.2 shows an example of a micromilled mould insert in a hardened tool steel with a smallest tool diameter of 100 µm. The positive aspects of this process include satisfying surface roughness of Rz ~ 0.3 µm, short manufacturing times and the high degree-of-freedom in geometry and workpiece material when milling with five axes. EDM can produce smaller structures than milling in harder materials and with very high aspect ratios. EDM is a rather slow process, limited to electrical conductive materials and produces a slightly worse surface roughness than milling. LBM can achieve small feature sizes (beam diameters in the range of a few microns), high aspect ratios in almost any material in quite a short time and in contrast to milling and EDM suffers no tool wear. The disadvantages of this process are: surface roughness and the flexibility of the geometry.

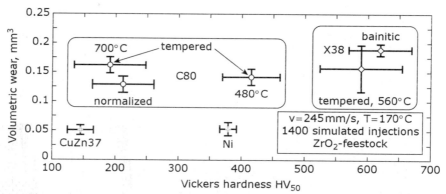

Figure 10.3 Volumetric wear of steels C80 and X38CrMoV5-1 versus Vickers hardness HV50 after 1400 simulated injections with ZrO2 feedstock [15].

In order to be able to process abrasive ceramic feedstocks and still achieve a substantial tool life, a tool material has to be chosen which is strongly wear resistant. Schneider *et al.* [15] have shown in a study, which compared unalloyed, low- and high-alloyed steels, electroplated nickel and brass matched with ZrO_2 and Al_2O_3 feedstocks, that hardness is not the most important material characteristic regarding wear resistance. The results rather indicated that the wear resistance of the tested mould materials depends in a higher degree on their microstructural parameters like homogeneity or work hardening and plastic deformation. Furthermore the wear behaviour was strongly influenced by the characteristics of the ceramic feedstock, i.e. shape and size of the ceramic particles as well as by the binder system. Figure 10.3 shows the volumetric wear of the different materials, which were investigated.

10.4 Cutting processes

10.4.1 Cutting with geometrically defined cutting edges

Figure 10.4 illustrates the applied cutting technologies with geometrically defined cutting edges. A short overview of these processes is given below.

10.4.2 Diamond cutting tools

In ultraprecision machining, cutting tools of monocrystalline diamond are almost exclusively used. Diamond has a very low friction coefficient and an excellent thermal conductivity which has a favourable effect on the cutting process. The main advantages of this cutting material are, however, its great hardness and the possibility of producing a cutting edge of an almost atomic sharpness. The production of extremely sharp cutting edges is one of the most important tasks in microcutting. Cutting edge sharpness in the submicrometre range allows the production of surfaces with roughness values of a few nanometres, providing them with an optical quality [16, 17].

Materials used in diamond machining include: aluminium, copper, brass, nickel-silver (as well as the so-called electroless nickel, an amorphous nickel–phosphor alloy) allowing the production of a particularly good surface and a low burr formation with a comparatively

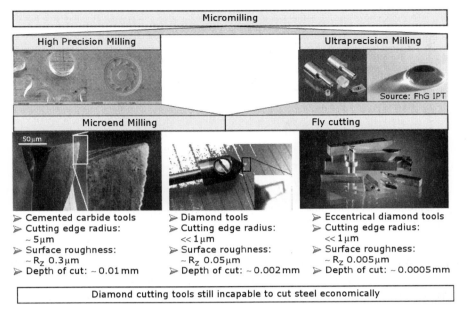

Figure 10.4 Classification of the various milling processes.

high hardness. The group of ferrous products, especially steels, is not yet suitable for diamond cutting. Because of the high process temperatures and due to the high affinity of carbon to iron the diamond is subject to a graphitization process leading to exorbitant tool wear. There are developments at various research facilities in Germany, Japan and USA to solve these problems and to establish a method which allows the cutting of steel with diamond tools. Presently, the approach of superimposing an ultrasonic excitation on the tool movement or nitriding the workpiece beforehand [18], seem to be most promising approaches to this problem.

10.4.3 Micromilling of steel

So far, microtechnology in general, but also microcutting has been limited to the structuring of silicon and nonferrous metals like electroless copper, brass and plastics [5]. For the future it is of great importance to be able to structure steel because of its material properties that range from hard-brittle to soft-ductile. In addition, silicon-based moulds are not as wear resistant as steel moulds when ceramic feedstock is employed.

Investigations regarding microcutting of steel have been performed particularly in Japan and Germany, since the middle of the 1990s [5, 19, 20–23], but are still at the research stage. The need for wear-resistant mould inserts is compelling for the economic efficiency of the moulding processes. The bending strength of the mould insert material, especially in the case of high aspect ratios of certain microstructures, can also be very important for the reliability of the replication process.

In contrast to the ultraprecision or microcutting of nonferrous metals, diamond cannot be used for the microcutting of steel. Here, hard metal milling cutters, which are widely used in conventional cutting, are of interest. Due to the single grains which appear in the form of micronotches at the cutting edge, cemented carbide tools are not suitable for obtaining optical surface qualities as in diamond cutting. The comparably low price and the possibility of machining steels in contrast to diamond tools, however, are the reasons for using cemented carbides as cutting materials for microcutting tools.

The formation of a sufficiently sharp cutting edge is crucial if a tool is to be suitable for microcutting. Since cemented carbides are sintered materials, in which carbide grains (mainly tungsten carbide) as abrasive material are bonded in a relatively soft cobalt phase, the size of the carbide grains determines the achievable cutting edge rounding. Current grain sizes are in the range 0.5–1.0 μm. Thus, cutting edge radii of 2–5 μm can be realized plus ~ 2 μm for tools when they are additionally coated by a special wear resistant layer, e.g. TiN.

Cemented carbide end milling cutters are quite commonly used in industry and are available from several manufacturers, as coated and noncoated tools and in diameters down to 0.02 mm. Tools in the experimental stages can reach diameters down to 0.005 mm. Regarding the microend milling of hard materials like steel, however, special attention also has to be paid to the process control and the bearing of the machine tool to prevent the occurrence of stochastic tool ruptures or premature wear. Furthermore, the process requires high cutting speeds. A minimum feed per tooth is also required to ensure a proper material separation.

10.5 Ablation processes

10.5.1 Electro-discharge machining processes

The industrial utilization of the electrodischarge effect began in 1954. First, minimum holes, e.g. injection nozzles, were produced on cavity-sinking EDM machines. The first investigations on micro-EDM began in 1967/69 [24, 25]. The process became more interesting in the 1980s with the propagation of silicon-based MEMS. At present the main fields of application are: holes for ink-jet printer heads, spinning nozzles, injection nozzles, turbine blades, electron tube grids, tools for punching of e.g. electronic components, microreactors, microtoothed wheels and mould inserts for injection moulding [26, 27].

According to the German standard DIN 8590, EDM belongs to the material removal processes, especially thermal removal by electric gas discharge with the spark as energy carrier. A spark is electrically generated, flashing over the electrode and the workpiece. This discharge has a material removal effect. Between the electrode and the workpiece is a dielectric which acts as an insulator. Another function of the dielectric is to take up the particles that have been removed and convey them out of the machining area. The sparks arising one after the other melt the electrode and the workpiece, throw parts of them into the dielectric and leave the rest of the melt on the material surface where it resolidifies. The result is a roughed surface with many small craters, depending on the process parameters. By using suitable parameters, the undesired material removal of the electrode, which has the negative contour of the form to be produced, can be kept much smaller than that of the workpiece. It is possible to remove a large volume of material if many removal pulses are used.

10.5.2 Micro-electrodischarge machining processes

The two most important variants of EDM are the wire-EDM and microcavity-sinking EDM. In the former process, a wire is used to cut out a contour from the workpiece and in the latter a cavity is produced by means of a negative form. Since all sides of the electrodes can be used as tool, it is possible to produce a multitude of forms.

For the production of complex 3D geometries, the EDM is used among other processes. Here, the form is generated over several paths using rotating, mostly cylindrical electrodes [28– 30], gradually removing layers which are up to 100 μm high from the workpiece until the final depth is reached [31].

The smallest electrodes produced have a diameter of 4.3 μm at a length of 50 μm. Depending on the electrode diameter, aspect ratios of the cavities of 10–50 are reached today. Sato has investigated holes of 15–300 μm which found application in the ink-jet nozzles of printers [32].

Due to its effective principle, the process allows workpieces to be machined independently of their hardness and strength. A certain minimum of electrical conductivity, however, is needed in order to enable EDM. Detailed research work on this has been undertaken by Reynaerts and his co-workers [33–35] and Masaki *et al.* [36].

10.5.3 Laser beam machining

LBM is based on the principle of using the high-energy electromagnetic radiation for the carefully directed machining of a workpiece. For this purpose, the radiation is guided in a defined way on the surface of the workpiece and formed to a desired power density and distribution by optical means. The absorption of the radiation in the workpiece or at its surface results in the desired influence on the material, in other words, the energetic effect on the material occurs thermally. The laser beam "tool" is not subject to wear and has a high flexibility. Depending on the laser beam source there are different machining mechanisms. With CO_2 and solid-state lasers it is possible to use two different mechanisms depending on the process control. In LBM, the material is transferred to a liquid phase by the energy input of the laser and then blown off from the machining spot by means of a gas flow. This method is superior due to high removal rates for medium surface qualities.

With higher beam intensities the material is directly vaporized and the removal is sublimed resulting in finer structures with better surface accuracies, however, the removal rates are lower. In the above-mentioned laser types or mechanisms the laser beam can be guided on to the workpiece surface either by a scanner or directly by an optic. In Excimer lasers the material removal is generated by electron band transitions and a mask projection process is used for the beam formation [37, 38].

For LBM, mostly pulsed laser beam sources are used, machining is possible over the whole spectrum of materials (metals, plastics, ceramics).

LBM is applicable in the following fields:

- lithography;
- boring and drilling;

- labelling;
- engraving;
- caving;
- structuring; and
- removal of surface layers/cleaning.

Especially in tool and mould manufacture, the laser is increasingly used as a machining tool because it is very flexible and independent of the mechanical properties of the material, like hardness or strength. Therefore, it is possible to machine materials which are difficult to cut as well as small structures. Two main fields of laser application for the tool manufacture are described below.

The first plants for LBM were introduced at the end of the 1980s, in which a CO_2 laser beam source was used [39, 40]. Since the CO_2 laser beam could only be guided to the machining spot via large-scale mirror systems and optics, the workpiece geometries to be machined were very limited and the plants very inflexible. In the process of laser caving the material is removed layer by layer. A characteristic feature is the pulsed operation with which it is possible to achieve beam densities of up to 1000 W/cm^2 with very short pulse times. In the case of materials that are difficult to machine, this leads to a sublimation removal without thermal effects on the adjacent material. Presently, it is basically possible to produce structures down to a size of 10 µm. Depending on the laser beam sources used and the machine design, the plants that are currently on the market allow the production of structures up to a size of 100 µm with aspect ratios of 50 and higher. The accuracy is currently at 0.01 mm. As this is not fully sufficient for micromachining some development is still needed in order to use laser beam sources as standard equipment for micromachining. Moreover, the surface machined by laser needs a further finishing to remove the residues of the excavated material [41].

Depth engraving is a commercialized process in tool and mould manufacture, which above all is used for the production of inscriptions. In this process, the power density of the laser beam is so high that the material vaporizes within a few nanoseconds during the machining. The surface and burr quality obtained by laser depth engraving can be compared with the results obtained by EDM. Using steel, removal rates of more than 10 mm^3/min can be reached [41]. Apart from the flexibility and the high speed, the laser engraving process stands out because it has zero wear when compared to conventional methods like EDM or high-speed milling.

Conclusions

As described above different production technologies are suitable for the manufacturing of micromoulds and cavities. The processes of cutting and material removal are able to meet the high requirements for moulds, especially for replication technologies with abrasive material (e.g. PIM). In nearly all cases of manufacturing EDM or LBM can achieve microstructures in hard material. For structures with aspect ratios below three and hardness of workpiece materials below 62 HRC cutting processes can also be used. As could be shown the selection of the most appropriate process technology has to be done carefully, because it is strongly dependent on the individual characteristics of the component that has to be fabricated e.g. material, details, accuracy and surface quality.

References

[1]   Byrne G., Dornfeld D. and Denkena B. **2003**. *Advancing Cutting Technology, CIRP Annals*, **52**(2), 483–507.

[2]   Wicht H. 2006. *Market Analysis for MEMS and Microsystems III*, 2005–2009, a NEXUS taskforce report, Henning Wicht, NEXUS Association, Grenoble, France.

[3]   Hesselbach J., Raatz A., Wrege J., Hermann H., Weule H., Buchholz C., Tritschler H., Knoll M., Elsner J., Klocke F., Weck M., von Bodenhausen J. and von Klitzing A. 2003. MicroPro Untersuchung zum internationalen Stand der Mikroproduktionstechnik. *Werkstattstechnik Online*, **93**, 119–128.

[4]   Anonymous. 2004. Rahmenprogramm Mikrosysteme, Bundesminsterium für Bildung und Forschug, Bonn, **Germany**.

[5]   Moriwaki T. and Shamoto E. 1991. Ultraprecision diamond turning of stainless steel by applying ultrasonic vibration. *Annals of the CIRP*, **40**(01), 559–562.

[6]   Ritzhaupt-Kleissl H., Binder J., Gietzelt T. and Kotschenreuther J. 2006. Net Shape Reaction Bonded Ceramic Micro Parts by Mechanical Microstructuring. In *Advanced Engineering Materials*, **8**(10), Wiley-VCH-Verlag, Germany. 983–988.

[7]   Walcher H. 2000. Spritzgiessformen für den Pulverspritzguss (PIM). In *Trends im Werkzeugbau, Konferenz-Einzelbericht Swiss Engineering Schweizerischer Technischer Verband STV*, 23 June, 2000.

[8]   Gastrow H. 1990. *Der Spritzgieß-Werkzeugbau*. Hanser, Vienna, Austria.

[9]   Buchholz C. 2005. Systematische Konzeption und Aufbau einer automatisierten Produktionszelle für pulverspritzgegossene Mikrobauteile. Fakultät für Maschinenbau, Universität Karlsruhe (TH), PhD Thesis.

[10] Menges M. 1999. *Mohren: Anleitung zum Bau von Spritzgießwerkzeugen*, Hanser, Vienna, Austria.

[11] Erstling: Plus Factors. Material properties of aluminium alloys are increasingly attractive for injection moulds. *Maschinenmarkt*, Heft **17**, 32–34, 37.

[12] Fleischer J., Weule H. and Buchholz C. 2005. Optimization of critical processes within an automated process-chain for the micro-PIM process. *Proceedings of 5th EUSPEN International Conference*, 8–11 May 2005, Montpellier, France.

[13] Fleischer J. and Dieckmann A-M. 2005. Entformkraftuntersuchung beim Micro-PIM-Prozess. In *wt Werkstattstechnik online Jahrgang 95*, Vol. **11/12**, 892–897.

[14] Hitachi. 2007. http://www.hitachi-tools.de (accessed 14 May 2007).

[15] Schneider J., Iwanek H. and Zum Gahr K. H. 2005. Wear behaviour of mold inserts used in micro powder injection moulding of ceramics and metals. *Wear*, **259**(7–12), 1290–1298.

[16] Moriwaki T. 1995. Experimental analysis of ultraprecision machining. *International Journal of the Japanese Society for Precision Engineering*, **29**(4), 287–290.

[17] Ohmori G. and Takada S. 1982. Primary factors affecting accuracy in ultraprecision machining by diamond tools. *International Journal of the Japanese Society for Precision Engineering*, **16**(1).

[18] Brinksmeier E., Dong J. and Gläbe R. 2004. Diamond turning of steel molds for optical applications. Proceedings of the 4th International Euspen Conference, 31 May–1 June, 2004, Glasgow, United Kingdom.

[19] Spath D. and Konold T. 1996. Mikrozerspanung – eine interessante alternative zur herstellung von mikrostrukturbauteilen. *Produktion und Management*, **86**(11/12), 579C582.

[20] Schmidt J., Hüntrup V., Tritschler H. 1999. Mikrozerspante formeinsätze aus stahl. *wt Werkstatttechnik*, **89**(11/12), 495–498.

[21] Peichl A., Schulze V., Löhe D., Tritschler H. and Spath D. 2000. Microcutting of steels – interaction of material properties and process parameters. In *Proceedings of the Micro.tec*, 25–27 September 2000, Hannover, Germany, 7–11.

[22] Klocke F., Rübenach O., Zamel S. 1996. Möglichkeiten und grenzen der spanabhebenden fertigung bei der herstellung von präzisionsbauteilen. In *Proceedings of Micro-Engineering 96*, 11–13 September, 1996, Stuttgart, Germany.

[23] Brinksmeier E. 1994. Herstellung von präzisionsbauteilen durch mikrozerspanung. *Industriediamantenrundschau*, **4**, 210–217.

[24] Kurafuji H. and Masuzawa T. 1968. Micro-EDM of cemented carbide alloys. In *Journal of the Japan Society of Electrical Machining Engineers*, **2**(3), 1–16.

[25] van Osenbruggen C. 1969. Micro sparc erosion. In *Philips Technisch Tijdschrift*, **20**, 200–213.

[26] Allen D. and Lecheheb A. 1996. Micro electro-discharge machining of ink jet nozzles: optimum selection of material and machining parameters. In *Journal of Materials Processing Technology*, **58**(1), 53–66.

[27] Michel F., Ehrfeld W., Lehr H., Wolf A., Gruber H. and Bertholds A. 1996. Mikrofunkenerodieren als strukturierungsverfahren in der mikrotechnik. In *Tagungsband zum 41. Internationalen Wissenschaftlichen Kolloquium*. Universitätsbibliothek der TU Ilmenau, Germany, 233–236.

[28] Bollen K. and van Dijck W. 1999. Development of machining strategies for complex (3D) surfaces with EDM milling. Master thesis. K.U. Leuven, Belgium.

[29] Kaneko T., Tsuchya M. and Fukushima T. 1989. Improvement of 3D NC contouring EDM using cylindrical electrodes. In: *Proceedings of 9th International Symposium for Electromachining (ISEM IX)*. April 1989, Nagoya, Japan, 49–52.

[30] Kruth J. P. and Bleys P. 2000. Machining curvelinear surfaces by NC electro discharge machining. In *Proceedings of 2nd International Conference on Machining and Measurements of Sculptured Surfaces*. 20–22 September 2000, Krakow, Poland, 271–294.

[31] Shinkai M. and Suzuki T. 1997. An electrical-discharge scanning machine. In *Mitsubishi Electric Corporation Catalogue*, 8–10.

[32] Sato T., Mizutani T. and Kawata K. 1985. Electro-discharge machine for micro hole drilling. In *National Technical Report*, **31**, 725–733.

[33] Reynaerts D., Meeusen W. and van Brussel H. 1998. Machining of three-dimensional microstructures in silicon by electro-discharge machining. *Sensors and Actuators A*, **67**, 159–165.

[34] Reynaerts D., Heeren, P. and van Brussel H. 1997. Microstructuring of silicon by electro-discharge machining (EDM)-part I: theory. In *Sensors and Actuators A*, **60**, 212–218.

[35] Heeren P., Reynaerts D., van Brussel H., Beuret C., Larsson O. and Bertholds A. 1997. Microstructuring of silicon by electro-discharge machining (EDM)-part II: applications. In *Sensors and Actuators A*, **61**, 379–386.

[36] Masaki T., Kawata K. and Masuzawa T. 1990. Micro electro discharge machining and its applications. In *Proceedings of IEEE Micro Electro Mechanical Systems*, 11–14 February 1990, Napa Valley, California, USA. S. 21–26.

[37] Eichler J. 2001. *Laser, Bauformen, Strahlführung, Anwendungen, 4. Auflage*, Springer Verlag, Berlin, Germany. 371–380.

[38] König W. and Klocke F. 2001. *Fertigungsverfahren, Abtragen und Generieren, 3. Auflage*, Springer Verlag, Berlin, Germany. 179–182.

[39] Anonymous. 1990. Das MAHO Lasercaving – ungeahnte Dimensionen im Formenbau. *Opto Elektronik Magazin*, **6**(1), 12–15.

[40] Eberl G., Hildebrand P., Kuhl M., Sutor U. And Wrba P. 1992. Neue Entwicklungen beim Laserabtragen. *Laser und Optoelektronik*, **4**, 44–49.

[41] Anonymous. 2001. Neue Laser im Formenbau. *Laser–Magazin*, Magazin Verlag, Hannover, Germany.

11

High-pressure microceramic injection moulding

V. Piotter and L. Federzoni

While injection moulding of microengineering products has become part and parcel of industrial-scale manufacturing for polymers, the corresponding injection moulding variants for metals and ceramics are still in an advanced stage of development.

Eventually, this so-called MicroPIM process aims at a large-scale fabrication technology for ceramic microcomponents with minimal geometrical details in the range of $10\,\mu m$ or even lower. Theoretical densities of up to 99% have already been achieved while the nominal sizes of the final parts exhibit a tolerance of ± 0.3–0.5%. Recent investigations into the parameter-geometry interdependency showed great promise for further increase of replication accuracy. Eligible materials are oxide and nitride or even electroconductive ceramics.

Specific variants are under development in an attempt to achieve functional combinations of materials with different physical/chemical properties. One of these improvements is micro two-component injection moulding which enables the fabrication of microcomponents consisting of two ceramic materials with different physical properties. Examples for the combination of different ceramics are alumina-zirconia combinations or ceramic mixtures with different mixing ratios.

11.1 Introduction

Fabrication of singular parts as well as components for microproducts by sophisticated variants of injection moulding represents an important future manufacturing technology and has already reached an industrially viable status in the case of polymeric materials.

Comparable to the macroscopic world the large number of application fields in MST demands the utilization of a wider range of different materials particularly towards inorganic materials. The actual choice is still dominated by silicon and thermoplastic polymers. An opportunity to overcome this lack is the adaptation of the industrially established PIM technology to microfabrication. It represents an economically viable process for complex shaped metal or ceramic parts to be manufactured in large-scale series. The technology itself can be used for metals as well as for ceramics. In the latter case the process is called microceramic injection moulding (MicroCIM).

For the realization of multimaterial devices as well as for the reduction of costs for joining and assembling, the development of micro two-component injection moulding represents a promising but also very ambitious challenge. The main technical problems are to adjust the sintering shrinkages and temperatures of different materials so that the internal stresses which occur are not critical during the manufacturing of the part as well as during the in-service use of this part. Further perspectives are the manufacturing of movable microjoints by using materials with different shrinkage ratios.

11.2 Micropowder injection moulding

11.2.1 Objectives and features

Although microinjection moulding is becoming more and more established for polymeric materials, many applications still require materials properties (mechanical, thermal, abrasion, magnetic, electrical etc.) that cannot be met by polymers. Some possible fields of application are listed in Table 11.1.

All these remarkable technical opportunities would be pointless if the economic demands could not be fulfilled, i.e. the demanded stronger materials have to be processed in an economically effective way. Therefore, extension of PIM to microfabrication purposes was an obvious way [1, 2] and attempts were carried out at various institutes all over the world. Just because of volume reasons two approaches, at the Commissariat à l'Énergie Atomique, Grenoble, France (CEA) and FZK, will be explained in more detail.

At CEA, developments on MicroPIM essentially concern the development of innovative feedstock formulations. These formulations comprise nanopowders particularly of ceramics (alumina, silicon nitride, SiALON, zirconia, …) for MicroCIM applications.

With such new innovative formulations, MicroCIM based on nanopowder feedstocks is now possible (solid content >50% typically). Hence, special projects regarding the development of the process of 2C-MicroCIM have led to the development of new types of feedstocks which allow the achievement of crack-free two-component parts.

At FZK development of high-pressure MicroPIM started in the late 1990s by using powders and binder systems which have already been in service for macroscopic applications. In further studies in-house produced feedstock systems specially tailored for MicroCIM were used. The latest component properties are given in Table 11.2.

Table 11.1 Examples of MST products requiring materials properties which are difficult to be met by silicon or by polymers.

Applications	Typical devices	Suitable materials
Micromechanical applications	Gear wheels, watches, surgical instruments	Steels, zirconia
Microchemical technology	Reactors, heat exchangers	Stainless steels, oxide ceramics
Microelectronics	Heat exchangers	Copper, Cu-alloys, aluminium nitride
Sensors and actors	Ultrasonic transducers	PZT ceramics
Optics	Ferrules	Alumina

Table 11.2 Technical data of MicroPIM as a function of the powder used.

Material	D_{50} (µm)	Max. aspect ratio	Min. structural detail (µm)	Density (% theo.)	Rmax (µm)
Al_2O_3	0.4–0.6	>10	<20	97	3
ZrO_2	0.2–0.4	>10	<3	99	<3
For comparison: 316L steel	4.5	10	50	97	8

As typical test structures, different kinds of gear wheels based on LIGA as well as on micromilled mould inserts were manufactured. Spectrometer test structures represent another kind of examples demonstrating the limitations of MicroPIM qualitatively. As shown in Figure 11.1, replication of structural details smaller than 1 µm is possible in principle. However, the performance of the whole part is determined by the surface quality rather than by the geometry. The dominant influence of the particle size of the powder is also shown in Figure 11.2.

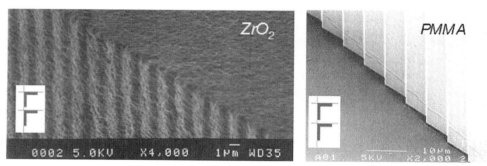

Figure 11.1 Spectrometer test structure in zirconium oxide (left) compared to a PMMA structure (right) injection moulded using the same LIGA mould insert. Considerable differences in surface quality are obvious.

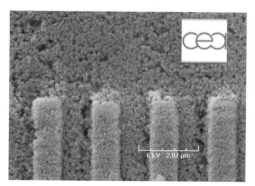

Figure 11.2 Surface quality achievement on a silica part using a silicon mould insert structured by lithography.

Table 11.3 Main characteristics of MicroPIM compared to other typical micromanufacturing techniques.

Parameter	MicroPIM	LIGA	Micromachining	Eximer laser	DRIE
Aspect ratio	10	100	14	<10	10–25
Minimal thickness	2 μm*	<20 nm	few μm	100 nm to 1 micron	2 μm
Accuracy	0.4–0.5% of the dimension	>1 micron	few μm	few μm	<1 μm
Need of a master?	yes	yes	no	no	yes

*These data correspond to a mean size (d_{50}) of 300 nm of a zirconia feedstock.

To estimate the technical performance of MicroPIM described above it should be compared to other micromanufacturing techniques for MST. Table 11.3 summarizes the main characteristics of these techniques.

As Table 11.3 shows, the MicroPIM process is particularly relevant for large-scale manufacturing as it is a replication technique, while the technical capabilities are nearly equal to that of other MST manufacturing techniques. An additional overview of the capabilities of MicroPIM is provided by Table 11.4.

11.2.2 Evaluations on the accuracy of the MicroCIM process

Geometrical accuracy of the final sintered parts is a critical issue in PIM. According to literature, tolerances of 0.2–0.5% can be achieved [3] which might be a sufficient level

Table 11.4 Characteristics and data for the current status of MicroPIM.

Preferred application	Large- and medium-scale fabrication (for prototyping preliminary series possible)
Typical powders	Zirconia, alumina, ZTA, silicon nitride etc.
Typical binder systems	Polyolefine/wax (+additives) polyacetal systems
Tool evacuation	Required if venting via parting plane or slits is not possible
Temperature control	Usually isothermal, in case of high aspect ratios also variothermal
Typical cycle time	10 s (isothermal) – 8 min (maximum for variothermal)
Tested mould insert materials	Steel, brass, nickel, (ceramics)
Machinery equipment	Standard industrial units with certain modifications (screw design, wear protection) for feedstock processing

of precision for many applications. On the other hand, for high value-added precision products process accuracy has to be improved. Thus various studies on the dimensional tolerances of zirconia single-mode ferrules have been performed to analyse the influence of process parameters on part quality [4].

The interdependencies between moulding tool and injection parameters on one side and the resulting dimensions of the green and sintered bodies on the other side have been comprehensively investigated. Injection velocity and tool temperature have been determined as the most important factors whereas other process parameters showed also significant but less obvious effects. Further detrimental effects like distortion or sink marks which occur during the cooling period after injection should not be neglected.

By comprehensive process analysis significant progress regarding considerably lower tolerances than the common values of approx. 0.5% has been achieved. In certain geometrical axes of the ferrules tolerances could be reduced to $\pm$ 0.1% [5].

11.2.3 Actual applications and further projects on MicroCIM

The large industrial potential of MicroCIM is demonstrated by the fact that some products have already entered the market as shown in Figure 11.3.

A special variant of PIM in macroscopic procurement is the merging with gas-induced moulding technology. This combination helps to avoid sink marks, reduces cycle times, saves weight and materials but, to date, no applications for microtechnology have been reported.

Another subvariant of MicroCIM is generated by the combination with inmould-labelling technology, which is also well-known from macroscopic plastic manufacturing. Such a merger should lead to a high-throughput process enabling promising kinds of material combinations and products of enhanced performance [6].

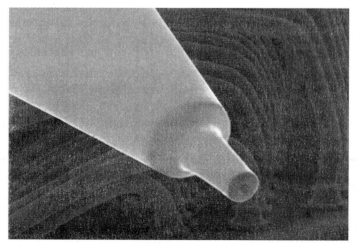

Figure 11.3 SEM picture of the top of an alumina die for fine pitch bonding with an inner diameter of 19 μm and outer diameter of 45 μm. (courtesy of SPT Roth Ltd., Switzerland.)

11.3 Micro-two-component injection moulding

11.3.1 General remarks

In obvious analogy to the further development of macroscopic injection moulding technology, new microreplication techniques promising to result in higher economic efficiency or new products, such as microassembly injection moulding or microinsert injection moulding, are currently being tested. Both variants facilitate relatively easy assembly and bonding by using insert parts and are expected to lead to a considerable reduction in total production costs [7].

A similar objective is pursued by micromulticomponent injection moulding, which additionally allows for the production of multicomponent and, hence, multifunctional microcomponents. This two- or multicomponent injection moulding in the micrometre scale would reveal advantages similar to using insert parts. There is no need to explain in detail that joining single microcomponents and assembling entire microsystems are by no means trivial processes, but involve a considerable expense. This is due to the small dimensions and the fact that microcomponents have to be handled quite differently from macroscopic components. For instance, in microdimensions, electrostatic forces become more important than the force of gravity, so that microparts may frequently stick to the handling systems instead of dropping. Accordingly, significant progress can be achieved by combining the moulding step of single components with a joining step, as in multicomponent injection moulding. Moreover, this allows multifunctional products to be generated which, sometimes, may combine different physical or chemical properties in one moulded specimen [8, 9].

There are a few other arguments for multicomponent CIM but there are, of course, also disadvantages as shown in Table 11.5.

Although the number of advantages is equal to that of disadvantages there are outstanding reasons (saving of mounting steps, function integration etc.) for applying multicomponent injection technique even in micromanufacturing. Therefore, the first attempts at developing multicomponent CIM had already started in the 1990s.

In this connection the research activities carried out at Cranfield University, UK, have to be mentioned. By using a sequential injection procedure two-component green bodies consisting of a core and a cover material were produced. Gear wheels with a green diameter of 60 mm acted as demonstrators. Although the experiments were carried out with metal feedstocks there is no obstacle in principle to using two-component injection technology for ceramic materials [10, 11].

Table 11.5 Advantages and disadvantages of multicomponent injection moulding in microfabrication.

Pro	Contra
Reduction of assembly expenditure	Longer cycle times
Creating functional material combinations	More complex tooling and process conduct
Higher degree of integration	Materials compatibility required (sintering behaviour)
Saving of tooling, machinery and plant expenditure	Higher difficulties to set up the process

Figure 11.4 Bicomponent SiAlON/Si$_3$N$_4$ ring
performed by two-component injection moulding.

Publications regarding two-component CIM are relatively rare in the literature. One exception is presented by the activities of ARBURG, Germany, which dealt with the coinjection of differently coloured ceramics e.g. for cups [12]. Due to the fact that the materials were identical, except for the colour, process conduct was relatively easy.

Further progress towards industrial application can be expected for the near future as currently running R&D projects will lead to considerable improvements, e.g. the EU-funded CarCIM project [13]. In this project, different combinations of biceramics are investigated for several industrial applications. Among them, combinations between SiAlON/Si$_3$N$_4$ (see Figure 11.4) or ATZ/ZTA are being investigated for industrial applications like brake disks, valve seats, gear wheels or glow plugs. All samples have been developed for applications in different fields of transport (automotive, railway).

The idea of two-component injection moulding leads to two different ways of realization, i.e. the generation of movable or immovable joints. Although the latter ones represent the majority of industrial utilization movable linkages of polymers have also found their way into practical application. Furthermore, first approaches for adapting two-compoment injection moulding to powder processing have already been carried out. On the other hand, to date there is no information about movable PIM joints for microapplications.

The question of how the different kind of joints can be generated depends strongly on the conduct of the process, mainly the sintering procedure, and the materials used. Table 11.6 shows the most important factors.

Table 11.6 Factors determining the generation of movable or immovable joints of MicroPIM components.

	Movable	**Immovable**
Binders	Preferably incompatible	Compatible
Powders	Different types and particle size	Preferably equal
Injection process	No mixing of feedstock	Merging advantageous
Debinding	Preferably two-step process	No special demands
Sintering	Different sintering temperatures and shrinkage ratios	Equal sintering temperatures and shrinkage ratios

11.3.2 Fixed connections with metals or ceramics

This variant of two-component-CIM is a combination of micropowder injection moulding and two-component injection moulding technologies.

The decisive hurdle to be overcome in this variant lies in the fact that not only two different materials must be joined in injection moulding, but that the bond must be preserved also throughout binder removal and sintering. For this purpose, the binder removal and sintering behaviour of the moulding compounds (feedstocks) and the powder must be adapted appropriately in terms of quantity, morphology, thermal and chemical behaviour.

Moreover, the powder moulding compounds must fulfil different, often opposite, functions such as low viscosity for complete filling of the mould, high stability of the green product for mould removal without deformation or crack appearance, brown parts from which the binder has been removed without causing any stress for warp-free sintering, and adhesive bonds, or bonds with targeted mobility, of at least two different ceramics and metals, respectively.

The example referred to here is the development of two-component-MicroPIM with ceramic feedstocks within a nationally funded project (BMBF 03N1080) which was carried out by the Fraunhofer Institute (IFAM Bremen) and FZK. Regarding the material combinations, two kinds of powder–metallurgical steels and two kinds of mixture ceramics which only differ in their mixing ratio have been merged.

To facilitate the injection moulding of two-component parts, a moulding tool and a special process control have been developed. This moulding tool allows the usage of various microcavities (Figure 11.5). One of these parts was determined for tensile and bending tests so that the bonding strength of the junction between two materials could be investigated.

An important feature is the possibility of evacuating of the moulding tool to avoid burns caused by compressed air. A further important feature of the tool is the integrated partition

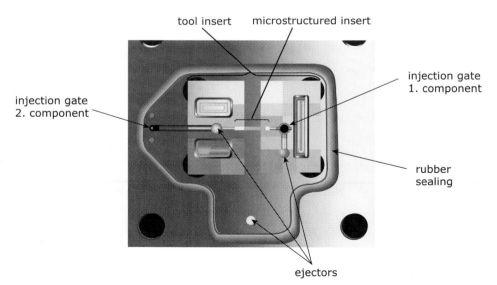

Figure 11.5 View on the parting plane of the micro two-component moulding tool with two runner gates and cavity for microstructured inserts in the centre.

dividing the cavity. This allows a sequential injection of both components: after the injection of the first material the partition can be pulled and the second component is injected onto the stationary front. It is also possible to work with an open partition. In that case the position of the junction can be adjusted by varying the speeds of simultaneous injections.

An interesting example, microheating elements based on a mix of electroconductive titanium nitride and insulating aluminium oxide will be explained here [8]. In this mixed ceramic material, aluminium oxide serves as the matrix material: the electric conductivity of the compound can be varied within wide limits via the fraction of percolated titanium nitride (TiN) powder in the matrix. The moulding compounds prepared in a twin-screw extruder have been processed in the two-component microinjection moulding process, among other samples, into mechanical test specimens and needle pin-shaped prototypes of ceramic microheaters.

In the case of the U-shaped heater elements, the curved section was moulded from a mixture with a low TiN-content, whereas in the two straight sections, the Al_2O_3 content was reduced. To determine the sintering parameters it has to be taken into account that that shrinkage of the TiN is nearly zero at the applied temperatures. To compensate, the powder content in the sections of higher TiN content has to be increased. It was found that the differences in powder content have to be in the range 4–6 vol% depending on the particular compositions of the feedstocks used. The sintering procedure itself was carried out at temperatures up to 1750°C with maximum heating rates of 10K/min. Using such parameter sets, two-material parts with a significant gradient in electrical conductivity were obtained. If the whole sintered part is set under current a glooming effect which is typical for resistivity heaters could be detected (Figure 11.6).

It should be noted that the position of the bonding seam can be set relatively accurately by the appropriate choice of injection moulding parameters, such as the injection rate, or by means of a slide valve.

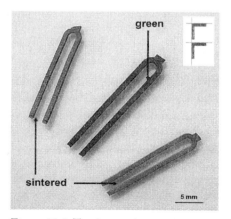
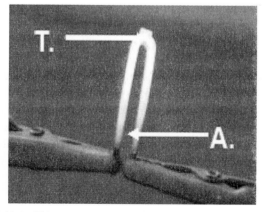

Figure 11.6 Test heater elements consisting of Al_2O_3/TiN ceramic with different electrical conductivity, sintered and green bodies (left). Glooming test of sintered heating element; section with reduced TiN-content are marked with T, sections with increased content are marked with A (right).

Figure 11.7 Microgear wheel/shaft combinations in the green state (left and right) and after sintering (middle). Dimensions of green bodies: outer diameter (gear wheel): 2.932 mm, height (gear wheel): 1 mm, hub diameter: 0.65 mm, shaft diameter: 1.65 mm.

The influence of the different sintering shrinkage rates on bonding surfaces and bonding seams is less pronounced in certain geometries of microcomponents than in macrocomponents. Differences, if any, in lateral sintering shrinkage can therefore be tolerated more readily at microdimensions than in macrotechnology.

A further example representing much more microscopic dimensions is the realization of micro gear wheel and shaft components as shown in Figure 11.7. While the gear wheel was moulded using a zirconia feedstock the shaft is made of aluminium oxide ceramic. Due to the special variation of shaft diameter hindering an axial movement of the gear wheel it is obvious that such a two-material component could not be produced by a simple mounting of the two parts.

11.3.3 Mobile connections

This variant imposes requirements nearly opposed to those mentioned above. First, one component must enclose the other either entirely or in part because, otherwise, the components would decompose during sintering. Moreover, the sintering shrinkage of the inner component must exceed that of the outer component. It is also important that compaction of the inner brown body starts earlier in order to prevent incipient sintering. Examples of such combinations of materials being employed are so far only known from the macroscopic world [14], while microscopic applications are currently under development as shaft/gearwheel composite projects at FZK. The geometries will probably be the same as those shown in Figure 11.7 whereas the mobility should be generated by a sophisticated choice of moulding and sintering parameters.

Conclusions

Microinjection moulding is being applied in an increasingly broader range of practical uses. The scope of examples, such as the increased use of special variants, additionally evidences the point.

For MicroPIM, this implies the tendency towards the utilization of finer powders, especially for metal materials. Development of ceramic feedstocks with larger ultrafine or even nanodispersed powder fractions in order to achieve smaller details and increased surface qualities has already started.

Accordingly, multicomponent microinjection moulding opens up entirely new perspectives to micro systems engineering, as not only the extent of assembly work is reduced in microsystems, but also new functional units can be manufactured. Thus, two-component microinjection moulding offers a clear economic and technical potential for future applications.

All variants of microinjection moulding are likely to experience advanced development and use of simulation programmes to include both micro- and PIM-specific aspects.

Acknowledgements

The research work described in this chapter was mainly sponsored by the Deutsche Forschungsgemeinschaft (DFG, SFB 499) and the Federal Ministry of Education and Science (BMBF). The authors also wish to thank their industrial partners and all colleagues at Forschungszentrum Karlsruhe (FZK) and CEA LITEN for their always helpful assistance.

References

[1] Rota A., Duong T.V. and Hartwig T. 2002. Micro powder metallurgy for the replicative production of metallic microstructures. *Journal of Microsystem Technologies*, **8**, 323–325.

[2] Ruprecht R., Gietzelt T., Müller K., Piotter V. and Haußelt J. 2002. Injection moulding of microstructured components from plastics, metals and ceramics. *Journal of Microsystem Technologies*, **8**, 351–358.

[3] German R.M. 2005. *Powder Metallurgy and Particulate Materials Processing: The Processes, Materials, Products, Properties, and Applications*. Metal Powder Industries Federation, Princeton New Jersey, USA.

[4] Beck M., Piotter V., Ruprecht R. and Haußelt J. 2006. Dimensional tolerances of micro precision parts made by ceramic injection moulding. In Dimov S. (Ed), *4M 2006: Proceedings of the 2nd International Conference on Multi-Material Micro Manufacture*, Grenoble 2006, Elsevier, Amsterdam, 135–38.

[5] Beck M., Piotter V., Ruprecht R. and Haußelt J. 2007. Effects of moulding conditions on the tolerances of precision parts in ceramic injection moulding. Paper presented at *10th European Congress and Exhibition on Advanced Materials and Processes (Euromat 2007)*, Nürnberg, Germany.

[6] http://www.vdivide-it.de/innonet/projekte/in_pp120_greentapim. pdf/view (accessed 17 November 2008).

[7] Michaeli W. and Opfermann D. 2006. Micro assembly injection moulding. *Journal of Microsystem Technologies*, **12**, 616–619.

[8] Örlygsson G., Piotter V., Finnah G., Ruprecht R. and Haußelt J. 2003. Two-component ceramic parts by micro powder injection moulding. *Proceedings of the Euro PM 2003*, 20–22 October 2003, Valencia, Spain, 149–154.

[9] Piotter V., Finnah G., Zeep B., Ruprecht R. and Hausselt J. 2007. Metal and ceramic micro components made by powder injection moulding. *Materials Science Forum*, **534–536**, 373–376.

[10] Alcock J.R., Logan P.M. and Stephenson D.J. 1996. Metal co-injection moulding. *Journal of Materials Science Letters*, **15**, 2033–2035.

[11] Alcock J.R., Logan P.M. and Stephenson D.J. 1998. *Surface engineering by co-injection moulding; Surface and Coatings Technology 105*, Elsevier, Amsterdam, 65–71.

[12] Schumacher C. 1999. Durch Pulverspritzgießen zu neuen Formen; Kunststoffe (Journal) 89, Hanser-Verlag, Germany. 70–72.

[13] http://www.carcim.de/ (accessed 17 November 2008).

[14] Maetzig M. and Walcher H. 2006. Assembly moulding of MIM materials. Proceedings of Euro PM 2006 – Powder Metallurgy Congress and Exhibition, 23–25 October 2006, EPMA, 43–48.

12

Low-pressure injection moulding

W. Bauer and R. Knitter

Low-pressure injection moulding (LPIM) is a variant of the PIM process, which is virtually identical with the conventional HPIM process. However, due to the use of low-viscous paraffin or wax binders the injection pressures can be reduced to less than 1 MPa. This allows the application of simple and inexpensive moulds making the process economical especially for small series production. The good flowability of the feedstocks and the ability to employ fine scaled powders also recommends LPIM for the moulding of microdimensional parts. The major drawback of LPIM is the low mechanical strength of the used binders, leading to demoulding problems with complex microparts. However, this problem can be overcome by customized tooling concepts, like soft tooling with silicone rubber moulds.

12.1 Introduction

During the last 20 years, the interest in CIM has increased as the demands of parts design have approached the limits of conventional powder processing routes. Nowadays CIM is a standard shaping method for the mass production of small ceramic parts with complex shapes. These parts can often not be fabricated by other ceramic shaping methods. Shaped by dry pressing or slip casting, the strength of the unfired ("green") compact is low. Thus, during the demoulding step, i.e. the withdrawal of the part from the mould, there is a high risk that fine structures are damaged. Mechanical tooling of ceramics is always an expensive issue, even for small workpieces. Due to the brittle nature of the material, machining can also introduce flaws which may lead to failure in use. In contrast to this, CIM is a near net-shape process for the manufacturing of complex shaped or miniaturized parts with dimensional accuracy. It requires little or no final machining and the high binder content of the injection moulded bodies offers sufficient green strength for a damage-free demoulding procedure. Injection moulding can be fully automated with proper process control, enabling the economic production of large series of parts.

There are two variants of CIM: the industrially established HPIM process described in Chapter 11 of this book and the less common LPIM. A comparison between these methods

Table 12.1 Comparison between HPIM and LPIM.

	HPIM	LPIM
Major binder components	Polymer, polymer/wax	Paraffin, wax
Viscosity range (Pa·s)	100–1000	2–20
Injection temperature (°C)	120–200	70–100
Injection pressure (MPa)	> 50	0.1–1
Tool costs	High	Low
Tool wear	High	Low
Green strength	Medium	Low
Application	Mass production	Prototypes and small series

is shown in Table 12.1. In HPIM, thermoplastic polymers are used as main binder components, whereas for LPIM paraffin or wax-based binders are used [1]. As a consequence, the pressures employed to LPIM are in the range of 0.1–1 MPa, whereas HPIM typically takes place at pressures above 50 MPa. Due to the different binder system, slight differences between HPIM and LPIM also occur for the feedstock preparation, the injection moulding machinery, and for the debinding process. Nevertheless, the methods are closely related as the basic principles of plastic shaping are common to both processes.

LPIM, which is sometimes also called hot moulding [2], was invented by Gribovski in the former Soviet Union [3], where the process gained a similar importance to its high-pressure counterpart in the USA, Western Europe or Japan. Recently, LPIM has been strongly propagated as a method for prototyping and small series fabrication as there is an increasing demand for individually manufactured and customized parts. In contrast to HPIM with its high costs of tooling that requires mass production for a return of investment, LPIM can work with simple and inexpensive moulds and is therefore economical even for a small number of parts. The good flowabilty of the low-viscous feedstocks and the ability to employ fine scaled powders also recommend LPIM for the injection moulding of microdimensional parts [4].

The major drawback of LPIM compared to HPIM is the lower mechanical strength of the used binders. This can lead to rupture of structures during the demoulding of the green compact, especially when fine particulars with high aspect ratio (height to width) must be removed from the cavity. For that reason, LPIM is normally limited to parts with lower complexity. Although this problem can be overcome by tailored tooling concepts, like soft tooling with silicone rubber moulds [5] or lost moulds [6], this still prevents the automation of the process for complex parts and restricts LPIM to smaller quantities.

12.2 Feedstocks for low-pressure injection moulding

CIM is a rather versatile shaping method, as almost every ceramic material is available as a powder. For nonclay ceramic powders a lack of plasticity is characteristic. Plasticity, however, is essential for the injection moulding process, so that the powders have to be blended with an organic vehicle to build a compound, called feedstock. The binder system usually consists of several organic components. The general range of feedstock properties is dominated by the main binder component, usually paraffin or wax in the case of LPIM. Dispersants improve the wetting of the particles and stabilize the dispersion against re-agglomeration. The amount of binder depends on the powder characteristics and covers a range from 20 to 50 vol% of the feedstock. In general, it is desirable to attain a high solid content in the mixture while maintaining a low feedstock viscosity. A solid content well below 50 vol% causes high shrinkage during the debinding and sintering process, leading to undesirable results like poor dimensional control or warping and cracking of the part. With decreasing particle size, the viscosity of the feedstock is increasing [7] and the mouldable solid content is decreasing [8]. For that reason, often relatively coarse powders with a mean particle size of more than 1 µm are preferred for injection moulding. However, in the case of microparts submicron powders have to be used to ensure a precise replication of the micron sized details and a good surface quality, as well as to achieve homogeneous properties on a micrometre scale. Fine powders also exhibit a high sintering activity allowing for a good densification and providing a high strength for the naturally fragile ceramic bodies. However, experience has shown that nanosized powders with a particle size below 100 nm cannot be compounded to a high solid content. Therefore, ceramic powders with a mean particle size in the range of 0.5 µm are preferred for the LPIM manufacturing of microparts.

Feedstock properties are determined by the powder [9], by the organic additives, and by the mixing conditions. Short-chained paraffin or wax binders enable the preparation of low-viscous feedstocks, where the viscosity is distinctly reduced, i.e. for several orders of magnitude in comparison to typical HPIM feedstocks made with long-chained thermoplastic polymers like polyethylene or polyoxymethylene [10]. For submicron or nanoscaled powder particles the use of binders with low molecular weight is beneficial as a higher solid loading of the feedstocks is possible due to the low viscosity of the unfilled material [11].

Binder systems based on paraffin or wax are inexpensive and nontoxic. They offer beneficial dispersion properties as they produce a liquid phase with a viscosity comparable to that of water. Typically, the melting point of paraffin is in the range 50–70°C, depending on the mean chain length of the binder. The thermal expansion of paraffin is quite high, which can be problematic as exact temperature control is required to manufacture parts with high accuracy. Also, the narrow melting range can cause distortion problems during the debinding process [12]. Besides that, the debinding behaviour of paraffin or wax is uncomplicated as they decompose without boiling. Their vapour pressure is negligible under normal conditions, unlike water, and therefore, constitution of the feedstock remains constant during the shaping process.

Due to the nonpolar nature of the paraffin, a surface active dispersant is necessary to improve the wetting of the polar oxide surfaces [13] that even exists on nonoxide powders. Nonionic dispersants have proven to be effective additives to paraffin-based binders. The polar functional group of the dispersant molecule couples to the surface of the ceramic powder; the nonpolar polymer chains extend into the paraffin and stabilize the dispersion by steric repulsion. Widely used dispersants include: fatty acids like oleic or stearic acid or surfactants like alkylphenol polyglycol ether, siloxane polyglycol ether or octadecylamine [2, 14, 15]. They differ in the kind of anchoring group towards the surface, but the number of carbon atoms in the chains is often between 16 and 18. The required amount of dispersant primarily depends on the specific surface area of the ceramic powder, but the binder/dispersant interactions may also influence the dispersant addition [15]. For a large number of ceramic powders, a recommended value in the range 0.5–2 mg dispersant per square metre powder surface has been found. However, although the flowability of feedstocks has reached an acceptable state, the effectiveness of most dispersants is not understood and efficient dispersants have still to be found by empirical testing. There is also a need for additives that improve the green strength of the feedstock without increasing the viscosity and also reduce the adhesion of the solidified part to the mould surface.

Special attention has to be paid to a careful drying of the used powder. Adsorbed water molecules have a strong affinity to the surface of oxide powders. They cover those surface sites where dispersant molecules are supposed to adsorb physically or chemically. As a consequence the dispersants are not effective in lowering the viscosity or in stabilizing the suspension if water is present [13]. The competition between dispersant and water also plays an important role in the long-term resistance against flocculation. In spite of the nonpolar character of the binder, the environmental humidity has a detrimental effect on the viscosity of the feedstock [16].

The achievement of a homogeneous feedstock mainly depends on the strength of powder agglomerates and on the shear stress imposed on the compound during mixing. Mixing time was found to be less important [17]. In LPIM feedstocks, high shear stresses can best be obtained by a fast moving mixer. A dissolver stirrer with a vane carrying disc is such a device which is well suited for the preparation of a small amount of feedstock. Heated three-roller mills [18] or ball mills [2] are alternatives for the homogenization and de-agglomeration of a larger amount of LPIM feedstock. Usually, feedstock mixing is performed at temperatures of 80–100°C for less than 1 h. To avoid trapping air bubbles during the stirring process, the use of a vacuum pump for the container is advisable, but one has to be aware that reduced pressure in combination with higher temperature can lead to partial evaporation of binder and dispersant components, which may affect the rheological properties of the feedstock.

12.3 Rheology of LPIM feedstocks

Some requirements on the feedstock rheology are contradictory to a certain degree. Good flowability is required for microcavities as it allows complete mould filling even at low moulding pressures. It simplifies degassing and removal of bubbles from the feedstock. On the other hand, a fluid-like feedstock behaviour promotes sedimentation of the powder particles during storage and processing, and it complicates shape retention in the binder removal step.

For that reason, feedstocks with shear-thinning behaviour and a distinct yield point are preferred. Unlike Newtonian behaviour, where the viscosity is independent of the shear rate, in the case of shear-thinning a decreasing viscosity is observed with increasing shear rate. Especially for the injection moulding of microparts this is an important feature, as small

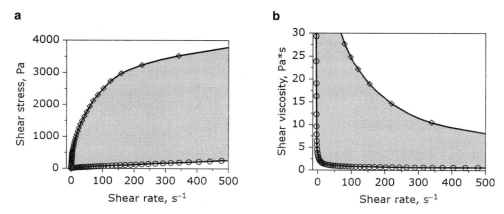

Figure 12.1 Typical working range for the rheological properties of paraffin-based LPIM feedstocks at 85°C; (a) shear stress versus shear rate; (b) shear viscosity versus shear rate.

cross-sections lead to high injection speeds and thereby to high shear rates. The opposite behaviour, shear-thickening, has to be avoided by all means, as filling of fine channels or micropatterns is nearly impossible when the viscosity of the feedstock increases with increasing shear rate.

Another important feature is the yield point [19]. When shear-thinning is accompanied by an adequate yield point, the risk of deformation for moulded parts is reduced during the thermal processing. Adequate means that a minimum force is required to cause the feedstock to flow. This force has to be high enough to prevent both sedimentation of particles and deformation under the influence of gravity and surface tension. On the other hand, the yield stress should be low enough to allow the release of enclosed air bubbles under vacuum treatment. Complete removal of air becomes increasingly important with decreasing feature size. With respect to microparts the resulting pores are in the dimension of the feature's cross-section and will lead to severe defects.

Examples of typical shear stress and shear viscosity curves of various submicron powder feedstocks are shown in Figure 12.1. Shear-thinning behaviour is recognizable for all shown feedstocks by decreasing shear viscosities with increasing shear stress in Figure 12. By using various additives, the yield point and flow behaviour at low shear rates can be adjusted to the desired value. To compare the rheological properties of different feedstocks the viscosity is often selected at a shear rate of 100 s^{-1}. Adequate mouldability for LPIM was found with viscosities of 2–20 Pa s at 100 s^{-1}.

12.4 Machines and tooling for low-pressure injection moulding

Although a standard injection moulding machine, which is used for HPIM, can also be charged with a LPIM feedstock, this combination provides some drawbacks as these machines are designed for a high-pressure regime and do not operate properly in the low-pressure range. Under high-pressure conditions the benefits of the low viscosity, e.g. inexpensive tooling, are lost. Additionally, at high pressure or high flow rates, respectively, a controlled mould filling is difficult, as jetting, splashing or air entrapment can take place [20]. For that reason, special LPIM machines were developed that can work at a moulding pressure of less than 1 MPa.

Two types of LPIM machines are commercially available: pneumatically driven machines and piston-type machines. Examples for the first category are the MIGL-machines (Figure 12.2(a) and (b)), manufactured by Peltsman Corp (Minneapolis, MN, USA), where the feedstock is pressed into the mould by compressed air. A built-in mixer prevents the segregation or sedimentation of low viscous feedstocks. Due to the simple concept, self-made constructions with a pneumatic injection principle do also exist in some laboratories and factories.

A piston-type machine is fabricated by GOCERAM AB (Mölndal, Sweden). The machine is designed for the medium pressure range as the use of a piston allows moulding pressures of up to 7 MP, but it is also suited for the low-pressure regime. In the standard machine, the feedstock is injected into the mould by a pneumatically driven piston (Figure 12.2(c)). Another model is equipped with an electrical servomotor to run injections with different rate profiles.

The machines can be used in principle for macro- as well as for microparts with only minor modifications. These modifications mainly concern the tooling concept. Thermal management of the tool must be adapted to prevent feedstock freezing in the small cross-sections of microparts and to handle the high thermal expansion of wax systems. Another microspecific feature is the evacuation of the tool so as to ensure complete filling of blind hole-type micropatterns.

a

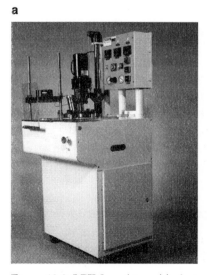

b

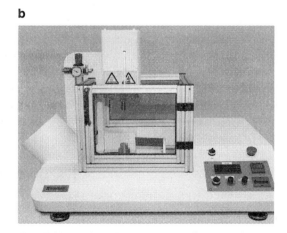

Figure 12.2 LPIM machines: (a) photograph of a MIGL-33 from Peltsman Corp. (courtesy of Peltsman Corp., USA); (b) photograph of a GOCERAM GC-MPIM-2-MA (courtesy of GOCERAM SA, Sweden).

A major problem arises from the low mechanical strength of the used binders requiring customized concepts for the demoulding of microparts. Typical feedstock temperatures are in the range 70–90°C at an injection pressure below 1 MPa. Therefore, LPIM feedstocks generate less wear and, although stainless steel is a standard material for LPIM tools, the tooling for LPIM can also be made from less expensive or easily machinable materials like brass, aluminium, or polymers. However, the low mechanical strength of the used binders can lead to rupture of patterns during the demoulding of the green compact, especially when fine particulars with high aspect ratio must be withdrawn from the cavity. For that reason, LPIM with metal tools is normally limited to parts with lower complexity.

An alternative material for LPIM tools is silicone rubber (soft tooling). In spite of the softness of silicone, it is possible to prevent distortion of the mould at a sufficiently low moulding pressure (usually less than 0.3 MPa). Tremendous benefits result from the elasticity of the material for the demoulding of micropatterned parts [21], as it becomes possible to remove parts with high aspect ratio, parallel side walls or even undercuts from the cavity. Whereas in noncompliant materials the demolding of such structures is challenging or hardly possible, with silicone tools a broad spectra of complex shapes can be moulded (see Section 12.6).

12.5 Thermal treatment

The conversion of the moulded green part to a dense ceramic part takes place in two process steps. Debinding, the removal of the organic constituents, is performed at temperatures below 600°C and usually takes place in an oven with circulated air. For the densification during sintering, significantly higher temperatures are required. While the sintering step is similar to other shaping processes, the debinding of LPIM parts offers some specific features.

Usually, binder removal is the most critical and time-consuming procedure in the CIM process and this step is frequently the origin of defects in the ceramic product. Depending on the solid content of the feedstock, 25–55 vol% of paraffin or wax-based binder must be removed from the green compact. A considerable fraction of the binder decomposes by pyrolysis with gaseous products producing an overpressure inside the bulk. With increasing wall thickness and with decreasing particle size this overpressure becomes more critical and the risk for the formation of defects increases [22]. Thus, the small dimensions of microparts promote the debinding of the green compact, whereas the continued trend to using finer powders has a contrary effect. Small part dimensions are also advantageous with regard to debinding time. While debinding times from days to more than one week are reported for macroscopic parts [23], typically an overnight debinding is possible for microparts.

During the first part of the debinding process a liquid binder phase is formed above the melting temperature of the binder. As LPIM binders have a narrow melting range, the complete binder becomes liquid, and deformation of the part can occur if a sufficiently high yield point is missing (Figure 12.3). While gravitational forces play a minor role for

a
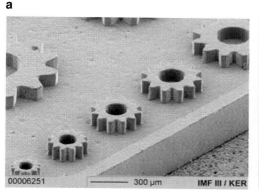
b
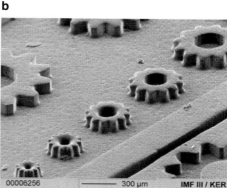

Figure 12.3 SEM micrographs of sintered zirconia micro gear wheels: (a) using a paraffin-based feedstock (solid content 50 vol%) with high yield point; (b) with lower yield point. Deformation took place during debinding.

microparts, the surface tension of the softened body leads to distortion that alters the shape of fine particulars by rounding initially sharp edges or by levelling surface patterns. To sustain shape integrity, the liquid phase has to be withdrawn for example by the capillary forces of a powder bed or by an absorptive support. For macroscopic parts a wicking powder bed is often used as mechanical support to prevent collapsing and to remove the excess binder by capillary forces. However, using a powder bed has the drawback of producing a higher surface roughness, as the powder bed is in contact with the molten binder phase, producing impacts and adhering particles. For large components a more rough-textured surface may not affect their applicability or it may later be removed by surface finishing. For micro parts that have to be applied as-sintered without finishing, a highly absorptive supporting carrier is essential for the debinding of LPIM microparts.

Figure 12.4 shows the representative thermoanalysis of a zirconia feedstock. Although significant mass loss does not start below temperatures of about 200°C, volatilization of a paraffin binder can be detected even at a temperature of 100°C. The highest decomposition rate is observed in the range 250–400°C. In this temperature interval a very slow heating rate is recommended. Above 400°C hardly any mass loss is observed, but the binder is not completely transferred into the gas phase. Reactions between binder components and their oxidation products, lead to cross-linked carbonaceous residues, preferentially located at the contact points of the powder particles [24]. These residues are beneficial as they provide a minimum stability which improves the handling of the so-called brown body. Figure 12.5 shows a representative profile for the debinding of LPIM microparts.

Besides the classic thermal debinding, a removal of the binder by solvents or by supercritical extraction is possible but will not be discussed in this chapter. Although these methods offer improved mechanical properties [25], they require further equipment and exact process control. They are more often used for HPIM feedstocks, but are less common for LPIM feedstocks than thermal processes.

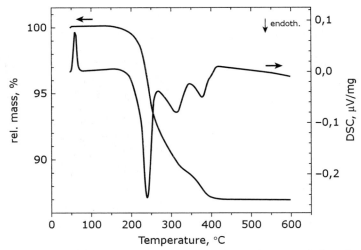

Figure 12.4 Thermoanalysis of debinding process for a paraffin-based zirconia feedstock (solid content 50 vol%); mass loss versus temperature, heating rate 5 K/ min, synthetic air atmosphere.

12.6 Examples

To date the application of direct rapid prototyping methods for ceramic microparts has been limited by restrictions in resolution and materials variety. A rapid prototyping process chain (RPPC), combining the fabrication of a polymer or metal master model by rapid prototyping methods like microstereolithography or by micromachining and the replication of this

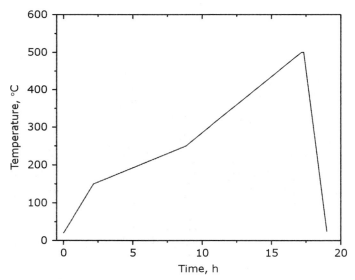

Figure 12.5 Representative temperature–time profile for thermal debinding of LPIM shaped microparts.

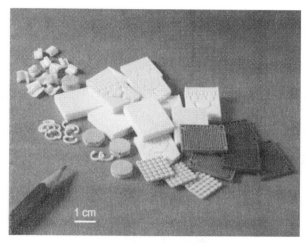

Figure 12.6 Examples of small ceramic parts made by
LPIM with silicone moulds. Materials used for the parts are
Al_2O_3, ZrO_2, PZT, $BaTiO_3$ and Al_2O_3/TiN.

model into the ceramic part by LPIM using silicone moulds, enables us to close this gap
[5, 26]. The master model is embedded into liquid silicone rubber that after curing can be
used as a tool for LPIM. With this process chain it is possible to produce ceramic prototypes
with full functionality down to the micron range within a few days or weeks. Some examples
and materials can be seen in Figure 12.6. A RPPC has also been developed at FZK for the
manufacturing of complex microsystems, e.g. the elements of a zirconia microturbine with
details down to 30 μm (Figure 12.7).

a   b

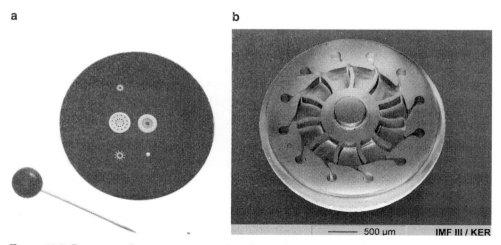

Figure 12.7 Prototype of a zirconia microturbine (outer diameter 3.2 mm): (a) individual parts,
made by LPIM in silicone moulds; (b) SEM micrograph of a partially assembled microturbine.
Channels in nozzle plate are 30 μm in width.

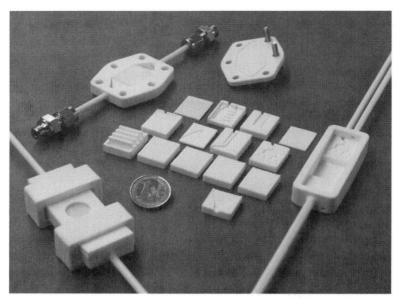

Figure 12.8 Modular ceramic microreactor system. Parts were made from same type of alumina feedstock, demonstrating applicability of LPIM process for macroparts as well as for micropatterns. ([27], reproduced with permission of the Royal Society of Chemistry.)

The benefits of LPIM can also be demonstrated by a ceramic microreactor system (Figure 12.8), which was developed and manufactured with the RPPC within a reasonable period of time and at acceptable costs [27]. Whereas metal microreactors have already been proved to work successfully, comparable ceramic systems for very high temperatures or corrosive conditions were still lacking. For the design development and due to the desired modular character of the reactor, a large variety of moulding tools was needed. As the final design could not be verified on design models, but only under operating conditions, the manufacturing of functional models was indispensable. With LPIM a fabrication technique was chosen that met the requirements for the moulding of the relatively large reactor housings as well as for the micropatterned details of the modular components.

Conclusions

LPIM is a versatile shaping method for the manufacturing of ceramic microparts. Generally, it can be used for a broad range of materials and allows for an economic fabrication from a few prototypes up to small and medium size lots. Thus, it covers the gap that cannot be economically filled by HPIM. With a resolution and an accuracy in the range of a few micrometres, the process is well suited to microfabrication.

Unfortunately, there are still deficits in machines and tooling concepts, which constrain a broader acceptance of the process in industrial applications. Without industrial support the scientific interest in LPIM declined during the last years. However, the development of new generations of HPIM machines for microparts also involved an improved compatibility with LPIM feedstocks. On this basis it can be expected that LPIM will play a major role in microfabrication in the future.

References

[1] Rak Z. S. 1998. New trends in powder injection moulding. *cfi/Ber. DKG*, **9**, 19–26.

[2] Lenk R. 1995. Hot Molding – An Interesting Forming Process, *cfi/Ber. DKG* **72**, 636–639.

[3] Gribovski P. O. 1961. *Hot Molding of Ceramic Parts*, GEI, Moscow-Leningrad, (In Russian).

[4] Knitter R., Bauer W., Göhring D. and Haußelt J. 2001. Manufacturing of ceramic microcomponents by a rapid prototyping process chain. *Advanced Engineering Materials*, **3**, 49–54.

[5] Bauer W. and Knitter R. 2002. Development of a rapid prototyping process chain for the production of ceramic microcomponents. *Journal of Materials Science*, **37**, 3127–3140.

[6] Elsebrock R., Makovica C., Meuffels P. and Waser R. 2004. Structured oxide ceramics by a sodium chloride moulding technique. *Journal of Materials Scinece*, **58**, 3348–3349.

[7] Weltman R. N. and Green H. 1943. Rheological properties of colloidal suspension, pigment suspensions and oil mixtures. *Journal of Applied Physics*, **14**, 569–576.

[8] Song J. H. and Evans J. R. G. 1995. The injection moulding of fine and ultra–fine zirconia powders. *Ceramics International*, **21**, 325–333.

[9] Bauer W., Bartelt G., Haußelt J. and Ritzhaupt-Kleissl H. J. 2001. Manufacturing of ceramic microcomponents by a rapid prototyping process chain. *Key Engineering Materials*, 206–213, 329–332.

[10] German R. M. 1990. *Powder Injection Molding*, Metal Powder Industries Federation, Princeton, New Jersey, USA.

[11] Song J. H. and Evans J. R. G. 1996. Ultrafine ceramic powder injection moulding: The role of dispersants. *Journal of Rheology*, **40**, 131–152.

[12] Mutsuddy B. C. and Ford R. G. 1995. *Ceramic Injection Molding*, Chapmann & Hall, London, United Kingdom, 368.

[13] Novak S., Vidovic K., Sajko M. and Kosmac T. 1997. Surface modification of alumina powder for LPIM. *Journal of the European Ceramic Society*, **17**, 217–223.

[14] Liu D. M. 1999. Effect of dispersants on the rheological behaviour of zirconia–wax suspensions. *Journal of the American Ceramic Society*, **82**, 1162–68.

[15] Zürcher S. and Graule T. 2005. Influence of dispersant structure on the rheological properties of highly–concentrated zirconia dispersions. *Journal of the European Ceramic Society*, **25**, 863–873.

[16] Novak S., Dakskobler A. and Ribitsch V. 2000. The effect of water on the behaviour of alumina–paraffin suspensions for low–pressure injection moulding (LPIM). *Journal of the European Ceramic Society*, **20**, 2175–2181.

[17] Song J. H. and Evans J. R. G. 1993. The assessment of dispersion of fine ceramic powders for injection moulding and related processes. *Journal of the European Ceramic Society*, **12**, 467–478.

[18] Leverkoehne M., Coronel-Hernandez J., Dirscherl R., Gorlov I., Janssen R. and Claussen N. 2001. Novel binder system based on paraffin–wax for low–pressure injection moulding of metal–ceramic powder mixtures. *Advanced Engineering Materials*, **3**, 995–998.

[19] Liu D. M. and Tseng W. J. 1998. Yield behaviour of zirconia–wax suspensions. *Materials Science and Engineering*, A254, 136–146.

[20] German R. M. and Hens K. F. 1991. Key issues in powder injection molding. *American Ceramics Society Bulletin*, **70**, 1294–1302.

[21] Bauer W., Knitter R., Bartelt G., Emde A., Göhring D. and Hansjosten E. 2002. Replication techniques for ceramic microcomponents with high aspect rations. *Microsystem Technologies*, **9**, 81–86.

[22] Evans J. R. G. and Edirisinghe M. J. 1991. Interfacial factors affecting the incidence of defects in ceramic mouldings. *Journal of Materials Science*, **26**, 2081–2088.

[23] M. A. Janney. 1995. Plastic forming of ceramics: extrusion and injection moulding. In *Ceramic Processing*, Terpstra R. A, Pex P. P. A. C., de Vries, A. H. (Eds), Chapman & Hall, London, United Kingdom. Chapter 6, 174–211.

[24] Zorzi J. E., Perottoni C. A. and da Jornada J. A. H. 2002. Hard–skin development during binder removal from Al_2O_3–based green ceramic bodies. *Journal of Materials Science*, **37**, 1801–1807.

[25] Chartier T., Ferrato M. and Baumard J. F. 1995. Influence of the debinding method on the mechanical properties of plastic formed ceramics. *Journal of the European Ceramic Society*, **15**, 899–903.

[26] Knitter R., Bauer W., Göhring D. 2003. Microfabrication of ceramics by rapid prototyping process chains. *Journal of Mechanical Engineering Science*, **217**(1), 41–51.

[27] Knitter R. and Liauw M. A. 2004. Ceramic microreactors for heterogeneously catalysed gas–phase reactions. *Lab Chip*, **4**, 378–383.

13

Fabrication of ceramic microcomponents by electrophoretic deposition

S. Bonnas and H. Elsenheimer

This chapter deals with electrophoretic deposition (EPD) as process engineering for the replication of ceramic microstructures. Since a stable suspension is required to use EPD as a reproducible shaping process an overview on stabilization mechanisms of colloid suspensions will be given. Interactions between particles in water (Derjaguin, Landau, Verwey and Overbeck theory) and electrokinetic effects, such as flow potential, sedimentation potential, electroosmosis and electrophoresis are also discussed followed by a presentation of the principle of EPD, different deposition setups, a summary of coagulation mechanisms and a mathematical description of deposition kinetics. Finally, various applications of EPD and some ceramic microstructures, fabricated by EPD are outlined thus demonstrating the potentials and limitations of the electrophoretic deposition as a replication process.

13.1 Introduction

At present the application of submicron and increasingly that of nanoscaled powders is gaining importance because of the demand for dimensional accuracy and smooth surfaces. In conventional shaping techniques like microinjection moulding or slip casting, nanosized powders are difficult to apply. For these powders EPD is a more suitable method [1, 2] because the deposition rate is independent of particle size as long as EPD is carried out perpendicularly to the sedimentation [3, 4].

EPD combines two processes: electrophoresis in a first step and deposition in a second step. In the first step particles suspended in a solvent move in the presence of an electric field to an oppositely charged electrode. In a second step the particles deposit on the electrode [1, 5–9] or on a parallel arrangement of an ion-permeable membrane [6, 10]. EPD is used for a wide range of applications such as: fabrication of coatings, laminated or graded materials, infiltration of porous materials or the manufacture of microstructures [5–7, 10, 11]. A wide range of materials and combinations can be employed [1, 12, 13]. Homogeneous deposits with low surface roughness and high mechanical strength can only be obtained by using well-dispersed suspensions [13]. Hence, knowledge of suspension stability is indispensable.

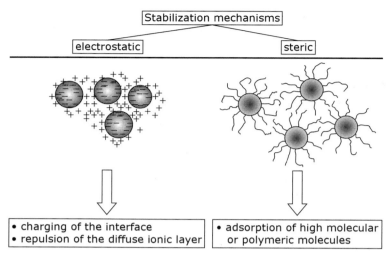

Figure 13.1 Schematic representation of the stabilization mechanisms electrostatic and steric stabilization.

13.2 Properties of suspensions for electrophoretic depositon

Generally, in liquid processing of ceramic materials the powders are dispersed, compacted to high densities and brought to a possible net shape, followed by removing the unrequested additives, and finally sintered [14]. Surface forces play a major role in dispersing and compacting. Colloidal stable suspensions can be fabricated by creating a high charge density on the particle surface, resulting in a strong repulsion of the electrical double layer enclosing the particles (electrostatic stabilization). A second possibility is the generation of repulsive forces by adsorption of polymers on the surface, induced by entropic repulsion of the polymers (steric stabilization) [15–18].

As a stable suspension of charged particles responding to an applied electric field is a prerequisite for EPD, only electrostatic and electrosteric stabilization are suitable for suspensions usable for EPD [2, 19].

13.2.1 Suspension stability

In general, the stability of a suspension can be defined in three different ways: as a state where no sedimentation of particles occurs [20–22], where a minimum of suspension viscosity is measured [23]; or in the case of a particle size distribution which is as fine and temporally invariant as possible [24, 25].

In the literature, four stabilization mechanisms are often defined and thus outlined in the following section. The nomenclature of the four mechanisms will be maintained, although electrosteric stabilization is a combination of electrostatic and steric stabilization. However, the fourth stabilization mechanism, which is depletion, will not be discussed here because it is irrelevant to EPD. Schematic representations of the stabilization mechanisms electrostatic and steric stabilization are shown in Figure 13.1.

Electrostatic stabilization

The mechanism preventing aggregation of particles is based on the presence of a repulsive electrostatic potential, partially enclosing each particle [20]. Dispersing a powder in a polar solvent induces the buildup of an electrical double layer surrounding the particles. This leads to a repulsive interaction between equally charged particles [17, 24]. The intensity of the interaction depends on pH, the concentration of specific adsorbed ions and ions in the suspension.

Steric stabilization

In this case an adsorbed nonionic polymeric layer prevents the aggregation of the uncharged particles [26]. The polymers must be tightly anchored on the particle surface in order to not desorb when collisions between particles occur. A steric stabilizing agent generally has tails, featuring a low solubility in the dispersing solvent, and exhibits a high affinity to the particle surface. Furthermore, there is a tail which is soluble in the surrounding solvent [17, 20, 24]. A full description of the mechanism is given in [26, 27].

Electrosteric stabilization

In this mechanism the aggregation of particles is prevented by combining electrostatic and steric stabilization [17, 20, 24]. This stabilization mode is usually associated with the adsorption of polyelectrolytes [20].

If the polyelectrolyte is adsorbed as a flat structure, the stabilization is mainly electrostatic with at the same time a weak steric repulsion. The steric contribution dominates as soon as the adsorbed layer increases and polymeric chains protrude in the solvent [17].

The stability of the suspensions is mainly governed by properties of the electrical double layer [26]. The concept of the electrical double layer describes the nonrandom ion assembly at an interface of two oppositely charged surfaces [20]. The interface charge on the surface develops because of ion exchange with the solvent. Due to this charge ions with an opposed charge are attracted, thus forming an electrical double layer, consisting of both, oppositely charged, ion species,[28, 29]. Different models exist to describe the structure of the double layer [20, 24].

13.2.2 Derjaguin, Landau, Verwey and Overbeck theory

Stabilization of particles is based on repulsive forces. However, there are also attractive van der Waals forces between particles. Thus suspensions become unstable under certain conditions. The pairwise interaction of repulsive and attractive forces between charged particles in a dielectric medium like water-based suspensions is described by the Derjaguin, Landau, Verwey and Overbeck (DLVO) theory [24, 30].

If two particles are approaching each other, their diffuse double layers overlap giving rise to a total force consisting of two components at any point between these particles [19]. The attracting component is named the van der Waals attraction, the other component the electrostatic repulsion. DLVO theory describes the simple addition of both energy potentials

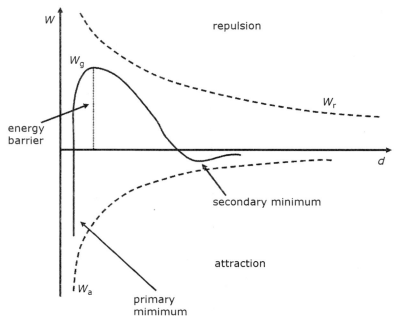

Figure 13.2 Total interaction energy between spherical particles as a function of particle distance.

[14, 31]. Thus it is possible to predict colloidal stability in polar solvents by considering the balance between these two potentials [14, 21].

Figure 13.2 shows the total interaction energy W between spherical particles as a function of particle separation d according to the DLVO theory for high ionic concentration.

At high ionic concentrations a secondary minimum can appear. Resulting from this a weak and potentially reversible particle agglomeration can occur [14, 20]. The stronger the ionic concentration the higher the energy barrier is and thus the suspension stability. Furthermore the maximum of total interaction energy shifts to shorter particle distances. So far, concerning nonaqueous solvents no theory has been validated. However, several papers give different examples concluding that DLVO theory is transferable to nonaqueous media [32–35].

13.2.3 Electrokinetic effects

The electrical double layer induces electrokinetic effects [36]. These effects are associated with the movement of charged species in an electric field. The field may be applied or created by motion of a liquid or particles due to an effective acceleration or pressure difference. As shown in Table 13.1, there are four different effects.

Table 13.1 Electrokinetic effects.

	Movement of media	Movement of particles
Absence of an electric field	Flow-potential 	Sedimentation-potential
Presence of an electric field	Electroosmosis 	Electrophoresis

The flow-potential results from the movement of the liquid caused by a hydrostatic pressure in a capillary or a stiff accumulation of particles as a green body filled with an electrolytic solvent. The electrical double layer is sheared from the solid phase by the movement of the liquid phase [36]. This induces an electric potential, the so-called flow-potential.

The sedimentation-potential is induced in the same way. However, the electrical double layer is sheared by the particle movement, caused by a gravitational or centrifugal force [36].

An electric field is applied to a capillary or fixed accumulation of particles, filled with an electrolytic medium. This generates a movement of ions from the double layer towards the electrode of opposite charge resulting in a movement of the liquid phase because of internal friction [36]. This effect is known as electroosmosis.

Electrophoresis is also generated by an electric field. The charged particles move to the electrode of opposite charge due to excess charge in the electrical double layer. This movement is contrary to the electroosmotic movement of the solvent [36].

Besides these classical electrokinetic effects there are further effects, induced by relative movements of the particle surface and the diffuse double layer. These are diffusiophoresis, electroaccoustic phenomena and electroviscous effects. A short overview of these effects has been given by Stahl and Weber [36].

EPD resulting from electrophoresis and the related mathematical description of the process will be described in Section 13.3.

13.3 Electrophoretic deposition

EPD was discovered in 1808 by Reuss. He observed the movement of clay particles in water because of the influence of an electric field [37].

13.3.1 Principles and models

The process of EPD consists of two parts. In a first step charged particles dispersed in a solvent move in presence of an electric field to an oppositely charged electrode. In a second step the particles deposit on the electrode [1, 5–9] or on an ion-permeable membrane which is parallel to it [6, 10]. The velocity of the particles during the movement towards an electrode is independent of their diameter [2, 38, 39].

Two kinds of parameters influence EPD. These are parameters affecting the suspension, such as solid content, zeta potential, viscosity and conductance and parameters concerning the experimental setup as geometry, electrode material, electric field and the time of deposition [9, 28].

Although the process of EPD has been well examined, the aggregation mechanism has not yet been clarified. There are several possible mechanisms which will be described below.

Movement due to electrophoresis

During the movement of a particle caused by an electric field the counterions in the diffuse double layer move in the opposite direction. Thus part of the dispersing medium is moved by viscous forces in the opposite direction, slowing down the moving particle. This effect is known as retardation [36]. As a result of movement of particles and counterions in the opposing directions the diffuse double layer is elliptically deformed along the semi-major axis in the direction of movement. While the particle is achieving a static condition the diffuse double layer is rearranged in a spherical form with the particle in the centre of the sphere. This rearrangement requires the so-called relaxation time. During the movement and the relaxation time there is a disequilibrium of charge behind the particle. This disequilibrium of charge induces a local electric field weakening the applied electric field. Thus the electrophoretic velocity of the particles is further reduced. This pseudo-force is known as the relaxation force [36]. In the case of high ionic concentration and, thereby thin diffuse double layers, the retardation and relaxation forces are both negligible and the primary breaking force is due to the frictional resistance of the dispersing medium [36].

A generally valid equation for the calculation of the electrophoretic velocity of spherical particles with the same conductance as the dispersing medium was formulated in 1931 by Henry [40].

$$v_{EPD} = \frac{\varepsilon_0 \cdot \varepsilon_r \cdot \zeta}{\eta \cdot E \cdot f(\kappa \cdot r)} \tag{13.1}$$

The factor $f(\kappa \cdot r)$ is dependent on particle radius and parameter κ (Debye–Hückel–parameter). This parameter describes the reciprocal thickness of the electrical double layer. There are two boundary cases for $f(\kappa \cdot r)$. These are $\kappa \cdot r \ll 1$ (thick double layer) for which

$f(\kappa \cdot r) = 2/3$ and $\kappa \cdot r >> 1$ (thin double layer) for which $f(\kappa \cdot r) = 1$. These boundary cases are called by the names of their inventors: Hückel ($\kappa \cdot r << 1$) and Smoluchowski ($\kappa \cdot r >> 1$). Henry found that for $\kappa \cdot r > 300$ Smoluchowski's equation is a good aproximation with an error of less than 1%. For $\kappa \cdot r < 0.5$ Hückel's equation is valid with the same error [40, 41].

The polarization of the double layer in an electric field (relaxation) has not been taken into account in an algebraic equation up to now. In 1978 O'Brien and White computed numerically the movement of a spherical particle under consideration of the retarding forces as friction, retardation and relaxation [42].

Mechanisms of coagulation during electrophoretic deposition

While EPD has been successfully used for a long time, the proper mechanism of coagulation is not yet exactly known. Different mechanisms have been postulated. Some of them could be disproved; others still have to be proved or disproved. Some of these models are explained below.

To explain the coagulation of particles during EPD Verwey and Overbeek [43] developed a concept based on DLVO-theory. By superpositioning attractive electrophoretic and van der Waals forces with repulsive forces due to the electrical double layer they calculated the electric fields necessary for coagulation. However, Brown and Salt [44] showed that the calculated fields differ significantly from the experimentally detected fields.

Sarkar and Nicholson [12] developed a model based on Verwey and Overbeek's concept to describe the coagulation during EPD. As shown in Figure 13.3 the double layer is deformed by fluid dynamics and the applied electric field. During the movement of the particle the double layer thins out at the front side and becomes thicker in the wake. The charged counterions move with the particle towards the electrode. The counterions on the border of the double layer tend to react with other ions outside the double layer. As a result of this chemical reaction the double layer becomes thinner and the next particle moving towards the electrode is able to approximate so closely, that the attractive van der Waals forces dominate. Figure13.3 shows this coagulation process. However, for this model the calculated fields also differ from the experimental data. This divergence may be due to a strong voltage drop over the electrode [12].

Wilson *et al.* [45] and Wang *et al.* [46] have investigated pairwise aggregation of particles caused by an electric field. They explained the aggregation by hydrodynamic forces. Solomentsev reasoned that convection around the particles, caused by electroosmosis near the electrodes, attracts further particles. The aggregation is explained by a domination of attractive van der Waals forces at very low distances between two particles [47].

A related model to describe coagulation of particles during EPD was proposed by Trau *et al.* [48]. Their theory indicates that electrohydrodynamic forces induce interaction between particles. These electrohydrodynamic forces are based on the passage of ionic currents through the suspension. They are induced when conformation of concentration polarization is disturbed by a colloidal particle. Nadal *et al.* [49] approved that coagulation of particles during EPD occurs because of electrohydrodynamic interaction. These attractive interactions compete against repulsive dipole interactions.

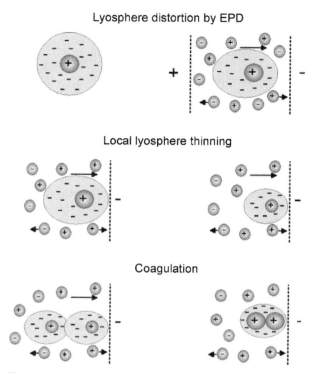

Figure 13.3 Coagulation mechanism during EPD proposed by Sarkar and Nicholson.

In the last years some simulations concerning the modelling of coagulation phenomena and deposition during EPD have been realized [50–53]. However, these calculations do not allow clear conclusions to be drawn about the mechanism.

A comprehensive overview of the different coagulation and deposition mechanisms is given in [12, 54].

Description of deposition rate during electrophoretic deposition

To control the deposition process during electrophoretic fabrication of protective coatings, functional graded and laminated coatings it is necessary to know the mathematic model for electrophoretic kinetics [55]. The exact correlation between the deposition rate and time response of the electric field during EPD is not yet known.

The first equation, describing the electrophoretic deposition rate dm/dt, standardized to the surface of the electrode A, was formulated by Hamaker [38] in 1940.

$$\frac{dm}{dt} \cdot \frac{1}{A} = a \cdot \mu \cdot E \cdot c(t) \tag{13.2}$$

The deposition rate is dependent on factor a, standing for the fraction of particles close to the electrode, contributing to the deposition, on electrophoretic mobility μ, electric field strength E and concentration of colloidal particles in the suspension.

Hamaker found that the deposited mass was to a certain degree proportional to the factors mentioned above. For low electric fields, short deposition time and low solid concentration there are variances between Equations (13.2) and experimental data [56]. Deposited mass is often considered to be independent of μ, E, c and A with respect to time. This is a strong simplification [13, 34]. However, Equation (13.2) cannot be solved algebraically without these simplifications because in reality electrophoretic mobility, electric field strength and particle concentration are time-dependent variables.

Experimental investigations lead to more than a linear increase of deposition rate if the solid concentration is raised. This is contrary to Hamaker's formula [57, 58]. Biesheuvel and Verweij [13] calculated a correction factor for Hamaker's formula considering the growth of the green body and the solid–liquid boundary.

In 1962 Avgustinik *et al.* [59] published a formula related to Equation (13.2) for the calculation of deposition rate for cylindrical geometries. In 1966 they extended their model by a factor considering decreasingly solid content of suspension due to EPD [60].

Further approaches to modelling the yield during EPD with constant current and constant voltage are given in [12].

Experimental investigations showed a S-shaped curve of yield as a function of the deposition time [61, 62]. Ferrari explained this phenomenon by a parallelism of two processes: the change of conductivity and the change of solid content in the suspension and developed a mathematical model [62].

The description of the deposition rate is complicated by possible inhomogeneities of the electric field. The field, calculated by voltage and distance of the electrodes does not necessarily correspond with the real applied field [5]. A dependence on location was shown, especially for EPD on a membrane.

Dispersing media

Most scientific studies concerning electrophoretic deposition use organic solvents as dispersing media, i.e. ethyl alcohol [15, 35, 63–68], acetyl acetone [69, 70], acetone [71, 72], isopropanol [73–76], acetic acid [77] or mixtures of different solvents [69, 78], partially with water [69, 79], or rarely with water [11, 80–82]. An overview has been given by Besra and Liu [82].

The great advantage of nonaqueous solvents over water is that high voltages can be applied. Thus, problems like electrolysis and Joule heating are avoided. Furthermore, organic media are preferred due to their good chemical stability and their lower conductivity [83]. However, compared to water, this results in higher costs, toxicity and combustibility and thus combined security precautions. Moreover, the waste management in aqueous suspensions is much less complicated. Due to the higher permittivity in water higher deposition rates are possible.

13.3.2 Applications

After the discovery of the principle of EPD only experiments to understand the procedure were realized for more than a century. Sheppard was the first to use the EPD for the commercial fabrication of rubber goods in 1927 [84]. The first patent on coating radiant bodies (i.e. electron discharge tubes or anticathode of X-ray tubes with tungsten or molybdenum) was taken out by Harsányi in 1933 [85]. In the following years investigations into the fabrication of china-like cups and dishes as well as the transfer to mass production were undertaken [86]. The shaping of Al_2O_3 [57] and ZrO_2 took place in the 1970s. A well-known application of EPD is the so-called elephant process which was used industrially to fabricate tiles [84].

However, in the industrial sector the most common use of EPD is electrophoretic lacquering [87], the so-called cataphoretic painting. Mainly in the automobile sector, car bodies and accessory parts are primed or lacquered with a film. In this process the workpiece is switched to be the cathode and the lacquer is deposited cataphoretically. The resulting homogeneous coating of all areas has advantages in comparison to other methods. Cataphoretic painting is used in device tool building, the manufacture of internal combustion engines and the bicycle industry [88].

Besides lacquering, EPD is commonly used as a fast and easy method to deposit coatings. Thus protective coatings for high temperature corrosion protection made of aluminium, aluminium–chromium and aluminium–silicon can be coated on a nickel-based super alloy [89].

Another application of EPD is in the fabrication of superconducting coatings of magnesium boride [90]. Suspensions made with $YBa_2Cu_3O_{7-x}$ powder were used to produce thick films for magnetic shielding of low frequencies magnetic fields [91]. A significant improvement of tribological properties was achieved [92] by the deposition of molybdenum sulphide.

Other investigations have dealt with the EPD of nontoxic ceramics like tungsten carbides for coating steel substrates in order to increase the hardness [93]. Similarly, films of silicon carbide and moulds have been deposited using organic and aqueous suspensions for applications in high temperature ranges [94, 95]. The fabrication of moulds made of aluminium nitride has been described [94, 96].

Moulds have also been electrophoretically deposited from oxide ceramics, e.g. tubes and plates have been made of ZrO_2 [2]. Suspensions tubes [74, 97, 98] and ogives have been fabricated [57] from Al_2O_3 suspensions.

In the field of medical equipment EPD is used to make hip joints [99]. Further applications include the coating of medical implants to increase the bioactivity [100, 101] or to improve their surface properties [102]. Bacteria can be immobilized on surfaces by means of EPD. This allows controlled fabrication of bacterial biofilms in biomedicine, i.e. protective layers on catheters [103] or possible biotechnological applications such as biosensors, bioreactors [103] or bioelectronics [104]. Another medical application of EPD is the fabrication of dental crowns [105].

Microtubes have also been electrophoretically deposited [106] from PZT suspensions in order to be used as piezoelectric actuating elements for micropositioning [107]. Functional graded thick films have been used as actuators [108].

For electronic applications dielectrics for thin-film capacitors were electrophoretically deposited from $BaTiO_3$ suspensions [109]. In this area the passivation is of great interest.

Thus high-voltage transistors were electrophoretically coated with zinc borosilicate glass [110]. Also Al_2O_3 was applied as isolation coating on tungsten heating coils of cathode ray tubes [111, 112]. Furthermore it has been deposited, for electrical isolation, on platinum wires used for sensor and instrument wiring in aircraft engines [113]. Lanthanum-doped PZT has been electrophoretically deposited on platinum to fabricate pyrosensors [114].

EPD has also been applied in the range of microoptic and optoelectronic applications. Layers of oriented diamond thin films for field emission displays have been fabricated [115]. Electrophoretically deposited nanostructured layers from fullerenes have been developed for optoelectronics, i.e. fabricate infrared filters [116]. Thick films from TiO_2 with large surface and high porosity have been deposited onto electric conducting glass to fabricate photovoltaic cells [117].

EPD is used in industry to manufacture shadow masks coated with ceramic insulators [112]. Phosphor has been electrophoretically deposited to fabricate monochromatic screens for information displays. Furthermore, by combining EPD and photolithography colour displays can be fabricated via selective deposition of different phosphors [118].

Several research works deal with the fabrication of fuel cells, especially high temperature fuel cells (solid oxide fuel cells (SOFC)). Yttrium-stabilized ZrO_2 is used as an electrolyte for fuel cells [119–122]. The fabrication of transparent ceramics is another field of great interest where most researchers are dealing with manufacturing transparent alumina. Two important application fields are: cover plates for erasable program memories and burner pipes for sodium discharge lamps [123]. However, so far the successful fabrication of transparent alumina has been very difficult [124]. Fabrication of transparent zirconia via EPD has not yet been possible whereas translucent ZrO_2 could be manufactured [125].

Besides dense moulds membranes were also fabricated. Thus electrophoretically dense ion exchange membranes were produced using zeolites [126]. Porous membranes were fabricated from Al_2O_3. The density of membranes was affected [127] by varying the applied electric field.

EPD is not only used for fabrication of layers and moulds, but also for infiltration of porous materials. With electrophoretic impregnation (EPI) nanoscaled second phases can be inserted into porous matrixes. In this way functionally graded materials (FGMs) can be fabricated [10, 128] and optical and electrical properties adapted [128]. During EPI, a green body with open porosity is impregnated with nanoscaled particles. The steepness of the gradient is dependent on the matrix porosity, ratio of particle and pore size, viscosity and zeta potential of the nanoscaled particle suspension and the applied electric field [10]. The amount of the secondary phase in the composite is limited by the matrix fabricated in the first processing step. Furthermore, the thickness of the graded layer is limited to a penetration depth of about 20 μm to several millimetres.

EPD research is dominated by fuel cell applications and the manufacture of composite materials. Manufacture of FGMs can be divided into two types: on the one hand the laminated stepwise gradients and on the other hand the continuous gradients. The stepwise gradient is fabricated by alternating electrophoretic deposition of multiple suspensions on the same substrate [129, 130]. A piezoelectric actuator was produced [131] by using several PZT compositions.

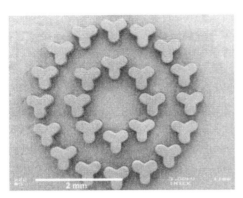 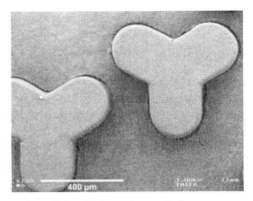

Figure 13.4 SEM pictures of an array of spinning nozzles and detail of a single nozzle (lost substrates made of C graphite-filled wax).

Fabrication of a continuous gradient was carried out by deposition from a suspension of one component gradually adding a suspension of another component [132]. In this way EPD of composite materials like tungsten carbide cobalt with a compound of titanium carbon nitrogen [133], Al_2O_3 with ZrO_2 or molybdenum silicide with Al_2O_3, respectively, ZrO_2 with nickel [7] was successfully carried out.

As a result of the colloidal properties of the different materials, different zeta potentials may result in varying deposition rates. Similarly heterocoagulation or segregation can occur. Establishing a perfect continuous gradient is only possible if these problems are solved. Complex calculations and an exact control of the process parameters are essential to predict and tailor the gradient in the layer.

13.4 Microstructures fabricated by electrophoretic deposition

EPD is used to fabricate layers and moulds and also for the fabrication of ceramic microstructures.

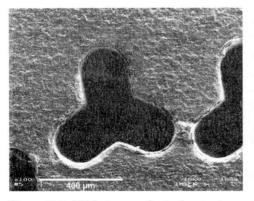 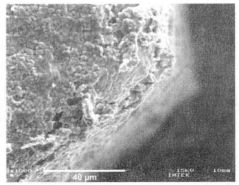

Figure 13.5 SEM pictures of a single spinning nozzle and detail of an edge of this single nozzle (sintered alumina).

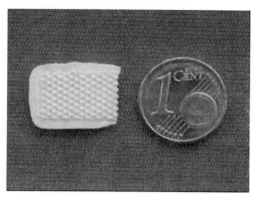

Figure 13.6 A triple mirror array.

Mould release is the greatest problem in manufacturing microstructures. There are different solutions to this problem: draft angles can be applied by simply striping off the substrate, or so-called lost substrates like hard plaster or compounds of graphite and wax can be used. Depending on the requirements, draft angles cannot always be employed. Therefore a lost substrate is the easier alternative. Some lost substrates are shown in Figure 13.4. The array of different forms of spinning nozzles made of a compound of graphite and wax is finally used as substrate for EPD [134, 135].

A lost substrate is thereby characterized by thermal or chemical removal. The compound was thermally removed and ceramics flawlessly sintered. In this way ceramic microstructures, as shown in Figure 13.5, were electrophoretically fabricated from Al_2O_3 suspensions.

A triple mirror array was also faultlessly electrophoretically deposited from a ZrO_2 suspension and sintered, as shown in Figure 13.6.

Trivial microstructures as dental crowns have been mentioned in Section 13.3.2. Details of the fabrication of such crowns have been published [36]. Figure 13.7 shows a SEM

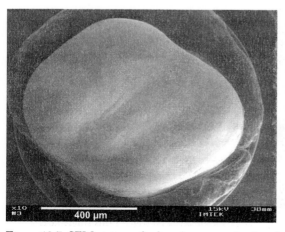

Figure 13.7 SEM picture of a dental crown occlusal surface.

picture of the occlusal surface of such a crown. The fitting accuracy is of great importance. In this case ZrO_2 was electrophoretically deposited on gypsum coated with a conductive lacquer. Gypsum was thermally removed and the dental crown flawlessly sintered.

Honeycomb structures made of PZT were deposited onto gold sputtered PMMA electrodes [137]. PMMA was thermally removed and the structures sintered. These microstructures can be used i.e. for sensor applications.

Conclusions

EPD is a versatile processing technique for the replication of ceramic microstructures and the coating of conductive devices. The EPD process consists of two parts: an electrical field induces a movement of charged particles in a solvent and in a second step the deposition on a substrate takes place. It is necessary to use stable suspensions for controlled formation of a deposit via EPD. EPD needs a charged particle surface, therefore two stabilization mechanisms are suitable: electrostatic and electrosteric stabilization.

DLVO theory describes electrostatic stabilization by simple addition of repulsive electrostatic and attractive van der Waals forces between two particles in aqueous media. Four electrokinetic effects are known to influence EPD. These are: flow potential, sedimentation potential, electroosmosis and electrophoresis. In the kinetic description of EPD only electrophoresis is considered.

There are two kinds of parameters influencing the kinetics of EPD. These are parameters affecting the suspension, such as solid content, zeta potential, viscosity and conductance and parameters concerning the experimental setup as geometry, electrode material, electric field and the time of deposition. The first simplified mathematical description of the deposited yield was given by Hamaker in 1939. This equation has since been adjusted by several scientists to consider additional effects neglected by Hamaker. Despite good kinetic modelling of the deposited mass, the mechanism of coagulation has not yet been clarified.

EPD has been employed to fabricate coatings and near net shape moulds consisting of various materials for use in electric and magnetic engineering, optic, superconducting, semiconducting and biomedical applications. Fabrication of ceramic microstructures is another field of application of EPD. In a first step a compound of graphite and wax was fabricated as a lost substrate. In a second step high-performance ceramic was deposited via EPD and finally sintered. Thus spinning nozzles and triple mirrors were fabricated.

Deposition onto a 3D conductive electrode is another possibility to manufacture devices of a simple geometry followed by a demoulding step. As mould release represents the major problem in manufacturing microstructures an adjusted electrode geometry or draft angles and an optimized demoulding technique are necessary.

References

[1]   Boccaccini A. R. and Zhitomirsky I. 2002. Application of electrophoretic and electrolytic deposition techniques in ceramics processing. *Current Opinion in Solid State & Materials Science*, **6**, 251–260.

[2] Harbach F. and Nienburg H. 1998. Homogeneous Functional Ceramic Components through Electrophoretic Deposition from Stable Colloidal Suspensions – I. Basic Concepts and Applications to Zirconia. *Journal of the European Ceramic Society*, **18**(6), 675–683.

[3] Bonnas S., Tabellion J., Ritzhaupt–Kleissl H. J. and Hausselt J. 2006. Systematic interaction of sedimentation and electrical field in electrophoretic deposition. In *Second International Conference on Multi-Material Micro Manufacture*, Grenoble, France, Menz W., Dimov S., and Fillon B. (Eds), Elsevier, Amsterdam. 187–190.

[4] Dushkin C., Miwa T. and Nagayama K. 1998. Gravity effect on the field deposition of two-dimensional particle arrays. *Chemical Physics Letters*, **285**(3–4), 259–265.

[5] Van der Biest O. O. and Vandeperre, L. J. 1999. Electrophoretic deposition of materials. *Annual review of Materials Science*, **29**, 327–52.

[6] Tabellion J. and Clasen R. 2004. Electrophoretic deposition from aqueous suspensions for near-shape manufacturing of advanced ceramics and glasses-applications. *Journal of Materials Science*, **39**, 803–811.

[7] Sarkar P., Datta S., and Nicholsn P. S. 1997. Functionally graded ceramic/ceramic and metal/ceramic composites by electrophoretic deposition. *Composites Part B-Engineering*, **28**(1–2), 49–56.

[8] Nagarajan N. and Nicholson P. S. 2004. Nickel-alumina Functionally graded materials by electrophoretic deposition. *Journal of the American Ceramic Society*, **87**(11), 2053–2057.

[9] Ferrari B. and Moreno R. 1997. Electrophoretic deposition of aqueous alumina slips. *Journal of the European Ceramic Society*, **17**(4), 549–556.

[10] Oetzel C., Clasen R., and Tabellion J. 2004. Electric field assisted processing of ceramics. *cfi-Ceramic Forum International*, **81**, 35–41.

[11] Simovic K. Miskovic-Stankovic V. B., Kicevic D and Jovanic P. 2002. Electrophoretic deposition of thin alumina films from water suspensions. *Colloids and Surfaces a–Physicochemical and Engineering Aspects*, **209**, 47–55.

[12] Sarkar P. and Nicholson P. S. 1996. Electrophoretic deposition (EPD): mechanism, kinetics and application to ceramics. *Journal of the American Ceramic Society*, **79**, 1987–2002.

[13] Biesheuvel P. M. and Verweij H. 1999. Theory of cast formation in electrophoretic deposition. *Journal of the American Ceramic Society*, **82**(6), 1451–55.

[14] Horn R. G. 1990. Surface forces and their actions in ceramic materials. *Journal of the American Ceramic Society*, **73**, 1117–1135.

[15] Widegren J. and Bergstrom L. 2000. The effect of acids and bases on the dispersion and stabilization of ceramic particles in ethanol. *Journal of the European Ceramic Society*, **20**(6), 659–665.

[16] Lowen H. 1995. Kolloide - auch für Physiker interessant? *Physikalische Blätter*, **51**, 165–168.

[17] Shojai F., Pettersson A. B. A., Mantyla T. and Rosenholm J. B. 2000. Electrostatic and electrosteric stabilization of aqueous slips of 3Y-ZrO$_2$ powder. *Journal of the European Ceramic Society*, **20**(3), 277–283.

[18] Fuerstenau D. W., Urbina Herrera R., and Hanson J. S. 1988. Adsorption of processing additives and the dispersion of ceramic powders. *Ceramic Transactions*, **82**(4), 333–351.

[19] Wei, M., Ruys A. J., Milthorpe B. K. and Sorrell C. C. 1990. Solution ripening of hydroxyapatite nanoparticles: Effects on electrophoretic deposition. *Journal of Biomedical Materials Research Part A*, **45**(1), 11–19.

[20] Hackley V. A. and Ferraris C. F. 2001. *The Use of Nomenclature in Dispersion Science and Technology*. National Institute of Standards and Technology, Washington, DC, USA.

[21] Hogg R. 1987. Flocculation phenomena in fine particle dispersion. In *Ceramic Powder Science*, Messing G. L., Mazdiyasni K. S., McCauley J. W. and Habers R. A. (Eds). American Ceramic Society Westerville, Ohio, USA. 467–482.

[22] Müller R. H. 1996. Zetapotential und Partikelladung in der Laborpraxis, Wissenschaftliche Verlagsgesellschaft mbH, Stuttgart.

[23] Briscoe B. J, Khan A. U. and Luckham P. F. 1998. Optimising the dispersion on an alumina suspension using commercial polyvalent electrolyte dispersants. *Journal of the European Ceramic Society* **18**,(14), 2141–2147.

[24] Lagaly G., Schulz O. and Zimehl R. 1997. Dispersionen und Emulsionen - Eine Einführung in die Kolloidik feinverteilter Stoffe einschließlich der Tonminerale. D. Steinkopff Verlag, Darmstadt.

[25] Nitzsche R. and Simon F. 1997. Bestimmung des Zetapotentials aus Messungen der elektrophoretischen Mobilität. *tm-Technisches Messen*, **64**(3), 106–113.

[26] Napper D. H. 1970. Colloid stability. Industrial and Engineering Chemistry Product Research and Development, **9**(4), 467–477.

[27] Napper D. H. 1977. Steric stabilization. *Journal of Colloid and Interface Science*, **58**(2), 390–407.

[28] Bouyer F. and Foissy A. 1999. Electrophoretic deposition of silicon carbide. *Journal of the American Ceramic Society*, **82**, 2001–2010.

[29] Stahl W. and Weber K. 1999. Verbundvorhaben: Verbesserung von Fest/Flüssig-Trennverfahren unter Zusatz elektrischer und akustischer Felder. Teilvorhaben: Elektrofiltration. Institut für Mechanische Verfahrenstechnik und Mechanik, Universität Karlsruhe.

[30] Evans D. F. and Wennerström H. 1999. The Colloidal Domain - Where Physics, Chemistry, Biology, and Technology Meet. Evans D. F., New York.

[31] Ottewill R. H. 1977. Stability and instability of disperse systems. *Journal of Colloid and Interface Science*, **58**(2), 357–373.

[32] Wang G., Sarkar P. and Nicholson P.S. 1999. Surface chemistry and rheology of electrostatically (ionically) stabilized alumina suspensions in polar organic media. *Journal of the American Ceramic Society*, **82**, 849–56.

[33] Wang G., Sarkar P. and Nicholson P. S. 1997. Influence of acidity on the electrostatic stability of alumina suspensions in ethanol. *Journal of the American Ceramic Society*, **80**, 965–72.

[34] Fukada Y., Nagarajan N., Mekky W., Bao Y., Kim H. S. and Nicholson P. S. 2004. Electrophoretic deposition- mechanism, myths and materials. *Journal of Materials Science*, **39**(3), 787–801.

[35] Wang G. and Nicholson P. S. 2001. Heterocoagulation in ionically stabilized mixed-oxide colloidal dispersions in ethanol. *Journal of the American Ceramic Society*, **84**, 1250–56.

[36] Stahl W. and Weber K. 2003. *Verbundvorhaben: Verbesserung der Filtrationskinetik von Filterpressen mit Hilfe elektrischer Felder*. Institut für Mechanische Verfahrenstechnik und Mechanik, Universität Karlsruhe.

[37] Adamczyk Z. and Weronski P. 1999. Application of the DLVO theory for particle deposition problems. *Advances in Colloid and Interface Science*, **83**, 137–226.

[38] Hamaker H. C. 1940. The influence of particle size on the physical behaviour of colloid systems. *Ransactions of the Faraday Society*, **36**, 186–192.

[39] Abramson H. A. 1931. The influence of size, shape, and conductivity on cataphoretic mobility, and its biological significance. *Journal of Physical Chemistry*, **35**(1), 289–308.

[40] Henry D. C. 1931. The cataphoresis of suspended particles. Part I. The equation of cataphoresis. *Proceedings of the Royal Society of London. Series A*, **133**(821), 106–129.

[41] Sumner C. G. and Henry D. C. 1931. Cataphoresis. Part II. A New experimental method, and a confirmation of Smoluchowski's Equation. *Proceedings from the Royal Society of London. Series A* **133**(821), 130–140.

[42] O'Brien R. W. and White L. R. 1978. Electrophoretic mobility of a spherical colloidal particle. *Journal of the Chemical Society-Faraday Transactions 2*, **74**, 1607–1626.

[43] Verwey E. J. W. and Overbeek J. T. G. 1948. Theory of the Stability of Lyophobic Colloids: The Interaction of Sol Particles Having an Electric Double Layer, Elsevier, New York.

[44] Brown D. R. and Salt F. W. 1965. The mechanism of electrophoretic deposition. *Journal of Applied Chemistry of the USSR*, **15**, 40–48.

[45] Wilson H. J., Pietraszewski L. A. and Davis R. H. 2000. Aggregation of charged particles under electrophoresis or gravity at arbitrary Peclet numbers. *Journal of Colloid and Interface Science*, **221**(1), 87–103.

[46] Wang H., Zeng S. L., Loewenberg M. and Davis R. H. 1997. Particle aggregation due to combined gravitational and electrophoretic motion. *Journal of Colloid and Interface Science*, **187**(1), 213–220.

[47] Solomentsev Y., Böhmer M., and Anderson J. L. 1997. Particle clustering and pattern formation during electrophoretic deposition: A hydrodynamic model. *Langmuir*, **13**, 6058–6068.

[48] Trau M., Saville D. A., and Aksay, I. A. 1997. Assembly of colloidal crystals at electrode interfaces. *Langmuir*, **13**(24), 6375–6381.

[49] Nadal F., Argoul F., Hanusse P., Pouligny B. and Ajdari A. 2002. Electrically induced interactions between colloidal particles in the vicinity of a conducting plane. *Physical Review E*, **65**(6), 061409.

[50] Perez A. T., Saville D., and Soria C. 2001. Modelling the electrophoretic deposition of colloidal particles. *Europhysics Letters*, **55**, 425–431.

[51] Cordelair J. and Greil P. 2004. Discrete element modeling of solid formation during electrophoretic deposition. *Journal of Materials Science*, **39**(3), 1017–1021.

[52] Greil P., Cordelair J. and Bezold A. 2001. Discrete element simulation of ceramic powder processing. *Zeitschrift Für Metallkunde*, **92**(7), 682–689.

[53] Hong C. W. 1997. New concept for simulating particle packing in colloidal forming process. *Journal of the American Ceramic Society*, **80**, 2517–24.

[54] Tabellion J. 2004. Herstellung von Kieselgläsern mittels elektrophoretischer Abscheidung und Sinterung. PhD thesis, University of Saarbrücken.

[55] Anne G., Van Meensel K., Vleugels J. and van Der Biest O. 2005. A mathematical description of the kinetics of the electrophoretic deposition process for AL_2O_3-based suspensions. *Journal of the American Ceramic Society*, **88**(8), 2036–2039.

[56] Hamaker H. C. 1940. Formation of deposit by electrophoresis. *Transactions of the Faraday Society*, **36**, 279–287.

[57] Andrews J. M., Collins A. H., Cormish D. C. and Dracass J. 1969. The forming of ceramic bodies by electrophoretic deposition. *Proceedings of the British Ceramic Society*, **12**, 211–229.

[58] Biesheuvel P. M. 2000. Porous ceramic membranes - suspension processing, mechanical and transport properties, and application in the osmotic tensiometer. PhD thesis, University of Twente.

[59] Avgustinik A. I., Vigderguaz V. S., and Zhuravlev G. I. 1962. Electrophoretic deposition of ceramic masses from suspensions and calculation of deposit yields. *Zhurnal Prikladroi Khimii*, **35**(10), 2090–2093.

[60] Avgustin A. I., Zhuravle G. I., and Vigderga V. S. 1966. Calculation of the yield of deposit in electrophoretic deposition. *Colloid Journal–USSR*, **35**(10), 379–383.

[61] Ma J. and Cheng W. 2002. Electrophoretic deposition of lead zirconate titanate ceramics. *Journal of the American Ceramic Society*, **85**(7), 1735–37.

[62] Ferrari B., Moreno R. and Cuesta J. 2006. A resistivity model for electrophoretic deposition. In *Proceedings of Conference in Electrophorectic Deposition: Fundamentals and Applications 2*, 29 May – 2 June 2005, Castelvecchio Pascoli, Boccaccini, A. R., Van der Biast O. and Claussen R, 175–180.

[63] Nicholson P. S., Sarkar P. and Huang X. 1993. Electrophoretic deposition and its use to synthesize ZrO_2/Al_2O_3 micro-laminate ceramic/ceramic composites. *Journal of Materials Science*, **28**, 6274–6278.

[64] Will J., Hruschuka M. K. M., Gubler L. and Gauckler L. 2001. Electrophoretic deposition of ziorconia on porous anodic substrates. *Journal of the American Ceramic Society*, **84**, 328–32.

[65] Sarkar P., De D., Yamashita K., Nicholson P. S. and Umegaki T. 2000. Mimicking Nanometer Atomic Processes on a micrometer scale via electrophoretic depostion. *Journal of the American Ceramic Society*, **83**, 1399–401.

[66] Wang G. and Nicholson P. S. 2001. Influence of acidity on the stability and rheological properties of ionically stabilized alumina suspensions in ethanol. *Journal of the American Ceramic Society*, **84**, 1977–80.

[67] Zhitomirsky I. and Petric A. 2000. Electrophoretic deposition of ceramic materials for fuel cell applications. *Journal of the European Ceramic Society*, **20**(12), 2055–2061.

[68] Hatton B. and Nicholson P. S. 2001. Design and fracture of layered $Al_2O_3/TZ3Y$ composites produced by electrophoretic deposition. *Journal of the American Ceramic Society*, **84**, 571–76.

[69] Wang Z. H., Shemilt J., and Xiao P. 2002. Fabrication of ceramic composite coatings using electrophoretic depostion, reaction bonding and low temperature sintering. *Journal of the European Ceramic Society*, **22**(2), 183–189.

[70] Ng, S. Y. and Boccaccini A. R. 2005. Lead zirconate titanate films on metallic substrates by electrophoretic deposition. *Materials Science and Engineering B Solid State Materials for Advanced Technology*, **116**, 208–214.

[71] Koura N., Tsukamoto T., Shoji H. and Hotta T. 1995. Preparation of various oxide films by an electrophoretic deposition method: A study of the mechanism. *Japanese Journal of applied Physics*, **34** (3), 1643–1647.

[72] Mizuguchi J., Sumi K. and Muchi T. 1983. A highly stable non-aqueous suspension for the electrophoretic eeposition of powdered substances. *Journal of the Electrochemical Society*, **130**(9), 1819–1825.

[73] Gutierrez C. P., Mosley J. R. and Wallace T. C. 1962. Electrophoretic deposition: A versatile coating method. *Journal of the Electrochemical Society*, **109**, 923–927.

[74] Harbach F. and Nienburg H. 1998. Homogeneous functional ceramic components through electrophoretic deposition from stable colloidal suspensions -II. Beta-alumina and concepts for industrial production. *Journal of the European Ceramic Society*, **18**(6), 685–692.

[75] De Beer E., Duval J. and Meulenkamp E. A. 2000. Electrophoretic deposition: A quantitive model for particle deposition and binder from alcohol-based suspensions. *Journal of Colloid and Interface Science*, **222**, 117–124.

[76] Maka K., Hadraba H. and Cihlar J. 2004. Electrophoretic deposition of alumina and zirconia - I. Single-component systems. *Ceramics International*, **30**(6), 843–852.

[77] Anne G., Van Meensel K., Vleugels J. and Van der Beist O. 2004. Influence of the suspension composition on the electric field and deposition rate during electrophoretic deposition. *Colloids and Surfaces a–Physicochemical and Engineering Aspects*, **245**, 35–39.

[78] Yamashita K., Nagai M. and Umegaki T. 1997. Fabrication of green films of single- and multi-component ceramic composites by electrophoretic deposition technique. *Journal of Materials Science*, **32**(24), 6661–6664.

[79] Lebrette S. 2002. Influence de l'éthanol sur la dispersion du rutile en milieu aqueux - Elaboration de couches céramiques par électrophorèse. PhD thesis, University of Limoges .

[80] Moreno R. and Ferrari B. 2000. Effect of the Slurry properties on the homogeneity of alumina deposits obtained by aqueous electrophoretic deposition. *Materials Research Bulletin*, **35**(6), 887–897.

[81] Ferrari B., Sanchez-Herencia A. J. and Moreno, R. 1998. Aqueous electrophoretic deposition of Al_2O_3/ZrO_2 layered ceramics. *Materials Letters*, **35**, 370–374.

[82] Tang F. Q., Uchikoshi T., Ozawa K. and Sakka Y. 2002. Electrophoretic deposition of aqueous nano-gamma-Al2O3 suspensions. *Materials Research Bulletin*, **37**(4), 653–660.

[83] Besra L. and Liu M. 2007. A review on fundamentals and applications of electrophoretic deposition (EPD). *Progress in Materials Science*, **52**(1), 1–61.

[84] Heavens S. N. 1990. Electrophoretic deposition as a processing route for ceramics. In *Advanced Ceramic Processing and Technology*, J. G. P. Binner, Noyes, Park Ridge, New Jersey, USA. 255–283.

[85] US Patent, 1897902. 1933. Harsanyi E. Method of coating radiant bodies.

[86] Entelis F. S. and Sheinina M. E. 1979. Electrophoretic forming of porcelain articles. *Glass and Ceramics*, **36**(11–1), 634–637.

[87] Berger, U. 1998. Elektrophoretische Klarlackierung - Anwendungsmöglichkeiten für die dekorative Galvanotechnik. *Metalloberglache*, **52**(9), 682–684.

[88] AG, Tisel Lackiertechnik. 2007. Personal communication 30 July 6, 2007.

[89] Fisch, H. A. Electrophoretic depostion of aluminide coatings from aqueous suspensions. *Journal of the Electrochemical Society*, **119**(1), 57–64.

[90] Ochsenkuhn-Petropoulow M. T., Altzoumailis A. F., Argyropoulow R. and Ochsenkhun K. M. 2004. Superconducting coatings of MgB_2 prepared by electrophoretic deposition. *Analytical and Bioanalytical Chemistry*, **379**(5–6), 792–795.

[91] Dusoulier L., Denis S., Vanderbemden P., Dirickx M., Ausloos M., Cloots R. and Vertruyen B. 2006 Preparation of $YBa_2Cu_3O_{7-x}$ superconducting thick films by the electrophoretic deposition method. *Journal of Materials Science*, **41**(24), 8109–8114.

[92] Panitz J. K. G., Dugger M. T., Pebbles D. E., Tallant D. R. and Hills C. R. 1993. Electrophoretic deposition of pure Mos2 dry film lubricant coatings. *Journal of Vacuum Science & Technology a*, **11**(4), 1441–1446.

[93] Put S., Vleugels J., Anne G. and van Der Biest O. 2003. Processing of hardmetal coatings on steel substrates. *Scripta Materialia*, **48**(9), 1361–1366.

[94] Wildhack S. 2003. Herstellung flüssigphasengesinterter Schichtkomposite aus SiC und AlN. PhD Thesis, University of Stuttgart.

[95] Tabellion J. and Clasen R. 2002. Electrophoretic deposition of SiC from aqueous suspensions. In *Electrophoretic Deposition: Fundamentals and Applications*, Boccaccini A. R,.van.Der. Biest O, and Talbots J. B.(Eds), The Electrochemical Society, Pennington, N. J. (USA), p. 102–109. 102–109.

[96] Moritz K. and Muller E. 2006: Investigation of the electrophoretic deposition behaviour of non-aqueous ceramic suspensions. *Journal of Materials Science*, **41**(24), 8047–8058.

[97] Powers R. W. 1975. The electophoretic forming of beta-alumina ceramic. *Journal of the Electrochemical Society*, **122**, 490–500.

[98] Powers R. W. 1986. Ceramic aspects of forming beta alumina by electrophoretic deposition. *American Ceramic Society Bulletin*, **65**, 1270–77.

[99] Anne G.,Vanmeensel, K.,Vleugels, J., and van Der Biest O. 2005. Electrophoretic deposition as a novel near net shaping technique for functionally graded biomaterials. In *Proceedings of Conference on Functionally Graded Materials VIII*, 11–14 July 2004, Leuven, Belgium, van Der Biest O., Goisk M., and Vleugels J. (Eds), 213–218.

[100] Wang R. and Hu Y. X. 2003. Patterning hydroxyapatite biocoating by electrophoretic deposition. *Journal of Biomedical Materials Research Part A*, **67A**(1), 270–275.

[101] Wang R. 2004. Pearl powder bio-coating and patterning by electrophoretic deposition. *Journal of Materials Science*, **39**, 4961–4964.

[102] Wei M., Ruys A. J., Milthorpe B. K., Sorrell C. C., and Evans J. H. 2001. Electrophoretic deposition of hydroxyapatite coatings on metal substrates: A nanoparticulate dual-coating approach. *Journal of Sol-Gel Science and Technology*, 21(1–2), 39–48.

[103] Poortinga A. T., Bos R. and Busscher H. J. 2000. Controlled electrophoretic deposition of bacteria to surfaces for the design of biofilms. *Biotechnology and Bioengineering*, **67**(1), 117–120.

[104] Brisson V. and Tilton R. D. 2002. Self-assembly and two-dimensional patterning of cell arrays by electrophoretic deposition. *Biotechnology and Bioengineering*, **77**(3), 290–295.

[105] USPatent, 6059949. 2000. Gal-Or L., Goldner R., Chernicak L., Sezin N., and Liubovich S. Method of electrophoretic deposition of ceramic bodies for use in manufacturing dental appliances.

[106] Yoo J. H. and Gao W. 2002. Electrical properties of PZT micro tubes fabricated by electrophoretic deposition. In *Electrophoretic Deposition: Fundamentals and Applications*, Pennington, New Jersey, USA, Boccaccini A. R., van Der. Biest O. and Talbots J. B. (Eds), The Electrochemical Society, 146–150.

[107] Glazounov A. E., Zhang Q. M., and Kim C. 1998. Piezoelectric actuator generating torsional displacement from piezoelectric d(15) shear response. *Applied Physics Letters*, **72**(20), 2526–2528.

[108] Chen Y. H., Li, T., and Ma, J. 2003. Investigation on the electrophoretic deposition of a FGM piezoelectric monomorph actuator. *Journal of Materials Science*, **38**(13), 2803–2807.

[109] Zhang J. P. and Lee B. I. 2000. Electrophoretic deposition and characterization of micrometer-scale BaTiO$_3$ based X7R dielectric thick films. *Journal of the American Ceramic Society*, **83**(10), 2417–2422.

[110] Shimbo M., Tanzawa K., Miyakawa M. and Emoto T. 1985. Electrophoretic deposition of glass powder for passivation of high-voltage transistors. *Journal of the Electrochemical Society* **132**(2), 393–398.

[111] Narisawa T., Arato T., Koganezawa N., Shibata M. and Nonaka Y. 1995. Formation of alumina insulation films for a CRT heater by constant-current electrophoretic deposition method. *Nippon Seramikkusu Kyokai Gakujutsu Ronbunshi-Journal of the Ceramic Society of Japan* **103**(1), 54–58.

[112] Verhoeckx G. J. and Leth N. J. M. 2002. Hamaker's law and throwing power in deposition of nano-sized alumina particles. In *Electrophoretic Deposition Fundamentals and Applications*, Boccaccini A. R., van Der Biest O. AND Talbots J. B.(Eds), The Electrochemical Society, Pennington, New Jersey, USA, 118–127.

[113] Kreidler E. R. and Bhallamudi V. P. 2001. Application of ceramic insulation on high temperature instrumentation wire for turbin engines. *Journal of Ceramic Processing Research*, **2**(3), 93–103.

[114] Sugiyama S., Takagi A., and Tsuzuki K. 1991. (Pb, La)(Zr, Ti)O$_3$ films by multiple electrophoretic deposition sintering processing. *Japanese Journal of Applied Physics Part 1-Regular Papers Short Notes & Review Papers*, **30**(9B), 2170–2173.

[115] Lee D. G. and Singh R. K. 1997. Synthesis of (111) oriented diamond thin films by electrophoretic deposition process. *Applied Physics Letters*, 70(12), 1542–1544.

[116] Barazzouk S., Hotchandani S. and Kamat P. V. 2001. Nanostructured fullerene films. *Advanced Materials* **13**(21), 1614–1617.

[117] Matthews D., Kay A., and Gratzel M. 1994. Electrophoretically deposited titanium-dioxide thin-films for photovoltaic cells. *Australian Journal of Chemistry* 47(10), 1869–1877.

[118] Talbot J. B., Sluzky E. and Kurinec S. K. 2004. Electrophoretic deposition of monochrome and color phosphor screens for information displays. *Journal of Materials Science* **39**(3), 771–778.

[119] Negishi H., Yanagishita H. and Yokokawa H. 2002. Electrophoretic deposition of solid oxide fuel cell material powders. In *Electrophoretic Deposition: Fundamentals and Applications*, Boccaccini A. R., van Der Biest O. and Talbots J. B.(Eds), The Electrochemical Society, Pennington, New Jersey, USA. 214–221.

[120] Basu R. N., Randall C. A. and Mayo M. J. 2001. Fabrication of dense zirconia electrolyte films for tubula solid oxide fuel cells by electrophoretic deposition. *Journal of the American Ceramic Society*, **84**(1), 33–40.

[121] Argirusis C., Damjanovic T. and Borchardt G. 2004. Electrophoretic deposition of thin SOFC-electrolyte films on porous La0,75Sr0,2MnO$_3$ –delta cathodes. In *Proceedings of 5th Conference of the Yugoslav Materials Research Society*, 15–19 September 2003, Herceg Novi, Yugoslavia, 335–342.

[122] Chen F. L. and Liu M. L. 2001. Preparation of yttria-stabilized zirconia (YSZ) films on La$_{0.85}$Sr$_{0.15}$MnO$_3$ (LSM) and LSM-YSZ substrates using an electrophoretic deposition (EPD) process. *Journal of the European Ceramic Society*, **21**(2), 127–134.

[123] Kertscher W., Bahn R. and Schott R. 1983. Transparente Al$_2$O$_3$-Keramik - Stand der Fertigung unter Beachtung des Rohstoffeinsatzes. *Silikattechnik*, **34**(12), 373–375.

[124] Braun A. 2005. Transparent polycristalline alumina ceramic by means of electrophoretic deposition for optical applications. PhD thesis, University of Saarbrücken.

[125] Wolff M. 2005. Untersuchungen zur Herstellung von transparentem Zirkonoxid. PhD thesis, University of Saarbrücken.

[126] Seike T., Matsuda M. and Miyake M. 2002. Preparation of FAU type zeolite membranes by electrophoretic deposition and their separation properties. *Journal of Materials Chemistry*, **12**(2), 366–368.

[127] Chen C. Y., Chen S.Y. and Liu D. M. 1999. Electrophoretic deposition forming of porous alumina membranes. *Acta Materialia*, 47(9), 2717–2726.

[128] Moritz K., Muller K., Tabellion J. and Clasen R. 2002. Elektrophoretische Abscheidung keramischer Feinst- und Nanopulver - Anwendungen als Formgebungsverfahren und Infiltrationsmethode. *Fortschrittsberichte Deutsche Keramische Gesellschaft*, **17**(1), 48–55.

[129] Kaya C. 2003. Al$_2$O$_3$-Y-TZP/Al$_2$O$_3$ functionally graded composites of tubular shape from nano-sols using double-step electrophoretic deposition. *Journal of the European Ceramic Society*, **23**(10), 1655–1660.

[130] Hadraba H., Maca K. and Cihlar J. 2004. Electrophoretic deposition of alumina and zirconia - II. Two-component systems. *Ceramics International*, **30**(6), 853–863.

[131] Li T., Chen Y. H. and Ma J. 2005. Characterization of FGM monomorph actuators fabricated using EPD. *Journal of Materials Science*, **40**(14), 3601–3605.

[132] Popa M., Moreno M., Hvizdos J. M. C., Bermejo R. and Anné G. 2005. Residual stress profile determinded by piezo-sprectroscopy in alumina/alumina-zirconia layers separated by a compositionally graded intermediate layer. In *Proceedings of conference on Fractography of Advanced Ceramics II*, Stara Lesna, 328–331.

[133] Put S., Anné S., Vleugels G. and van Der Biest O., 2004. Advanced symmetrically graded ceramic and ceramic-metal composites. *Journal of Materials Science*, **39**(3), 881–888.

[134] Von Both H., Dauscher M. and Haubelt J. 2004. Elektrophoretische Herstellung keramischer Mikrostrukturen. *Keramische Zeitschrift*, **56**(5), 298–303.

[135] Von Both H., and Hasselt J. 2002. Ceramic microstructures by electrophoretic deposition of colloidal suspensions. In Electrophoretic Deposition: Fundamentals and Applications, Boccaccini A. R., van Der Biest O. and Talbot J.B.(Eds), *The Electrochemical Society*, Pennigton, New Jersey, USA, 78–85.

[136] Bonnas S. 2003. Elektrophoretische Abscheidung von ethanolischen Zirkonoxid-Suspensionen. Diploma thesis, University of Freiburg.

[137] Ritzhaupt-Kleissl H. J., Bauer W., Gunther E., Laubersheimer J. and Haubelt J. 1996. Development of ceramic microstructures. *Microsystem Technologies*, **2**, 130–134.

14

Tape casting in micro- and nanofabrication

E. Carlström and L. Palmqvist

Tape casting is a well-established method for mass production of ceramic components based on thin layers. The major part of industrial tape casting production consists of single substrates for electronic purposes and multilayer capacitors with integrated printed metal electrodes. It is possible to extend these methods to new ranges of materials as well as a much wider range of component structures. Older technology requiring the use of solvent-based systems has been replaced by water. The high precision and the possibility of using nanoscaled powders makes the technology well suited for the production of new types of microsystems. The challenges of using finer powders, creating small-scale structures and integrating new materials as well as examples of how these challenges can be overcome are described in this chapter.

14.1 Introduction

Tape casting is a forming method for ceramic powders. The powder is dispersed in a liquid together with an organic binder. The powder dispersion is spread onto a moving carrier film (or steel belt) with a doctor blade. The dispersion dries forming a thin ceramic layer that can be separated from the carrier film. The ceramic tape is cut in sheets, heated to burn away the organic binder and heated further to sinter the ceramic powder to a dense material (see Figure 14.1).

Tape casting was developed in the 1950s at Bell Labs as a method to make ceramic substrates for integrated electronic circuits. The basic method is still in use and silicon chips for integrated circuits are mostly mounted on a substrate of tape cast ceramic. Alumina was the first material to be tape cast and it is still the dominant electronic substrate material [1]. Alumina is an electric insulator but is also a good heat conductor.

Beryllia has even better thermal conductivity than alumina but it forms a very toxic hydroxide (beryllium hydroxide) and is avoided today for that reason. If a thermal conductivity higher than that of alumina is required the alternative is aluminium nitride. Aluminium nitride substrates are generally more difficult to fabricate but are used in high-end applications where the heat dissipation of alumina is not sufficient.

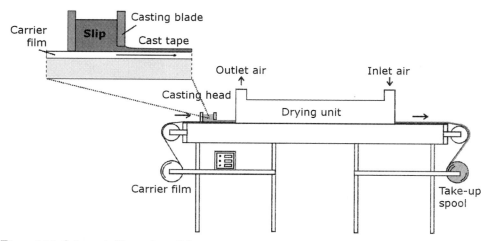

Figure 14.1 Schematic illustration of the tape casting operation.

Tape casting was developed using organic solvents. In order to minimize the fire hazard chlorinated solvents such as carbon tetrachloride were used initially. Chlorinated hydrocarbons are hazardous materials and have therefore been replaced by less toxic solvents over time. The next logical step is to totally replace organic solvents by water. Research has shown that this is possible in most cases but industrially there is still not much use of water-based systems.

The first application of tape casting needed tapes of approximately 0.5–1.0 mm thickness. Over time thinner and thinner tapes have been developed. The introduction of tape cast ceramics as dielectrics in multilayer capacitors (MLCs) have driven the development towards very thin tapes. Tape cast tapes in the micrometre range have been produced by tape casting for the production of multilayer capacitors. At present MLC technology is limited by retaining a number of grains over the thickness of the layer rather than by the ability to form thin tapes [2]. In order to have a large number of grains in a thin layer there is a need to go to nanosized powders where the main challenge is to disperse these extremely fine powders at high enough volume concentration and low enough viscosity [3].

Currently there is a trend towards using tape cast ceramics to manufacture a much wider range of products. It is possible to produce ceramic filters and membranes with tape casting. Both the anodes and electrolytes for fuel cells can be produced by tape casting.

By printing electrically conductive metallic patterns on the ceramic tape, stacking tapes, laminating and sintering it is possible to produce complex electronic substrates. This technology is used to produce electronic circuit boards with integrated passive components that are suitable for applications at high frequencies (microwave) and in rough environments (vehicle and industrial electronics). This technology was initially developed using alumina substrates. The high sintering temperature (1400–1600°C) required refractory metals such as platinum or molybdenum as conductors. The technology, known as HTCC (high temperature cofired ceramics) for electronic applications has to a large extent been replaced by the low temperature cofired ceramics (LTCC) technology (see Figures 14.2 and 14.3). In LTCC technology the ceramic substrate sinters at a lower

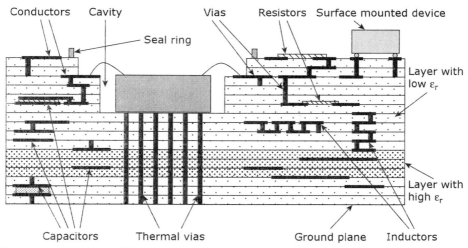

Figure 14.2 Schematic LTCC circuit.

Figure 14.3 LTCC test circuit.

temperature (900°C). This makes it possible to use silver alloys as the conducting material with lower resistive losses.

Today there is a trend towards producing microsystems using tape casting. Tape casting is an efficient mass production method for all kinds of devices that can be built up with a layered structure. One example is the development of the multilayered piezoceramic actuator [4]. This development has made it possible to introduce these actuators in the fuel injection system of smaller car engines. Small tape cast actuators are also starting to be used in micromotors [5], in active vibration control [6] and in energy harvesting for mobile power generation [7].

Most microsystems are built with polymers and metals but ceramics are increasingly used in high temperatures and harsh environments and for special applications where the ceramic behaves as for example a sensor, an actuator or a catalyst.

14.2 Microsystem applications of tape casting

14.2.1 Micro-actuators

Monolithic piezoceramic actuators require high driving voltages. A field-strength of $1000\,V/cm$ is a common value for the maximum deflection of a ceramic actuator. An actuator stack consists of a number of thin ceramic layers divided by electrode layers. Because every layer is thin, even a low driving voltage can result in a high field over the ceramic layer. Multilayered actuators are produced today by tape casting thin ceramic sheets ($10–100\,\mu m$) screen printing a metal electrode on the ceramic tape, followed by stacking and laminating the tapes and sintering [8].

Apart from problems with high driving voltage a piezoceramic actuator has a very small deflection. A typical maximum deflection is in the range of 0.1%. In order to create actuators with larger deflections the bimorph structure was invented. It consists of two electroded ceramic layers. The polarity is arranged so that one layer expands while the other contracts. This will cause the bimorph to bend with a much larger deflection than a monolithic actuator can achieve [9].

Jing and Luo [10] developed a micro-actuator using tape casting with screen printed electrodes. The micro-actuator was developed to improve the tracking of a hard disk drive. The piezoceramic was a lead niobate zirconate ($(Pb(Mg_{1/3}Nb_{2/3})O_3$-$Pb(Zr_{0.52}Ti_{0.48})O_3$) electroded with an Ag/Pd electrode. The ceramic material was tape cast, screen printed with the electrode paste and cofired. When a U-type micro-actuator with two 10-layer piezoelectric elements was used an actuation stroke of $1.496\,\mu m$ was realized with an alternating voltage of ± 20 V.

Hooley et al. [11] have proposed a new concept to use piezoactive transducers in a phased array digital sound projector. Wagner et al. [12] have developed a helical shaped piezoelectric actuator for use in a sound projector. The actuator was fabricated from ceramic that could be tape cast. The piezoceramic tape was electroded and laminated as a bimorph. A strip of the tape was bent in a helical shape and sintered (Figure 14.4).

14.2.2 Fuel cell membranes

Solid oxide fuel cells (SOFCs) have high efficiency, low cost and can be used with a wider range of fuels compared to proton exchange membrane (PEM) fuel cells. The disadvantage of the ceramic fuels cells is the high operating temperature (900°C range) that creates material and design problems. Newer developments include the application of nanostructured materials, which can decrease the operating temperature dramatically (300–400 °C).

Figure 14.4 Sintered 3D helical shaped PZT structures based on cast tapes (top, 10 active layers of 90 μm thickness, 12 mm inner diameter) [12].

Work has been done both with replacing the Ni anode with porous Ni/ceria composites or replacing the electrolyte by nanoscaled ceria doped materials [13–15]. Tape casting is the preferred production method for SOFCs with a planar design [16]. The requirements for a SOFC are a very thin electrode layer and a porous anode layer. This can be achieved with tape casting where pore formers are added to the anode composition to control the amount and size of the pores. Pore former are particles that break down and evaporate during sintering leaving porosity that remain after sintering. Organic pore formers such as starch particles or inorganic particles such as ammonium oxalate have been used [17].

14.2.3 Microcombustion chamber

An example of the possibilities of fabricating microsystems using tape casting is a micro catalytic combustion chamber developed by Okamasa and Lee [18]. A schematic sketch is shown in Figure 14.5. The combustion chamber was developed for various mobile power generation systems such as the gas turbine engine, thermoelectric generator or thermophotovoltaic generator. The advantage of the combustion-based systems compared to fuel cells is their higher electrical generation density. The design by Okamasa and Lee uses a catalyst to reduce the combustion temperature but the resulting temperature is high enough to require a ceramic material in the combustor.

14.2.4 Photonic bandgap structures

Photonic bandgap (PBG) materials are periodic dielectric structures featuring electromagnetic waves that cannot propagate in certain frequency ranges and directions. By using tape casting combined with screen printing, Li et al. [19] were able to create such structures. Arrays of either rectangular or hexagonal structures were screen printed with metallic paste on the ceramic tape. Via holes are drilled and filled with a metallic paste providing electrical contact between the layers. Because of their effective surface wave suppression, these novel

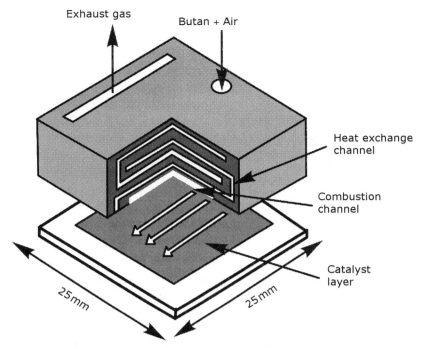

Figure 14.5 Microcatalytic combustion chamber [16].

PBG materials could be used as high-performance antenna substrates to enhance broadside gain of patch antenna devices. These materials may also have potential applications as waveguides or filters in the communication industry, serving as a high-impedance surface.

14.3 Aqueous tape casting

Even if organic solvents still dominate industrial tape casting the new developments in tape casting are mostly done in aqueous (water-based) systems. The arguments for using water-based system are cost, environmental effects, work environment, and development time.

Water is a cheaper solvent and fire hazard protections are costly measures. Chlorinated solvents can reduce the fire risk but are not allowed because of their carcinogenic effects. Reducing emissions of carbon dioxide to the environment and saving nonrenewable oil resources favour water-based systems. Organic solvents are toxic and require measures to protect the workforce from fumes during handling. Working with water-based systems is simpler and faster in a lab environment and this saves development time.

If there is a well-functioning industrial production it is a big step to switch to a new system. But as new systems are developed there are very few arguments left for using organic solvents.

The initial problems with water-based systems were slower casting speeds and problems with cracking of thicker tapes. These problems were solved by switching to latex binders

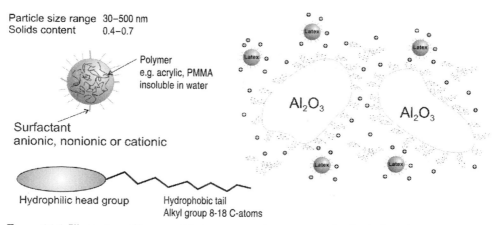

Figure 14.6 Illustration of latex particle in a tape casting system with adsorbed surfactants.

(acrylic emulsions). These binders made it possible to use higher concentrations of particles in the ceramic suspension without increasing the viscosity. This made it possible to cast with higher speeds in continuous casting and to cast thicker tapes without drying cracks [20]. Figure 14.6 shows schematically the characteristics of latex-binder dispersions.

However, latex systems can cause porosity in the sintered tapes, but that can be avoided by using a nonionic latex instead of an anionic latex [21]. There can also be problems regarding the release from the carrier film. This can also be solved by proper choice of latex amount and type and by using carrier films pretreated with a release agent.

Another problem with water-based tape casting relates to ceramic powders that are slightly soluble in water. One example of this is barium titanate ($BaTiO_3$) where the Ba^{2+} ions dissolve from the surface changing the surface chemistry and interact with the dispersant making the dispersion more difficult. In spite of these difficulties it is possible to tape cast barium titanate in water by careful selection of dispersant and pH [22]. Barium titanate compositions are commonly used as dielectrics in multilayer capacitors. Another example is aluminium nitride, a powder that reacts slowly with water releasing gases (hydrogen and nitrogen). The released gas bubbles are difficult to remove from the suspension and cause porosity in the sintered material.

Powders that react in water can, however, be tape cast after passivation of the powder surface. Passivation can be achieved by surface-active substances that adsorb or react with the surface creating a layer that protects it from direct contact with the water [23]. The effect of dissolving ions on dispersants can be avoided by choosing dispersants that are insensitive to dissolving ions.

A comparison of aluminium nitride tape cast in a water-based system with the same material cast using a system with organic solvents shows small differences. The water-based system showed a small increase in oxidation of the powder. This could be seen as a minor decrease in thermal conductivity in the final material. The water-based systems showed a lower viscosity and thus better casting properties [24, 25].

14.4 Dispersion of fine powders for tape casting

In order to sinter a ceramic to a dense material there is a need to use fine powders. Most high performance ceramics are fabricated from powders with particle sizes mainly below $1\,\mu m$ with a typical mean particle size of $0.1\,\mu m$. The recent advances in nanotechnology have increased the interest in going to even finer powders. One reason for this is to use the added functionality of very fine grain size materials as catalysts or sensors. Another reason is to decrease the sintering temperature. The last is especially interesting in cosintering of several materials to make microsystems where the sintering temperature has to be adapted to the material that is most sensitive to high temperatures. Another reason to use very fine powders when tape casting microsystems is that fine powder can create smother surfaces with a potential for closer dimensional tolerances

Fine particle systems can only be dispersed in water if the attractive van der Waals force is balanced by a repulsive force. The repulsive force can be electrostatic repulsion of particles with a surface charge or the steric repulsion caused by adsorbed surface-active molecules.

Good dispersants are available for ceramic systems that are used industrially in tape casting. Polyacrylic acid-based dispersants (with trade names such as Dispex, Darvan C and Dolapix) can disperse many ceramic systems that have moderate or high isoelectric points (such as alumina, zirconia and silicon nitride). Ceramic materials that have low isoelectric points such as silicon carbide, silica or tungsten carbide can be dispersed using polyethylene imines. Both of these types of dispersants are polyelectrolytes that mainly disperse by electrostatic stabilization. These dispersants are mainly used in water.

Dissolution of soluble ions increases the ionic strength of the suspension, which tends to flocculate the dispersed power. Electrostatic stabilizers are very effective at moderately high ionic strengths, but may loose some of its electrostatic repulsion ability at high salt concentrations due to counter ion screening of the charges. This problem can be overcome by using a dispersant with a mainly steric stabilization mechanism. An example of such a dispersant type is comb copolymers. The comb copolymers have a charged backbone, for example a polyacrylic acid, with grafted brushes of a nonionic chain, for example polyethylene oxide. Nonionic comb copolymers are relatively insensitive to high ionic strengths, and may be efficient stabilizers in highly concentrated ceramic suspensions. The different structures of polyacrylic dispersants and comb copolymer dispersants are shown in Figure 14.7.

Comb copolymers are also attractive as dispersants in solvent mixtures, where polyacrylic acids are poor dispersants. For example, alcohols with low molecular weights are interesting as solvents for ceramic powders that react with water, as they are virtually nontoxic, have low surface tension and medium polarity. Partial substitution of water with alcohol can suppress the reactivity and improve the long-term stability of ceramic suspensions. When alcohols are used, electrostatic stabilization is less effective due to low permittivity of the solvents. For these systems, steric stabilizers such as comb copolymers may be attractive dispersants. Particularly at high alcohol concentrations, comb copolymers have shown superior performance to polyacrylic acids (see Figure 14.8). The reason for this is attributed

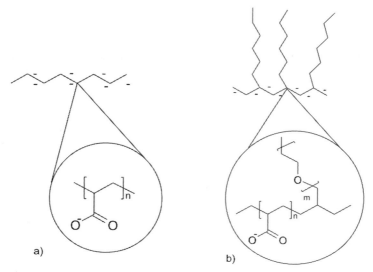

Figure 14.7 (a) Polyacrylic dispersant; (b) Comb copolymer dispersant with polyacrylic backbone and oxyethylene chains.

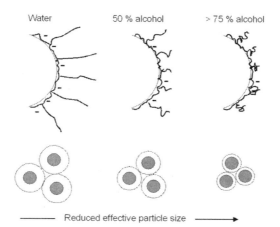

Figure 14.8 Effect of solvent on effective particle size.

to a balance between effective particle volume and dispersion efficiency. The solubility of the comb copolymer is high enough in alcohols to give a stabilizing layer on the particle surface, but at the same time the effective particle diameter is kept small since the chains are not fully extended, as they would be in water.

The balance between effective particle volume and dispersion efficiency from extended chains becomes even more important for nanosized particles. Particle size reduction to nanometre dimensions will be needed in the microsystems industry to make dense sheets for multilayer devices as the thickness of the sheet approaches that of conventional powder sizes. Ceramic particles in the nanometre range have a high sinterability, which is required for densification at reduced sintering temperatures. Lowered sintering temperatures will save energy, cost and time, and are necessary for cofiring of ceramics and metal pastes. However, nanoparticles are also difficult to disperse at high volume concentrations. This is because of their tendency for network formation, which increases with decreasing particle size due to more particle–particle contacts. When commonly used dispersants are adsorbed on nanosized particles, the thickness of the dispersant layers will be in the same size range as the particles themselves. Hence, the dispersants will constitute a considerable part of the total volume fraction of the system. This leads to demands for more effective, smaller and more insensitive dispersants. Comb copolymers with short graft chains are promising candidates for the stabilization of nanosized powders, since short graft chains give a smaller increase in effective volume than conventional dispersants [26].

Very fine particles will be more susceptible to dissolution in water due to high specific surface areas. Solvent mixtures that suppress dissolution and comb copolymers as dispersants may then be a better choice [27].

14.5 Optimization of dispersants and binders for tape casting systems

In a tape casting process it is important to have a low viscosity suspension and a high particle concentration. The principles and mechanisms acting in a ceramic–solvent suspension are shown in Figure 14.9. The dispersant concentration and the pH in a water suspension are crucial parameters that influence viscosity as a function of the particle concentration.

The optimum dispersion is achieved with a monolayer of the dispersant adsorbed on the powder particles. If insufficient dispersant is added to cover the particle surfaces the suspension will tend to flocculate, which increases the viscosity. If too much dispersant is added the free dispersant in the solution or the thicker adsorbed layers will create a resistance to flow and also increase the viscosity.

The pH of the suspension will influence the function of a polyelectrolyte dispersant. The pH influences the charge of the particles and the charge regulates the adsorption of the polyelectrolyte. A pH below the isoelectric point will create a powder with positive charged

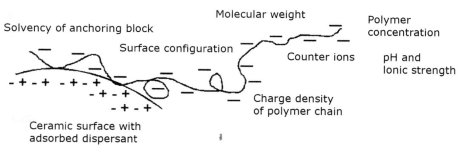

Figure 14.9 Principles for dispersion in ceramic systems.

surface. A pH above the isoelectric point will create a negatively charged surface. At the isoelectric point the surface charge is low and no electrostatic forces are available to enhance adsorption.

Polyacrylic acid is a very common type of dispersant that is negatively charged when the acid is dissociated. To adsorb the polyacrylic acid a positively charged surface is needed. This means that a pH below the isoelectric point is needed. But if the pH is too low (below the pK_a of the acid) the polyacrylic acid will not dissociate. The nondissociated molecule has no charge and will not be adsorbed by electrostatic attraction and the electrostatic repulsion of the dispersant is also lost. This means that there is an optimum pH where adsorption occurs and where the polyacrylic acid works as an electrostatic dispersant. Similar mechanisms are valid for all types of polyelectrolyte dispersants.

The surface charge behaviour of the powder is mainly influenced by its main constituents. However, minor surface impurities can have a strong influence of the charge behaviour of the powder in water as a function of pH. The variation in particle size and specific surface can also cause variations in the dispersion properties of the powders. Some powders will have soluble species that can release ions to the suspension. Ions in the solution can react with the dispersant and increase the demand for dispersant concentration.

This means that there will always be an optimization problem that has to be solved with each new ceramic powder. This optimization is best solved by viscosity measurements to find the lowest viscosity for a specific powder as a function of dispersant concentration and pH. Care also has to be taken with time-dependent effects, as slightly soluble ions will dissolve slowly over time.

The binder also influences the viscosity of the suspension. A charged binder can adsorb on the powder, thus competing with the dispersant. A charged binder with an opposite charge to the powder can flocculate the powder by neutralizing the powder charge. Some binders also react with soluble ions from the powder. Latex binders that are stabilized with nonionic surfactants have the advantage of not causing any electrostatic interactions in water.

The binder will also influence the release from the carrier film and the plasticity of the cast green tape. More binder will create a stronger and more plastic tape while there can be problems with release from the carrier film. The wetting properties of the binder play an important role in the release from the carrier film. Problems with the release from the carrier film will cause deformation of the released tape and make it more difficult to keep the required dimensional tolerances of very thin tapes or can even lead to tape fracture. A high concentration of binders can also influence the green density and decrease the final sintered density.

Because of the possible interactions between binder, dispersant and powder and the interactions of the binder with the carrier film it is important both to choose a compatible binder system and to optimize the binder concentration.

14.6 Precision and shrinkage in tape casting

A crucial issue in tape casting for microsystems is control over shrinkage and the possibility to create structures with high precision. Traditionally this has been an area where problems were solved by practical measures in the industry. Common methods to avoid warpage are placing weights on tapes during sintering or annealing weight-loaded sintered tapes to a temperature where warpage can be removed by creep.

Recently these phenomena have been subject to scientific studies something that should eventually should lead to better dimensional control. Besendörfer and Roosen [28] have studied the influence of the amount of solvent and casting speed on the dimensional tolerances of LTCC green tapes. Li and Lannutti [29] have studied the curvature evolution in LTCC tapes and laminates during cofiring. Their work showed that the burn out prior to sintering was as important for the curvature of the final material as the sintering itself.

Sintering of a ceramic on a solid surface leads to constrained shrinkage. This constraint may lead to cracking but even if no cracking is observed it still leads to anisotropy of the sintered body. The constraining stresses produce both anisotropy of the grains (minor effect) and anisotropy of the pores (major effect) [30]. Ravi and Green [31] have developed a method to measure the deformation of the layers during sintering using cyclic loading dilatometry and also developed a model to calculate the resulting distortion.

14.7 Templated grain growth

A recent technique to create ordered microstructure with enhanced properties is templated grain growth. This technique is based on adding seed crystals to the tape casting suspension. The flat seed crystals are oriented in the casting direction by the flow during casting. During sintering the seed crystals influence the grain growth, thus creating an oriented microstructure.

Templated grain growth has been used to create ceramic laminates with extremely high fracture toughness [32]. It has also been used to create piezoceramics with a very high piezoelectric coefficient due to the oriented microstructure [33].

14.8 New materials and composite structures

Tape casting was traditionally used for a limited range of ceramic materials with main application for the electronic industry. However, it is starting to be used for a wider range of material and applications such as hydroxy apatite [34] or calcium phosphates [35] for biomedical applications. It can also be used to fabricate composites such as carbon composites [36] and even metal. There are more efficient production methods for metal components but tape casting of a metal powder followed by sintering is an established process for the production of porous metals and has been used for porous titanium [37].

It is also possible to use tape casting to create layered composites. Composites with silicon carbide layers interspersed with carbon layers have shown a high resistance to cracks and thermal shock and have been proposed as materials for combustions chambers [38, 39]. Composites based on one material with alternating porous and dense layers is another way to create materials with excellent thermal shock properties. Figure 14.10 shows a picture of the porous–dense layer structure with a crack. The crack propagation will be much lower as the porous layer will prevent the crack propagation in the material. A crack in ZrO_2 will propagate through the entire layer at a thermal shock of 400°C while in a porous–dense material it will only propagate through 10% of the layer thickness (see Figure 14.11). [40, 41].

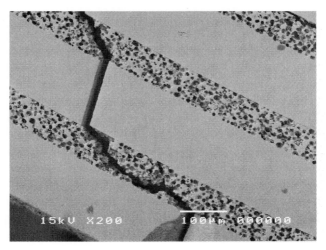

Figure 14.10 Microstructure with dense and porous layers showing crack deflection in porous layers.

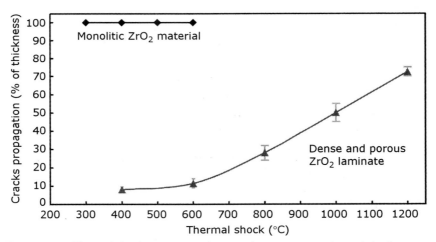

Figure 14.11 Thermal shock experiment showing the resistance to thermal shock damage in a material with alternating dense and porous layers.

These types of composites can also be used in sensors and actuators to change the elastic modulus in order to match with the acoustic impedance [42].

Conclusions

Tape casting is a ceramic forming process that is well suited for the fabrication of microsystems. It can be used to form layers in the range 5–500 μm thickness for a great palette of materials. The tape casting technology can also tune the materials properties within a wide range. The industrial production of multilayer capacitors has shown that it is a cheap and efficient industrial process even for the generation of thin layers.

Recent developments make it possible to proceed from older methods requiring organic solvents to water-based processing systems. This makes the method well suited to tomorrow's cheap mass production of microsystems where ceramics are required. By integrating printed metallic structures and vias it is possible to build complete multilayer structures with 3D interconnections. There are also other possibilities for forming more complex structures than plain layers. By using the features of the forming process it is possible to create unique material properties such as improved thermal shock resistance or oriented microstructures with anisotropic properties.

In order to be able to fully utilize tape casting for microsystems very fine (nano) powders need to be dispersed. Such powders make it possible to reduce sintering temperatures and thus cosinter materials that would otherwise be incompatible. Nanosized powders are important to realizing the very fine microstructures that are necessary for microscale precision. Nanostructured materials are also important in creating new or enhanced functional properties. However, there is some fundamental difficulty in dispersing nanosized powders to reach the high volume concentrations needed for tape casting. Nevertheless, new dispersants such as comb copolymers offer possibilities and show advantages that could be used to solve this problem.

References

[1] Mistler R.E. 1990. Tape casting the basic process for meeting the needs of the electronics industry. *American Ceramic Society Bulletin*, **69** (6) 1022–1026.

[2] McCauley D E., Chu M.S.H. and Megherhi M.H. 2006. PO$_2$ Dependence of the diffuse-phase transition in base metal capacitor dielectrics. *Journal of the American Ceramic Society*. **89** (1), 193–201

[3] Vinothini V., Singh P. and Balasubramanian M. 2006. Optimization of barium titanate nanopowder slip for tape casting. *Journal of Materials Science*. **41** (21), 7082–7087

[4] Takahashi S., Ochi A., Yonezawa M., Yano T., Hamatsuki T. and Fukui I. 1983. Internal electrode piezoelectric ceramic actuator, *Ferroelectric*, **50** (181).

[5] Bexell, M. and Johansson S. 1999. Microassembly of a piezoelectric miniature motor. *Journal of Electroceramics*, **3** (1) 73–83.

[6] Hanselka H. 2001. Adaptronics as a key technology for intelligent lightweight structures. *Advanced Engineering Materials*, **3** (4), 205–215.

[7] Lesieutre G.A., Ottman G.K. and Hofmann H.F. 2004. Damping as a result of piezoelectric energy harvesting. *Journal of Sound and Vibration*, **269**, 991–1001

[8] Lubitz K., Schuh C., Steinkopff T. and Wolff A. 2002. Material aspects for reliability and life time of PZT multilayer actuators. In *Piezoelectric Materials in Devices*. Setter N. (Ed), 183–194. Ceramics Laboratory, EPFL Swiss Federal Institute of Technology, Zurich, Switzerland.

[9] Colla E.L. and Morita T. 2002. Piezoelectric technology for active vibration control pp. In *Piezoelectric Materials in Devices*, Setter N.(Ed), 123–154. Ceramics Laboratory, EPFL Swiss Federal Institute of Technology, Zurich, Switzerland.

[10] Jing Y. and Luo J. 2005. Structure and electrical properties of PMN-PZT micro-actuator deposited by tape-casting process. *Journal of Materials Science: Materials in Electronics* **16**, 287–294.

[11] World Intellectual Property Organization Patent, 2000. Hooley A., Pearce D.H., Lenel U.R., McKevitt G. and Shepherd M.R. Electro-active devices. WO01 47041.

[12] Wagner M., Roosen A., Oostra H., Höppener R. and de Moya M. 2005. Novel low voltage piezoactuators for high displacements. *Journal of Electroceramics*, **14**, 231–238.

[13] Ziao-Feng Y., Bo Huang, S.R., Wang, Z.R., Wang, L., Ziong, T. and Wen, L. 2007. Preparation and performance of a Cu-CeO_2 -ScSZ composite anode for SOFCs running on ethanol fuel. *Journal of Power Sources*, **164**, 203–209.

[14] Esposito V. and Traversa, E. 2008. Design of electroceramics for solid oxides fuel cell applications: Playing with ceria. *Journal of American Ceramic Society*, **91** (4), 1037–1051.

[15] Zhu B., Rundgren K. and Mellander B.E. 1997. Ceramic membranes – potential uses for solid state protonic conductors. *Solid State Ionics*, **97**, 385–391

[16] Luo L.H., Tok A.I.Y. and Boey F.Y.C. 2006. Aqueous tape casting of 10 mol%-GdO_3 -doped CeO_2 nano-particles. *Materials Science and Engineering A*, **429**, 266–271.

[17] Haslam J., Pham A.Q., Chung B.W. DiCarlo J.F. and Glass R.S. 2005. Effects of the use of pore formers on performance of an anode supported solid oxide fuel cell. *Journal of the American Ceramic Society*, **88** (3) 513–518.

[18] Okamasa T. and Lee G.G. 2006. Development of a micro catalytic combustor using high-precision ceramic tape casting. *Journal of Micromechanics and Microengineering*, **16**, 198–205.

[19] Li B., Zhou Ji and Li L. 2006. Microwave bandgap in multilayer ceramic structures, *Journal of the American Ceramic Society*, **89** (3), 1087–1090.

[20] Kristoffersson A. and Carlström E. 1997. Tape casting of alumina in water with an acrylic latex binder. *Journal of the European Ceramic Society*, **17**, 289–297.

[21] Engebretsen C. and Carlström E. 2002. Binder induced porosity in tape casting. *Ceramic Engineering Science Proceedings*, **23** (4), B:253–259.

[22] Song Y.L., Liu X.L., Zhang J.Q., Zou X.Y. and Chen J.F. 2005. Rheological properties of nanosized barium titanate prepared by HGRP for aqueous tape casting. *Powder Technology*, **155** (1), 26–32.

[23] Ferreira J.M.F., Olhero S.M.H. and Oliveira M.I.L.L. 2005. Processing of aluminium nitride. *Ceramic Transactions*, **166**, 71–84.

[24] Luo X., Li J., Zhang B., Li W. and Zhuang H. 2006. High thermal conductivity aluminium nitride substrates prepared by aqueous tape casting. *Journal of the American Ceramic Society*, **89** (3), 836–841.

[25] Luo X., Li J., Zhang B., Li W. and Zhuang H. 2005. Comparison of aqueous and non-aqueous tape casting of aluminium nitride substrates. *Journal of the American Ceramic Society*, **88** (2), 497–499.

[26] Palmqvist L., Lyckfeldt O., Carlström, E., Davoust P., Kauppi A. and Holmberg K. 2006. Dispersion mechanisms in aqueous alumina suspensions at high solids loadings. *Colloids and Surfaces A- Physicochemical and Engineering Aspects*, **274** (1–3), 100–109.

[27] Lyckfeldt O., Palmqvist L. and Carlström E. 2008. Stabilization of alumina with polyelectrolyte and comb copolymer in solvent mixtures of water and alcohols. Journal of the European Ceramic Society, **10**, 1016.

[28] Besendorfer G., Roosen A., Modes C. and Betz, T. 2007. Factors influencing the green body properties and shrinkage tolerance of LTCC green tapes. *International Journal of Applied Ceramic Technology*, **4** (1), 53–59.

[29] Li W. and Lannutti J.J. 2005. Curvature evolution in LTCC tapes and laminates. *IEEE Transactions on Components and Packaging Technologies*, **28**, (1), 149–156.

[30] Guillon O., Krauss S. and Rödel J. 2007. Influence of thickness on the constrained sintering of alumina films. *Journal of European Ceramic Society*, **27**, 2623–2627.

[31] Ravi D. and Green D.J. 2006. Sintering stresses and distortion produced by density differences in bi-layer structures. *Journal of the European Ceramics Society.* **26**, (1/2), 17–25.

[32] Park D.S., Kim C.W. and Park C. 1998. Self-reinforced silicon nitride composite containing unidirectionally oriented silicon nitride whisker seeds. *Ceramic Engineering and Science Proceedings*, **19** (3), 97–104.

[33] Messing G.L., Trolier-McKinstry S., Sabolsky E.M., Duran C., Kwon S., Brahmaroutu B., Park P., Yilmaz H., Rehrig P.W., Eitel K.B., Suvaci E., Seabaugh M. and Oh K. S. 2004. Templated grain growth of textured piezoelectric ceramics. *Critical Reviews in Solid State and Materials Sciences*, **29** (2), 45–96.

[34] Tian T., Jiang D., Zhang J. and Lin Q. 2007. Aqueous tape casting process for hydroxyapatite. *Journal of the European Ceramic Society*, **27**, 2671–2677.

[35] Tanimoto Y., Hayakawa T. and Nemoto K. 2007. Characterization of sintered TCP sheets with various contents of binder prepared by tape-casting technique. *Dental Material* s, **23** (5), 549–555.

[36] Geffroy P.M., Chartier T. and Silvain J.F. 2007. Preparation by tape casting and hot pressing of copper carbon composites films. *Journal of the European Ceramics Society*, **27** (1), 291–299.

[37] Rak Z.S. and Walter J. 2006. Porous titanium foil by tape casting technique. *Journal of Materials Processing Technology*, **175** (1–3), 358–363.

[38] Clegg W.J., Andrees G., Carlström E., Lundberg R., Kristoffersson A., Meistring R., Menessier E. and Schoberth. A. 1999. The properties of ceramic laminates. *Ceramic Engineering Society Proceedings*, **20** (4), 421–426.

[39] Carlström E. and Kristoffersson A. 1995. Solid-state-sintered silicon carbide by water-based tape casting. *Fourth Euro Ceramics Vol 1*, Gallassi C.(Ed), Faenza Editrice, Italy. 367–374.

[40] Vandeperre L.J., Kristoffersson A., Carlström E. and Clegg W.J. 2001. Thermal shock of layered ceramic structures with crack-deflecting interfaces. *Journal of the American Ceramic Society*, **84**(1), 104–110.

[41] Davis J.B., Kristoffersson A., Carlström E. and Clegg W.J. 2000. Fabrication and crack deflection in ceramic laminates with porous interlayers. *Journal of the American Ceramic Society*, **83** (10), 2369–2374.

[42] Palmqvist L., Lindqvist K. and Shaw C. 2007. Porous multilayer PZT materials made by aqueous tape casting. *Key Engineering Materials*, **333**, 215–218.

15

Low temperature cofired ceramic-processing for microsystem applications

W. Smetana

LTCC technology may be considered as a further development of thick-film technology. This technology was originally developed for the realization of multilayer circuits of high reliability, especially for radio frequency applications. But LTCC technology has found new applications since it became evident that complex 3D structures could easily be realized. It covers areassuch as: microfluidics, integrated device packaging, bioreactors, sensors and transducers. Its fields of application are similar to those of MEMS but are realized in a mesostructure performance. The process technique for a 3D microsystem architecture with LTCCs has been developed from a standardized process which has to be modified to a large extent for most unconventional applications. New procedures also have to be implemented. This chapter outlines aspects and specific processing methods for the fabrication of 3D microstructures. It deals with the different micromachining methods for structuring LTCC tapes, lamination techniques, bonding of tapes, the application of sacrificial materials for the creation of cavities and channel structures, and the assembling of LTCC modules.

15.1 Introduction

LTCC technology is based on the application of tapes, which are cast from slurries of ceramic-filled glass systems mixed with an organic vehicle. They are produced by commercial suppliers at different thicknesses varying from 30 μm to 350 μm. The terms "low temperature" and "cofired" refer respectively, to the firing at temperatures below 900°C and the simultaneous firing of LTCC tapes and thick-film pastes of electronic components. LTCC technology has been proving to be a valuable development of thick-film technology that has established new application areas. It is an ideal platform for the realization of high-speed signal circuitry due to the low dielectric constants and losses [1] of the tape materials. LTCC technology has also been successfully applied for the fabrication of microsystems because of its inherent features (e.g. chemical inactivity, biocompability, hermeticity, matching of the CTE with silicon, high temperature stability, and 3D structuring) [2–4]. The implementation of complex channel structures, integration of electronic and optoelectronic components (light sources, photodetectors, optical fibres)

209

a
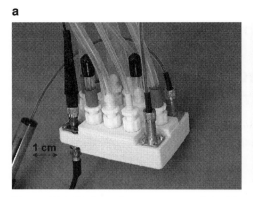

b
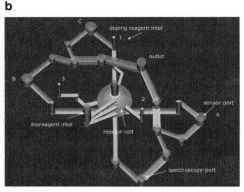

Figure 15.1 Biological monitoring module: (a) completely assembled with sensor-, inlet-, outletports and sockets for optical fibres, (b) schematic of fluid duct network [4].

in one LTCC module (Figure 15.1) are the main advantages of this technology over silicon, glass and polymers.

Fields of application are similar to that of MEMS but only realized in mesostructure dimensions (the mesosize range varies typically from ten to several hundred microns) [2]. LTCC technology allows high flexibility thanks to the availability of tapes of varied physical characteristics (e.g. permittivity, permeability, elasticity etc.) and thickness [5]. The technique for making a 3D architecture with LTCCs is rather simple and standardized. Device production in LTCC technology covers the machining, punching or laser drilling of vias and channels on individual layers. For electronic circuits via holes are filled with a conductive paste by using a stencil printing process and electrical features are screen printed onto the layers using the appropriate pastes. The individual layers are collated and laminated in a heated platen press or in a heated isostatic press. The laminated stack is then exposed to the firing cycle. That is a somewhat critical process where heating rate, holding time at burnout temperature and total firing cycle time have to be matched to the thickness of the ceramic stack.

15.2 Materials

LTCC tapes are glass–ceramic composite materials. LTCC materials are based on either crystallizable glass or a mixture of glass and ceramics, e.g. alumina, silica or cordierite. The properties of the ceramic tape can be modified by using materials with different electrical and physical properties (e.g. piezoelectric or ferroelectric characteristics, varistor performance etc.). The CTE can be matched to alumina, gallium arsenide or silicon. For example a formulation containing 65 vol% cordierite ($2MgO_2Al_2O_35SiO_2$) and 35 vol% mullite ($3Al_2O_32SiO_2$) yields a tape system that shows a CTE matching with Si. The added glass frit binder lowers the sinter temperature of the tape as well as provides compatibility with thick-film pastes. A third component of the composite is an organic vehicle for binding and viscosity control of the tape before sintering [6]. Green ceramic tape is produced by

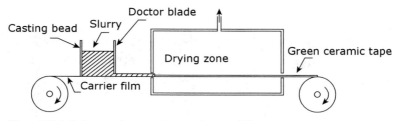

Figure 15.2 Scheme of tape casting equipment [6].

tape casting methods [7]. Once the tape-casting slurry has been properly dispersed it is deposited onto a moving plastic carrier sheet. This is usually done by passing under a doctor blade as shown in Figure 15.2. In order to control the thickness of the sheet it is necessary to control a range of parameters like the blade gap, carrier film speed, or the height of the fluid level in the slurry tank. The solvents start to evaporate and the tape cures as it passes through a drying zone. The tape is then inspected for defects and rolled onto spools. Polyethylene terephtalate (PET) film is usually used as carrier film. Additionally, a silicone release agent is often applied to improve the peelability of the tape from the carrier film. Most fired tapes show a uniform shrinkage in all directions as a result of sintering. The lateral shrinkage typically varies in the range 13–15% and the shrinkage in thickness is in the range 15–20%.

But there are also tapes and procedures available which provide only a shrinkage in thickness where the lateral dimensions remain nearly unchanged (i.e. lateral shrinkage is 0.1–2% and shrinkage in thickness is 43–47%). In the case of constrained sintering, several versions have been developed. One approach is the so-called "tape-transfer" technology. This procedure prevents shrinkage by laminating and firing each layer of tape onto a dense substrate made of alumina, BeO or AlN. The tape adheres to the substrate and does not shrink in the x- or the y- direction. Upon firing it shrinks only in the z-direction [8]. Another procedure which provides zero shrinkage uses two additional tape layers (sacrificial layers), which have to be colaminated to the top and the bottom of the LTCC stack. These two additional tape layers are fabricated of a material (e.g. alumina) with a much higher sintering temperature compared to the LTCC. Hence, these tapes do not shrink or sinter themselves but maintain a uniform frictional contact with the surface of the LTCC stack constraining the lateral shrinkage of the LTCC during firing. After firing the additional sacrificial layers have to be removed mechanically (e.g. abrasive-jet cleaning). The resulting shrinkage is about 0.1–0.2%. By contrast, the application of self-constrained LTCC tapes requires no additional processing steps. The "self-constrained" LTCC tape is actually a compound of tapes with different sintering characteristics which alternately constrain lateral shrinkage during sintering and results in a defined residual lateral shrinkage of only 0.2% ± 0.03%. Its density increases by shrinking primarily in the z-axis.

An advantage of most constrained sintering processes is that the fired part tends to be free from warping [9].

15.3 General processing

One of the important features of green glass–ceramic tape technology is the possibility of fabricating 3D structures using multiple layers of unfired tapes [9]. The processing of the green ceramic tapes is usually done in three basic steps: first, patterning of individual layers with via holes and applying resistor, conductor, and dielectric pastes, depending on the specific application; secondly, collation and lamination of the tapes under pressure and at elevated temperatures; and finally, cofiring of the laminated stack to sinter the material.

Each layer is fabricated in the green (unfired) state adapted to the requirements of the overall performance of the 3D structure. Via holes and registration holes are machined into the tapes. Each tape is then aligned on the vacuum chuck of a screen printer made up of a porous metal or ceramic plate. The via holes are filled with a conductive paste using a stencil and electrical features are screen printed onto the layers using the appropriate pastes for the creation of the internal electrical elements. After printing, the cavities are machined using an automatic punch or laser. The finished sheets are collated in a mould and aligned by registration pins. Successive layers are rotated by 90° to compensate for the texture (preferential orientation) induced by the fabrication of green tape. It must also be ensured that all the sheets consistently face with the same side up, i.e. the "shiny" sides (side originally attached to the carrier tape) must always either face up or down. The lamination of the stack of tapes is carried out in a heated platen press or in an isostatic press (typical laminating parameters are 200 bar at 70°C for 10 minutes). The laminated LTCC-tapes are cofired in a two-step process. The first step at about 350°C burns out the organic components while the second step consists of the viscous sintering process. In this second step the ceramic material densifies within a temperature range of 850–950°C depending on the composition of the LTCC. The optimal heating rates and dwell times depend strongly on the number of tapes forming the stack. After cofiring additional thick-film structures can be applied on the top and bottom surfaces in a postprocessing step using a conventional thick-film process. Afterwards the LTCC structures are singularized using a dicing saw, ultrasonic cutting or laser cutting and additional active or passive components can be added. The stacking technique is useful for creating solid 3D parts but the stacking of planar layers inevitably leads to a staircase pattern and the vertical smoothness is limited by the minimum layer thickness and alignment accuracies.

Another attractive aspect of this technology is demonstrated by the creation of 3D shell structures with single layers of LTCC tape [10]. Shell structures with curved geometry may be easily placed on a corresponding macrosized mould. Since the tape can be easily moulded to shape and machined in the green state, it can be processed into 3D curved shells. The tape is laid on top of a mandrel, which is a metal mould of the desired shape. If the shell is based on a developable surface such as a cylinder one piece of tape is adequate. If it is not a developable surface such as a sphere multiple pieces must be used (Figure 15.3). The mandrel is made from brass, which has a melting point above the maximum temperature of 850°C of the firing process and is easy to machine. Prior to laying down the tapes on the mould, a boron nitride aerosol spray (Advanced Ceramics Corporation) is applied on the metal surface of the mandrel. It prevents the tapes from sticking to the mandrel even if

Figure 15.3 Hemispherical LTCC shells, unfired diameter: 45 mm [10].

brass oxidizes slightly during the firing step. Thus, the shell can easily be removed from the mandrel when the sintering has been completed.

15.4 Machining

In order to use LTCC multilayers as microsystem modules channels, holes and cavities with defined geometry and structures need to be fabricated. Fortunately, the unfired tapes are easy to machine. Suitable mechanical processing procedures are punching, drilling, embossing, and cutting. Laser cutting and laser ablation are useful laser machining technologies are.

15.4.1 Computer numerically controlled micromachining of low temperature cofired ceramic tapes

The milling of tapes is critical if channels of small dimensions and adjacent to each other have to be manufactured since it is not possible to hold the LTCC layer adequately with a vacuum chuck during processing. Additionally, when lamination is attempted, the channels shift and become occluded due to the delicate structure of the machined layer. Manufacture of channels using the hogging method is a viable approach. "Hogging" is a well-known term and process in computer numerically controlled (CNC) milling where layers are laminated together and features such as channels are then milled in steps, increasing the depth of the mill with each pass until the required depth is obtained [11].

CNC milling methods have been used to fabricate feature sizes in the range of 100 μm. The dimensional tolerances of fabricated structures are comparable to those attained with a general purpose punching machine.

15.4.2 Jet vapour etching

A technique for etching the organic component of the binder in various LTCC substrates has been presented by Espinoza-Vallejos *et al.* [12]. Using a collimated solvent jet, it is possible to remove organic material and with the fluid flow momentum the ceramics particles of the filler can be removed at the same time. This technique makes it possible to fabricate a

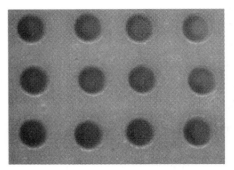

Figure 15.4 Punched via holes, entry-side on Mylar backing, diameter: 55 µm (courtesy of KMS Kemmer Technology Centre GmbH, Dresden, Germany).

wide variety of shapes. The liquid solvent is heated, thereby increasing its vapour pressure. In a nitrogen stream the acetone acting as selected solvent is atomized and ejected through a microfabricated nozzle. The acetone vapour is able to dissolve the organic binder and the gas mixture momentum removes the filler. This etching technique has been successfully used to drill round holes of approximately 25 µm in diameter in LTCC tapes.

15.4.3 Punching

Punching of via holes is traditionally used in LTCC mass production [13]. As a punched via is actually an imprint of the punch pin itself, punched vias look very uniform. This is especially true for the punch pin entry side. At the punch pin exit side, the edges tend to look more like a fringe or chipped. The geometry of punch pin and die, especially from the perspective of tool wear, and the cutting clearance determine the quality of the punched hole. Punching is usually done with Mylar on top (see Figure 15.4). From the top view, the holes look uniform and clean. Figure 15.5 shows a corresponding exit (tape) side. While the hole geometry on the entry

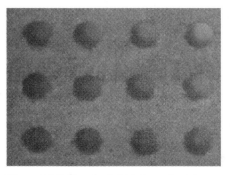

Figure 15.5 Punched via holes, exit-side on tape, diameter: 55 µm (courtesy of KMS Kemmer Technology Centre GmbH, Dresden, Germany).

side is almost entirely determined by the punch pin, on the exit side it is determined by pin and die.

Via holes with a diameter of 30 µm have been already fabricated in laboratory by punching but are still not state-of-the-art.

15.4.4 Laser machining

Laser cutting allows very flexible structuring of green tape, like cutting of cavities and complex shapes. But as the laser cut is the result of absorption of the laser beam in the material, the cutting quality depends considerably on the specific material. Nd-YAG-, excimer- and CO_2-lasers can be utilized to machine the tapes. The tape is either moved under the laser beam in a computer controlled x–y movement or the laser beam is deflected by a galvosystem in order to machine complex shapes. In general, it seems to be hard to avoid a zone along the laser cut becoming impaired/damaged. This impairment can be detected as a colour change or vitrification of the tape material. This has been attributed [14] to the Gaussian shape of the laser energy density distribution. It can be avoided by applying metal masks. Changes in the chemical composition of that zone have been described [15].

CO_2-lasers (midinfrared, wavelength 10.6 µm) are widely used for technical applications. The standard spot size of CO_2-lasers is around 70 µm [6], which makes them unsuitable for fabrication of via holes smaller than this without the use of masks. Nd-YAG-lasers (near infrared, wavelength 1.064 µm) have a well-established position in thick-film technology as trim lasers and tools for cutting and scribing, their spot size is around 50 µm [14]. Generally laser drilled via holes exhibit a tapered shape, where the taper mainly depends on the laser focus depth, via size, and tape thickness (i.e. aspect ratio). Hole diameters of around 80 µm at the exit side of the tape, with larger diameters at the entry side, depending on the tape thickness, can be achieved with CO_2- and Nd-YAG-lasers. A spot size of around 15–20 µm is attained using UV-lasers (e.g. a tripled Nd-YAG, wavelength 355 nm) which make them predestined for the fabrication of very small structures and via holes. As the ablation process in the UV spectrum is partly photochemical and partly photothermal, UV lasers allow, at least partly, a cold ablation of organic material. Hole diameters of approx. 55 µm and 80 µm at the exit and entry side of the tape can be easily achieved with UV-lasers (Figure 15.6). Nevertheless it must be pointed out that laser machined via holes in green LTCC tape cannot be as small and uniform as punched holes.

The so-called water jet guided laser can be used for the machining of fired LTCC substrates [16]. The laser beam is focused through a quartz window into the water jet nozzle by means of a standard fibre coupling and is then contained inside in the water jet by total reflection at the air–water interface. Pure de-ionized and filtered water at a pressure of 50–500 bar is used to form the water jet. The nozzles are made of sapphire or diamond in order to generate a stable water jet with diameters in the range 25–100 µm. Compared to classical laser cutting the water jet guided laser technology minimizes the heat damage in any kind of sample. Because of its high momentum, the water jet expels the molten material from the cut and at the same time the remaining particles are flushed away.

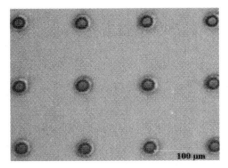

Figure 15.6 Via holes, entry-side on tape, diameter: 35 µm. Machined with frequency converted diode pumped laser (Nd-YVO4-Laser, 532 nm) (courtesy of KMS Kemmer Technology Centre GmbH, Dresden, Germany).

15.4.5 Embossing

The embossing technique makes it possible to replicate very accurately complex 2D and 2.5D patterns onto ceramic surfaces. The lamination and firing of embossed tapes makes it possible to build up complex 3D structures like multilayer channel systems. Different manufacturing methods may be used for the fabrication of embossing tools, among others micromachining and direct laser machining. The method selected depends on the required structural resolution of the tool. A specific embossing tool fabrication method is presented here which is based on the selective laser structuring (for instance by means of a femtosecond-laser) of a thick photoresist layer which is spin coated on a glass wafer [17]. The well-defined threshold for the ablation with ultrashort laser pulses provides a process parameter window where the resist is selectively removed from the underlying substrate without affecting the substrate. Thus it is possible to generate 2.5D structures where the height is solely determined by the thickness of the resist layer. After a cleaning procedure a 50 nm thick Ni-coating is sputtered on the laser structured resist. Subsequent microelectroplating generates a 100 µm thick metallic tool which is bonded to an embossing tool holder (see Figure 15.7). Ceramic tape

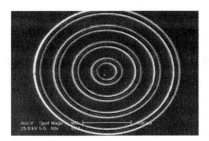

Figure 15.7 Embossing tool carrying concentric circle structure [17].

Figure 15.8 Embossed structure of concentric circle structure with enlarged detail [17].

samples, tools as well as the parts of the machine in contact with the samples and tools, are preheated to 70°C and the temperature is kept constant throughout the embossing process. This allows easier "flow" and deformation of the ceramic. At higher temperatures ceramic tape sticks to the tool while at lower temperatures the tape has a more rigid consistency which often results in the formation of cracks. Samples are exposed to embossing forces in the range 0.5–5 kN with steps increasing by 0.1 kN. The required force is usually reached after a maximum time of 80 s. The dwell time at peak pressing force is about 15 s (Figure 15.8). The quality of structures obtained by microembossing techniques depends on the following parameters: tool and tape temperature; embossing pressure; the embossing depth in relation to the tape thickness; and the spacing of the grooves (ridge breadth). If the line-to-line distance is 100 μm and the embossing pressure is 100 MPa, a groove can only be embossed into the tape for up to 33% of its thickness [18]. This conclusion may not be generalized. It concerns only the results attained with the ferritic tape AHT03-003 made by Heraeus. Results with other tapes may differ since the embossing performance depend strongly on the specific material characteristics of the tape (e.g. tape viscosity).

15.4.6 Photolithographic patterning

This approach opens the possibility of patterning features smaller than 10 μm for tapes with optimal grain size distribution. For this technique a dry photoresist is laminated at 70°C to a partially sintered tape [2]. The patterns are transferred using contact UV photolithography. The dry photoresist is developed using sodium carbonate and UV postexposure hardens the mask against etchant attack. This procedure allows to etch the glass grains to be etched using hydrofluoric acid. Finally the sintering of tape has to be completed.

15.4.7 Photoformable low temperature cofired ceramic tapes

The photoformable ceramic consists of a glass–ceramic composite with a conventional nominal composition except that in the organics a negative photoresist is included besides the plasticizers and antiflocculants. A sodium carbonate spray shower is used [2] to develop the exposed tape. After the developing process the tape is laminated and fired. The partial etching depth can be controlled by modifying the process parameters like shower pressure and developing time. Features smaller than 30 μm in size may be achievable with this type of process.

15.5 Lamination

Sagging of suspended or laminated structures is a common problem in the processing of LTCC materials. It has been proved that the lamination of a LTCC tape with holes of diameters beyond 400 μm results in a deformation of the adjacent tapes above and below the cavity. For smaller diameters the deformation is negligible. There are different methods to prevent deformation of LTCC structures during firing [12]. These can be classified as passive and active methods, which are based on an adequate adaption of lamination conditions and utilization of sacrificial layers, using etchable as well as fugitive materials.

15.5.1 Deposition of thick films

In the case of bridging structures, a screen-printed thick-film layer may be applied which avoids tape sagging during sintering [16]. The effectiveness of this procedure for deformation prevention is due to the shrinkage of a selected paste that is matched to fired alumina substrates, which have a higher CTE than the glass–ceramic tape. Consequently, the thick-film layer exerts a tensile force on the interface structure, preventing the sagging.

15.5.2 Use of sacrificial materials

15.5.2.1 Etchable

For suspended or bridging tapes, a lead bisilicate composition has been used as a sacrificial material [19]. After deposition and firing, the lead bisilicate glass is etched in buffered hydrofluoric acid in order to remove the material beneath a suspended or bridging tape. This method offers some problems since etching rate is different inside cavities from in open space or due to limited solubility of the glass frit, it is difficult to remove all the material. LTCC tapes which comprise glass–ceramic composites are also attacked by the etchant. Thus, the removal of the etchable paste, which is a common problem in closed structures, requires special treatment in order to improve the quality of the fabricated devices.

15.5.2.2 Fugitive phases

The use of fugitive phase materials which are exhausted during firing is another approach to supporting bridging and channel structures [19]. Several of these strategies have been explored. Some authors are using carbon black as the fugitive phase [20–23]. The carbon can be applied as tape or paste. Additionally, it does not interact with the LTCC tape during firing. Malecha and Golonka [23] reports about fabrication of channels with varying width (between 100 μm and 5 mm). After machining channel structures the ceramics foils are stacked together in proper order and initially pressed by an isostatic press with a pressure equal to 20 atmospheres. Then channels are filled by sacrificial material (carbon-black paste or cetyl alcohol) and structures are laminated a second time with an increased pressure equal to either 50, 100 or 200 atmospheres. The LTCC samples that are laminated at lower pressures (50 and 100 atmospheres) show poor bonding between ceramics foils. In structures laminated at a standard pressure of 200 atmospheres there is complete fusion of the individual tape layers without any trace of the junction. The insertion of sacrificial material and the two-step lamination process improve the performance of cross-sections of channel structures for narrow as well as for wide channels and avoid deformation effects such as sagging, contraction etc. Sagging and delamination are typical of relatively wide channels (width equal to 500 μm or more) whilst contraction is characteristic of narrow channels (width equal to 200 μm and less).

It has been proved that the granularity of graphite powder is a critical parameter since the temperature of total burnout depends strongly on the particle size (770°C for 2 μm particles and 865°C for 15 μm particles) [20]. Although sagging and delaminating are avoided for all membrane structures, a new type of defect, the swollen membrane surface, was observed in those prepared with "coarse" powder-based sacrificial paste. The swelling in the membranes is ascribed to the burnout characteristics of the powder. It has been observed that any gas producing reaction following the closure of open porosity in the LTCC tape (at 785°C) results in swelling. However, the burnout of glass ceramic tape occurs up to 865°C, which results in entrapment of the gas products which evolve beyond 785°C. This increases the pressure inside the membrane and causes swelling. In the case of "fine" powders, the burnout is already completed prior to the open porosity elimination, thus swelling of membranes does not occur.

In addition, other materials that are part of the organic constituents of tapes may be applied as a fugitive phase that exhausts free of residues at the burnout temperature of tape [24, 25]. Best results have been attained with PMMA (see Figures 15.9 and 15.10). The organic sacrificial material is used in powder form. The channels and cavities are filled by a vacuum sucking technique (see Figure 15.11). It has been found that the main task of the sacrificial material is to provide a uniform pressure distribution within the LTCC tape stack during lamination. Additional support from cavity structures during firing is not required since the LTCC material shows adequate strength and stability throughout the firing process. It has only to be ensured that the PMMA starts to pyrolize slowly which

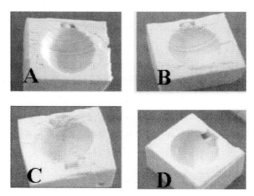

Figure 15.9 Sections of spherical cavities (volume: appr. 1 cm³) prepared with different organic sacrifical materials: A: polystyrene, B: hydroxyethylcellulose, C: without filler, D: PMMA [24].

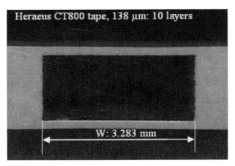

Figure 15.10 Cross-section of channel, PMMA filled (aspect ratio H/W = 0.43) [25].

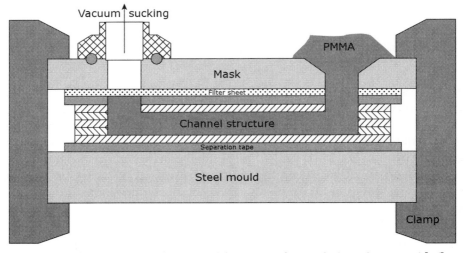

Figure 15.11 Schematic of vacuum assisted filling setup for sacrifical powder material [25].

enables the gaseous decomposition products to be expelled via the channels. In contrast, a rapid decomposition of the organic filler induces a sudden intensified production of gaseous burnout products which cannot escape adequately via the channels and finally results in the destruction of the LTCC module. The burnout phase of the firing schedule has to be adapted to these requirements. An adequate dwell time at the critical degradation temperature of the filler also has to be provided.

The burnout of the organic fugitive phase can be accomplished at lower temperatures as has been described for graphite pastes or foils [20] as already described in this section.

15.6 Bonding and assembling

It is possible to bond LTCC tapes to other materials, like, glass, ceramics, metals and silicon, using a cofired or postfired scheme. Glass is bonded using a cofired scheme or by application of postfired glass frits at lower sintering temperatures. For ceramics in particular, one can bond using cofiring frits and glazes, epoxy or brazing techniques. Silicon can be cofired or postfired using metallic die-attaching pastes.

15.6.1 Tape joining

Tape in the green state may be joined with pressure sensitive adhesives e.g. poly(2-ethyl-2-oxaline) or poly(2-ethyl-2-oxazoline) [26]. This provides a low pressure bond that allows very large stacks to be fabricated without high-pressure lamination. Full pressure lamination should be avoided in order to prevent the collapse or deformation of cavities during the process. Alternatively this material may be used as lamination aid that burns all its volatiles before 300°C is reached.

For the formation of shell structures it has been found convenient to use a silver-based paste (e.g. DuPont 6141 paste) to stick the different pieces together in the desired position before they are placed on a mandrel [10].

15.6.2 Bonding of fired low temperature cofired ceramic-stacks

In order to join two already fired LTCC segments permanently together they are covered with tape, and a thin layer of glass encapsulant paste is then printed over this. The two fired segments are then aligned and pressed together and allowed to dry for approximately one hour. When the glass encapsulate paste has dried, the entire unit is run through a modified furnace profile, in order to form one functional unit from two already fired segments [26].

The [27] bonding of a valve set comprised of sapphire onto a LTCC module using an encapsulant glass (DuPont 9615) usually used for coating of thick-film circuits has been described [27].

15.6.3 Component integration in low temperature cofired ceramic module

Several attempts have been made to embed optical quartz fibres in a LTCC stack by means of coprocessing [28]. Fibres which are positioned in a LTCC module, laminated and cofired with the whole structure result in delamination. If fibres are placed in a channel inside the LTCC module, laminated together, and cofired their positions are not stable due to the deviations of the sintering shrinkage. Finally, the attempt to position the fibre on a glass film inside the channel within the LTCC module and use a conventional processing steps has failed. The fibres are easily broken during processing. Optical fibres have been successfully integrated in the LTCC module by gluing them inside specially made microchannels. Nevertheless, the realization of buried sapphire fibres in multichip modules fabricated in LTCC technology is possible if a self-constrained sintering tape (e.g. HL 2000 of Heraeus) acting as a distortion-free substrate is applied (see Figure 15.12) [9].

Encapsulation of Si-dies into LTCC layers has also been investigated. Knowing the exact shrinkage for the $x-y$ plane, Si pieces are fitted inside the LTCC, laminated and sintered together. either the Si-die nor the LTCC broke, and the device is fixed inside just by mechanical clamping [2, 29].

Figure 15.12 Top: Cofired optical sapphire fibre: A: in a conventional and B: in a self-constrained LTCC-tape. Bottom: illuminated 25 µm diameter optical fibre embedded in self-constrained tapes (Heralock, HL 2000) [9].

Sapphire and alumina tubes have been successfully connected to LTCC for fluid connection. Different diameters of openings in a LTCC stack have been prepared and the quality of connection tightness has been checked. During firing, the shrinkage of the LTCC provides a tight seal around the sapphire tube if the drilled hole corresponds to the tube diameter. Devices have been constructed and successfully tested to 414 kPa. In contrast, the opening corresponding to a scale up factor of 15% (a value exceeding the actual tape shrinkage of 12.7%) provides only a loose connection [22].

Petersen *et al.* [30] have demonstrated sapphire lenses imbedded into LTCC stacks. Several prototypes using 6.35 mm diameter lens were built up using the three different methods illustrated in Figure 15.13. The shrink fit method involves drilling a 6.35 mm hole in several layers of green LTCC tapes and inserting the lens in the hole. The device is then laminated and fired. The assembled device shows excessive warping. Because the lens does not shrink during firing, the upper and lower halves of the device densify to different extents. In the clearance method the drilled hole size is increased from 6.35 mm to 7.14 mm. The ~11% difference results in less impediment to shrinkage of tape by the sapphire lens resulting in less distortion of the fired part. There was leakage occurring between the lens and the wall of the hole. The pocket method involves capturing the sapphire lens between layers of LTCC with smaller diameter holes. The sapphire lens is in a clearance hole of 7.14 mm diameter. Pressure testing was accomplished by immersing the devices under water, applying pressurized air and observing the bubbles that form from the leaks. The clearance and pocket methods were more efficient than the shrink fit method but yield only low pressure (<7 kPa) seals.

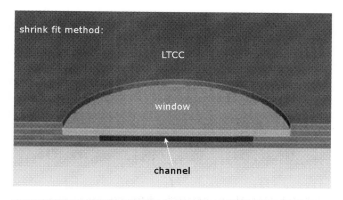

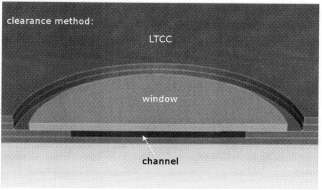

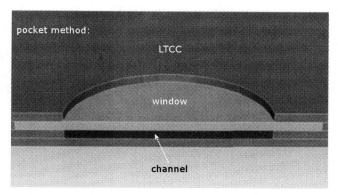

Figure 15.13 Methods for imbedding sapphire windows [22].

15.6.4 Anodic bonding

The LTCC tape material LTCC 951 from DuPont contains 50% alumina, the rest is mainly borosilicate glass with a CTE of 5.6 ppm/K, has been used for anodic bonding trials [29]. This material is not appropriate for direct bonding to silicon (mainly due to the CTE). A suitable glass composition for anodic bonding has been selected and deposited on the LTCC substrate in the requested pattern by either screen or stencil printing. The waviness of the LTCC substrates can be compensated to some extent by using a thicker glass print.

The thickness obtained varied in the range 35–100 µm. But the surface of the fired glass layer shows, depending on the applied screen printing process, a certain surface roughness. Thus, the surface has to be polished to obtain the required flatness and uniformity required for anodic bonding. The samples are cleaned in H_2O_2 + H_2SO_4 (1:4), rinsed in de-ionized water and successfully anodically bonded at 420°C and an applied voltage in the range 130–500 V. Fracture occurs when the samples are cooled down after bonding. The fracture seems to occur in the interface between the printed glass and silicon as glass residues are found on silicon, and silicon residues are detected on the printed glass on the LTCC substrate [31]. In order to accomplish a hermetic bonding, the glass paste needs to be modified for lower temperature bonding.

A new LTCC material has been developed by HITK and VIA Electronic (Germany) that has a lower CTE (3.6 ppm) [29, 32]. This new material already contains the required alkali ions, which permits direct anodic bonding to silicon. The alkali ions can be in the bulk of LTCC or only at the surface. This type of LTCC substrate with bulk alkali ions has been bonded to silicon by standard anodic bonding at 420°C, 800 V in less than a minute. No voids could be seen. When lower bonding voltage and/or confined electric field are required, the LTCC substrate with metal electrodes embedded in the substrate (placed below the surface) should be used.

Conclusions

LTCC-technology shows attractive features for the fabrication of complex 3D microsystems.

Examples of different processing procedures which have been already applied to the production of 3D microsystems are highlighted. Nevertheless, many of these processes are still being developed and need to be pursued in order to strengthen the position of LTCC technology as a competitive alternative method for the fabrication of microsystems.

References

[1] Müller J., Perrone R., Rentsch S., Hintz M. and Stephan R. 2006. Improved RF Performance for embedded passives in LTCC by fine line structuring methods. In *Proceedings of the 30th International Conference of IMAPS*, 24–27 September 2006, Cracow, Poland, 69–77.

[2] Gongora-Rubio M. R., Espinoza-Vallejos P., Sola-Laguna L. and Santiago-Aviles J. J. 2001. Overview of low temperature cofired ceramics tape technology for meso-system technology (MsST). *Sensors and Actuators A*, **89**, 222–241.

[3] Golonka L. J. 2006. Technology and applications of low temperature cofired ceramic (LTCC) based sensors and microsystems. *Bulletin of the Polish Academy of Sciences, Technical Sciences*, **54**(2) 221–231.

[4] Smetana W., Balluch B., Stangl G., Gaubitzer E., Edetsberger M. and Köhler G. 2007. A multi-sensor biological monitoring module built up in LTCC-technology. *Microelectron Engineering*, **84**(5–8), 1240–1243.

[5] Feingold A.H., Heinz M. and Wahlers R.L. 2003. Dielectric and magnetic materials for integrated passives. In *Proceedings of the Conference on Ceramic Interconnect Technology*, Denver, Colorado, USA., 189–194.

[6] Imanaka Y. 2004. *Multilayered low temperature cofired ceramics (LTCC) technology*. Springer, New York, USA. 105–117.

[7] Mistler R.E. and Twiname E.R. 2000. Tape Casting. *American Ceramic Society*, Westerville, Ohio, USA.

[8] Modes C., Neidert M., Herbert F., Reynolds Q., Lautzenhiser F. and Barnwell P. 2003. A new constrained sintering LTCC technology for automotive electronic applications. In *Proceedings of the 14th European Microelectronic and Packaging Conference*, 23–25 June 2003, Friedrichshafen, Germany, 118–122.

[9] Lautzenhiser F. and Amaya E. 2002. HeraLockTM 2000 self-constrained LTCC tape. In Proceedings of the 2002 IMAPS Advanced Technology Workshop, on Ceramic Materials, September 2002, Providence, Rhode Island, USA.

[10] Li J. and Ananthasuresh G K. 2002. Three-dimensional low-temperature co-fired ceramic shells for miniature systems applications. *Journal of Micromechanics and Microengineering*, **12**, 198–203.

[11] Moeller K., Besecker J., Hampikian G., Moll A., Plumlee D., Youngsman J. and Hampikian J. M. 2007. A prototype continous flow polymerase chain reaction LTCC chain. *Materials Science Forum*, **539–543**, 523–528.

[12] Espinoza-Vallejos P., Sola-Laguna L. and Santiago-Aviles J.J. 1998. Jet vapor etching: a chemical machining technique for green LTCC. In *Proceedings of IMAPS, 98 Fall Meeting*, 2–4 November 1998, San Diego, California, USA.

[13] Hagen G. and Rebenklau L. 2006. Fabrication of smallest vias in LTCC Tape. *Proceedings of Electronics Systemintegration Technology Conference*, 5–7September 2006, Dresden, Germany, 642–647.

[14] Drüe K.H. 2005. Precise drilling and structuring of LTCC materials using a 355 nm YAG Laser. In *Proceedings of European Microelectronics and Packaging Conference(EMPC)*, 12–15 June 2005, Brugge, Belgium.

[15] Wolter K.J. et alt. 2005. Via formation in LTCC tape: a comparison of technologies. In *Proceedings of ACerS/IMAPS First International Conference on Ceramic Interconnect and Ceramic Microsystems Technologies*, 10–13April 2005, Baltimore, Maryland, USA.

[16] Wagner R., Spiegel A. and Richerzhagen B. 2004. Electronic packaging: new results in singulation by laser microjet. In *Proceedings of the SPIE*, **5339**, 494–499.

[17] Andrijasevic D., Smetana W., Zoppel S. and Brenner W. 2007. An investigation on development of MEMS in LTCC by embossing technique. Proceedings of the *Third international conference on multi-material micro manufacture (4M 2007)*, 2–5 October, Borovets, Bulgaria, 143–146.

[18] Kallenbach M., Bartsch H., Hintz M. and Hoffmann M. 2007. High-inductive small-size microcoil with high ampacity. In Proceedings of Smart Systems Integration 2007, 27–28 March 2007, Paris, France, 505–507.

[19] Espinoza-Vallejos P., Zhong J., Gongora-Rubio M., Sola-Laguna L. and Santiago-Aviles J.J. 1998. The measurement and control of sagging in meso electromechanical LTCC structures and systems. *Materials Research Society Conference Proceedings*, **518**, 73–79.

[20] Birol H., Maeder T. and Ryser P. 2006. Processing of graphite-based sacrificial layer for microfabrication of low temperature co-fired ceramics (LTCC). *Sensors and Actuators A*, **130–131**, 560–567.

[21] Birol H., Maeder T., Jacq C., Straessler S. and Ryser P. 2005. Fabrication of LTCC Micro-fluidic devices using sacrificial carbon layers. *Journal of Applied Ceramic Technology*, **2**(5), 345–354.

[22] Youngsmann J., Marx B., Schimpf M., Wolter S., Glass J. and Moll A. 2006. Low temperature co-fired ceramics for micro-fluidics. In *Proceedings of 56th Electronic Components and Technology Conference*, 30 May–1 June 2006, San Diego, California, USA. 699–704.

[23] Malecha K. and Golonka L.J. 2007. Microchannel fabrication process in LTCC ceramics. *In Proceedings of 31st International Conference of IMAPS*, 23–26 September 2007, Rzeszow-Krasiczyn, Poland.

[24] Smetana W., Balluch B., Stangl G., Gaubitzer E., Edetsberger M. and Köhler G. 2007. Set-up of a biological monitoring module realized in LTCC technology. In *Proceedings of the SPIE*, **6465**, 646509-1–646509-10.

[25] Wang X.S., Balluch B., Smetana W. and Stangl G. 2008. Optimization of LTCC-channel abricaion for biomedical micro fluidic 3D-structures. In *Proceedings of the 31st International Spring Seminar on Electronics Technology*, 7–11 May 2008, Budapest, Hungary.

[26] Jaques B., Weston H., Plumlee D.G. and Moll A.J. 2006. Advanced fabrication techniques or an ion mobility spectrometer for low-temperature co-fired ceramics. In *Proceedings of the MAPS/ACerS 2nd International Conference and Exhibition on Ceramic Interconnect and Ceramic Microsystems Technologies*, 25–27 April 2006, Denver, Colorado,USA.

[27] Rodrigues da Cunha C., Ketzer S. and Gongora-Rubio M.R. 2006. Numerical and experimental analysis of microfluidic diode manufactured in LTCC technology. In *Proceedings of BERSENSOR 2006*, 27–29 September 2006, Montevideo, Uruguay.

[28] Golonka L J., Zawada T., Radojewski J., Roguszczak H. and Stefanow M. 2006. LTCC microfluidic system. *International Journal of Applied Ceramic Technology*, 3(2), 150–156.

[29] Rusu C., Persson K., Ottosson B. and Billger D. 2006. LTCC interconnects in microsystems. *Journal of Micromechanic Microengineering*, **16**, 13–18.

[30] Peterson K.A., Walker C.A., Patel K.D., Turner T.S. and Nordquist C.D. 2004. Microsystem Integration with new techniques in LTTC. In *Proceedings of IMAPS Interconnect Technology Conference*, 27–28 April 2004, Denver, Colorado, USA.

[31] Bergstedt L. and Persson K. 2001. Printed glass for anodic bonding–a packaging concept for MEMS and system on a chip. In *Proceedings of IMAPS Nordic 38th Conference*, 23–26 September 2001, Oslo, Norway, 10–14.

[32] Müller E., Bartnitzek T., Bechtold F., Pawlowski B., Rothe P., Ehrt R., Heymel A., Weiland E., Schroeter T., Schundau S. and Kaschlik K. 2005. Development and processing of an anodic bondable LTCC tape. In *Proceedings of the 15th European Microelectronics and Packaging Conference*, 12–15 June 2005, Brugge, Belgium, 313–318.

16

Different ceramic materials for micromilling of high aspect ratio microstructures

T. Gietzelt and G. Bissacco

Micromilling is a very flexible and universal technique with a high material removal rate. The achievable surface roughness is in the submicron range. There are almost no limitations on design and 3D structures can easily be machined. Additionally, compared to other techniques like LIGA or EDM a very wide range of materials is accessible. Ceramic materials offer a wide range of favourable properties compared to plastics and metals, such as: high temperature stability, extreme hardness, high wear resistance and biocompatibility. This chapter discusses and compares different processing routes, which exhibit specific advantages and disadvantages.

16.1 Introduction

By using micromilling for the fabrication of microstructures, a high material removal rate can be realized. Compared to other techniques like LIGA (only suitable for materials that can be deposited by electroforming) or EDM (electrically conductive materials) a wide range of materials like polymers, metals and alloys, and ceramics can be microstructured for different applications. Ceramic materials in particular offer some interesting properties compared to metallic alloys like extreme hardness, good corrosion resistance, high wear resistance and biocompatibility. Of course, the machining of such materials needs some special adaptation of machining processes. For example, to avoid machining of ceramics with extreme hardness such materials can be machined in a presintered state with a subsequent heat treatment to achieve nearly full density. On the one hand cutting forces and tool wear can be reduced by some orders of magnitude. On the other hand machining parameters have to be adapted to avoid faults at the edges of microstructures. Additionally, the shrinkage related to the sintering process has to be taken into account in advance, making it difficult to meet very tight tolerances.

To avoid damage to the sensitive parts of the machine by the abrasive dust of the ceramics, for example, the feed drives and guideways have to be protected by blocking with compressed air.

Table 16.1 Overview of machinable ceramic materials [1].

Machining State	Final state	Pressed	Presintered	Before ceramization
Material	Graphite MACOR™ aluminiumsilicate	Reaction bonded ceramic (see Chapter 7)	ZrO_2 Al_2O_3 aluminiumsilicate	Glassy carbon
Advantage	No thermal treatment, easy to clean	No sinter shrinkage after reaction sintering, green equal to sintered dimension	Good chemical resistance	Low tool wear, good chemical resistance, joining possible
Disadvantage	Tool wear (graphite), chemical resistance? (MACOR™)	due to binder no cleaning in US and organic solvents in green state, limited wall thickness	Shrinkage during sintering	Very brittle, limited wall thickness due to degassing during pyrolysis

16.2 Ceramic materials for micromilling

Table 16.1 gives an overview of ceramic materials accessible for micromilling. There are several kinds of materials available with different processing routes exhibiting specific advantages and disadvantages.

16.2.1 Ceramics for machining in the final state

Microparts can very easily be made from ceramics that can be structured in the final state like graphite and MACOR™ (Figure16.1).

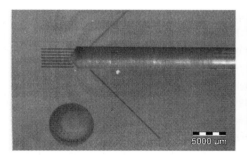 

Figure 16.1 Mould for microcasting made of fine grain graphite R8710 by SGL Carbon. Microchannels are 0.2 mm in square (left) and comb-like microstructure (right) of a microheat exchanger made of MACOR. Width of channels 0.4 mm, depth 0.6–2.9 mm (AR more than seven), thin walls 0.2 mm (AR almost 15).

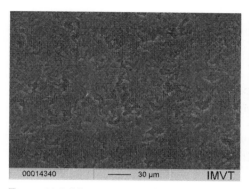 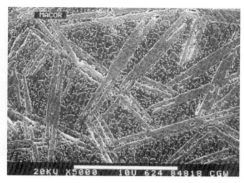

Figure 16.2 Microstructure of fine grain graphite-type R8710 supplied by SGL Carbon (left) [2] and MACOR™ by Corning International Europe (right) [3].

However, the mechanical properties of graphite are limited due to pores. The fine grain graphite R8710 supplied by SGL Carbon has an open porosity of 10% and a grain size of 3 μm. Additionally, the chemical resistance of MACOR™ is not as good as for alumina or zirconia due to the chemically different phases (Figure 16.2). The glassy ceramic consists of a matrix of approximately 45% borosilicate glass with 55% of embedded mica crystals causing microcracks when machining.

16.2.2 Ceramics for machining in the pressed state

To achieve a sufficient strength of the green blank, a certain amount of additives (e.g. polymers or other compaction aids) have to be added to the ceramic powder particles to reduce the friction between the particles as well as supporting the adhesion between the particles. Typically, uniaxial pressing in a die or cold isostatic pressing (CIP) for a more uniform green density allocation is used for the fabrication of the blanks. Well-known materials processed this way are alumina and zirconia where shrinkage during sintering must be taken into account and hence the achievable tolerances are not very tight.

Another interesting example for a ceramic machinable in the pressed state, however, is the shrinkage free ceramic based on the intermetallic phase $ZrSi_2$. During reaction sintering $ZrSi_2$ is oxidized and transformed into $ZrSiO_4$ (see Figure 16.3).

By using a low loss binder with a high ceramic yield for compaction and zirconia as an inert phase, the sinter process can be adjusted so no shrinkage of the final part occurs (Figure 16.4).

A prerequisite for this is a sufficient access of oxygen. Hence, the wall thickness is limited to a maximum value. This ceramic provides better mechanical and chemical properties than graphite and MACOR™. The big advantage is that shrinkage does not have to be taken into account. If the composition and the compaction of the blanks is well balanced, very tight tolerances can be achieved [4].

16.2.3 Ceramics for machining in the presintered state

Alumina and zirconia exhibit premium properties in terms of hardness, chemical resistance and mechanical properties. However, micromilling of these materials is only possible in the

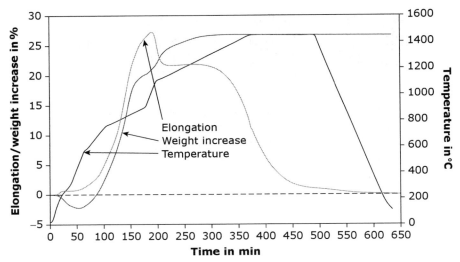

Figure 16.3 Reaction sintering of shrinkage free ceramic based on ZrSi$_2$.

Figure 16.4 Columns made of shrinkage free ceramic in green (left) and sintered state (right). Height 1 mm, trench width 0.5 mm, AR of two.

presintered or pressed state taking into account the subsequent shrinkage during sintering. If the material is presintered, it means that the formation of sinter necks and compaction has already started but the mechanical strength due to the remaining pores is low enough for milling. However, as the materials are subjected to shrinkage when sintering to full density (see Figure 16.5), it is difficult to meet very tight tolerances and tool wear cannot be neglected.

In the final state such ceramics can only be grinded with diamond tools causing 60–80% of the total costs of the ceramic part [5].

16.2.4 Machining of ceramics before ceramization

The micromilling of cross-linked resins, the preliminary product of glassy carbon, is very easy. A wide range of polymeric materials can be used. A prerequisite is highly cross-linked,

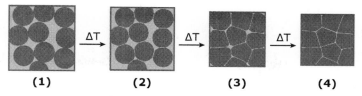

(1) loose bonded powder particles with low strength
(2) formation of so-called sinter bridges
(3) formation of a communicating pore skeleton
(4) sphericity and elimination of pores, closed porosity

Figure 16.5 Schematic characterization of the four states of densification during the sintering process.

3D aromatic polymers like polyphenylene, polyimides, aromatic epoxy formulations as well as phenol- and furan resins where pyrolysis takes place in the solid state. A high carbon yield is favourable to limit shrinkage. During pyrolysis up to 1000°C, volatile constituents from the polymer backbones are removed. The result is a twisted microstructure consisting of carbon ribbons with nanoporosity in between. Shrinkage of about 25% occurs. During a high temperature treatment up to 2200°C the part expands by approximately 5%. Afterwards the glassy carbon has excellent chemical and corrosion resistance; however, it is very brittle and can only be finished by grinding. The material is biocompatible. For example heart valves have been made from glassy carbon [6]. Like for the reaction sintering of the shrinkage-free ceramic, a certain maximum wall thickness has to be met or venting channels must be added.

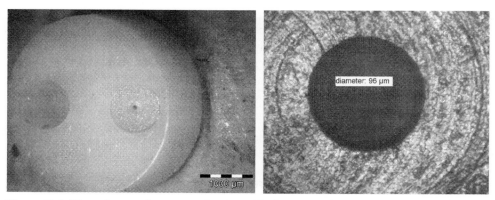

Figure 16.6 Dispersion nozzle in sintered state made of presintered zirconia with two holes with 100 μm diameter (AR of seven) under an angle of 60°.

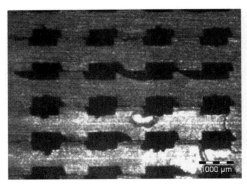 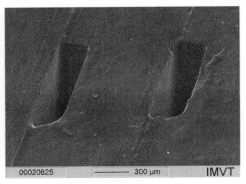

Figure 16.7 V-mixer made of several layers of microstructured and cross-linked resin sheets.

The chipping of these materials is comparable with the machining of polymers. The tool wear is negligible.

It is possible to build complex apparatuses of glassy carbon by stacking and joining several precured and microstructured layers using resin as glue and subsequent cross-linking of the resin at moderate temperatures of 60–80°C. An example is the V-mixer shown in Figure 16.7.

To prove the corrosion resistance of the material the V-miver was stored in concentrated sulphuric acid (95–97%) at a temperature of 100°C for two weeks. No traces of corrosion could be observed.

16.3 Micromilling of ceramic microstructures

Machining of ceramics is not as challenging as chipping of materials which are difficult to machine e.g. the nickel-based alloys, titanium or tantalum used in process engineering. These materials are chosen because of their good corrosion resistance but tend to weld with the cutting edge of the tool. Compared to ceramic materials, however, sometimes the corrosion resistance is much lower.

For most ceramic materials the tool life is reasonable and the cutting forces are moderate. Graphite, however, can be very abrasive. A sharp cutting edge is favourable to achieve sharp edges without faults at the microstructures. Hence, uncoated tools should be preferred in relation to more wear resistant coated tools but with rounded cutting edges. Machines must be equipped with appropriate solutions to avoid damaging the feed drives and guideways, e. g. using pressured air inside the encapsulated area. For dry machining, the dust of the removed ceramic material should be sucked off, preferably near the point of origin or emulsion should be used.

The infeed and feed can be about one order of magnitude larger for ceramics than for steel. Hence, the productivity of ceramic parts by micromilling is much higher than for metallic parts and offers the additional advantage of the enhanced properties in terms of hardness and corrosion resistance.

Unfortunately, in mechanical engineering, ceramics are generally considered to be brittle materials, thus avoiding exploitation of their favourable properties. Especially in MST

Figure 16.8 Ceramic test components made of alumina and zirconia joint with Hastelloy C-22 by thermal shrinking.

where corrosion often limits the lifespan of microchemical devices made of metallic alloys, ceramics are favourable. The reason is that the grain size of ceramic materials is often in the submicron range and hence one order of magnitude lower than for metallic materials. Consequently, the impact of intercrystalline corrosion of the functionality is much lower.

Of course, special design rules should be considered for ceramics. Rounded corners should be used instead of sharp edges to avoid the concentration of tensions. Joining or adaptation of ceramic components is often a problem. For example, using glassy solder will limit the chemical or thermal properties considerably. By thermal shrinking, ceramic components can be combined with metallic housings. An example is given in Figure 16.8.

Conclusions

There is a wide range of very different ceramic materials with different properties available for mechanical micromilling.

Ceramic materials are very interesting for MST and offer interesting properties. Especially for chemical and high temperature application as well as for applications subjected to wear ceramics are well suited. However, the design has to be adapted compared to metallic parts according to the typical characteristics of ceramics like brittleness and crack propagation. New technologies to improve joining have to be developed. Laser welding of alumina has been reported [7]. A large area of the material is preheated to limit the temperature gradient and thermal tensions using a CO_2 laser. The welding process is carried out using a Nd:YAG-laser. In this way, a vacuum-proof joint free of cracks could be obtained. However, the process has to be adapted very carefully to each type of material and test welding experiments have to carried out.

Acknowledgments

The authors acknowledge the funding of this research within the framework of the EU-Network of Excellence Multi-Material Micro Manufacture: Technologies and Applications (4M).

All ceramic microstructures displayed in this paper were machined by the members of the group "Materials for Micro Process Engineering" of the Institute of Micro Process Engineering.

References

[1] Gietzelt T. and Eichhorn L. 2007. Micro milling of high aspect ratio micro structures in ceramics. Proceedings of *Third International Conference on Multi-Material Micro Manufacture (4M 2007)*, 2–5 October 2007, Borovets, Bulgaria, 191–194.

[2] http://www.sglgroup.com/export/sites/sglcarbon/_common/downloads/ products/product-groups/gs/cut-to-size-graphite-for-electrical-discharge-machining-edm/Marken_EDM_d.pdf (accessed 17 November 2008).

[3] http://www.corning.com/docs/specialtymaterials/pisheets/Macor.pdf (accessed 17 November 2008).

[4] Binder J., Schlechtriemen N., Jegust S., Pfrengle A., Geßwein H. and Ritzhaupt-Kleissl H. 2007. Green machining of net-shape reaction-bonded ceramics: materials and applications. *Ceramic Forum International: cfi/Berichte der Deutschen Keramischen Gesellschaft*, **84**(6), E57–E60.

[5] Warnecke G., Eichgrün K. and Schäfer L. 2000. Zuverlässige Serienbauteile aus Hochleistungskeramik. *Ingenieur-Werkstoffe*, **2**(2), pp 42–45.

[6] Dübgen R. and Popp G. 2004. Glasartiger Kohlenstoff SIGRADUR®–ein Werkstoff für Chemie und Technik. *Materialwissenschaft und Werkstofftechnik*, **15**(10), 331–338.

[7] http://laz.htwm.de/3_forschung/50_keramik/1_keramikschwei%Dfen/default. Asp?content=%2 F3%5Fforschung%2F50%5Fkeramik%2F1%5Fkeramikschwei%25Dfen%2F02%5Fverfahren%2 Ehtml (accessed 17 November 2008, only in German).

17

High precision microgrooving on ceramic substrates

G. Bissacco, D. Andrijasevic, H.N. Hansen and T. Gietzelt

The realization of microgrooves is an important task in microproduction. At the same time the generation of high precision shaped grooves in ceramic materials is very cumbersome. Established and flexible processes like grinding and diamond cutting face problems when dealing with small concave features. Alternative possibilities are the machining of ceramics in the green or presintered state by means of dedicated shaped tools. This chapter presents the development of the process chain for the generation of high precision shaped microgrooves on ceramic substrates for applications in the high precision microbonding process.

17.1 Introduction

The generation of high precision shaped microgrooves in ceramic materials is very cumbersome. Simple V-shaped or square-shaped grooves can be produced by means of established processes. However, when the geometry of the microgroove cross-section is not elementary, such processes can no longer be applied. Ceramic materials in the sintered state can be machined by means of diamond grinding processes and diamond cutting processes. Micro-grinding shows strong limitations in generating very small features due to the wheel geometry and dimensions, furthermore the diamond wheel changes profile during machining due to wear, with consequent loss of accuracy. Fly cutting using diamond tools can be used for accurate machining, but only low aspect ratios can be realized for short grooves. Micro-milling can be used for the machining of ceramic materials in the presintered, pressed and green state. However the minimum groove width is equal to the tool diameter used. At present the smallest available micro-end mills have diameters down to $30\,\mu m$. Hence, the smallest groove width is $30\,\mu m$. However, it is limited to an aspect ratio of 1.5 due to stability reasons. The aspect ratio for micro end mills depends strongly on the diameter. For micro-end mills of $100\,\mu m$ in diameter, a maximum aspect ratio of 10 can be realized. However, the geometry of the groove is always strictly rectangular.

An alternative possibility for the generation of microgrooves in ceramic materials is a grooving process, where a nonrotating tool, provided with linear motion, is pushed into the

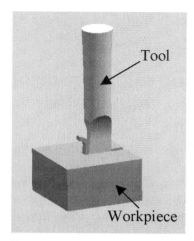

Figure 17.1 Schematic of the microgrooving process.

material, removing part of it as chips. A schematic of the process is shown in Figure 17.1. In such a process the tool can be made of hardened tool steel or tungsten carbide. Due to the limited tool hardness, microgrooving can be used in connection with the same materials machinable by micromilling, namely presintered, pressed and green ceramics. An important advantage of microgrooving in comparison to other processes for microstructuring of ceramics is that it can easily be applied to the realization of shaped microgrooves whose cross-section contains small details. In fact, the geometry of the tool is replicated in the workpiece material. The application of microgrooving processes requires the capability to design a functional tool and to manufacture it. Additionally, precautions have to be taken in order to ensure high accuracy of the groove's depth.

17.1.1 Shaped grooves and process chain

Figure 17.2 shows a shaped groove's cross-section, required for applications in the hot stream gas bonding of optic fibres. Due to its thermal stability and the capability for easy integration in electrical circuits, low temperature cofired ceramics (LTCC) tape is used as a substrate material for the grooves necessary for alignment of the optic fibres. Mechanical machining of the material in the fired state is not possible, thus machining in the green state was chosen. The dimensions in Figure 17.2 refer to the unfired ceramic. The profile consists of a 90° V shape with depth of 200 mm and a 40 μm wide rectangular feature underneath. The depth of the V part of the groove in the fired ceramic, taking into account 21% shrinkage, is 165 μm [1]. For the proper alignment of the optic fibres, the required accuracy on the groove's depth is 3 μm and the side wall roughness must be Ra < 3 μm.

A dedicated process chain was developed [2] for the realization of the recommended shape of the grooves. The first step was the realization of a 4 mm single edge tool to be used in orthogonal cutting for the generation of a reference surface on the ceramic substrate. The

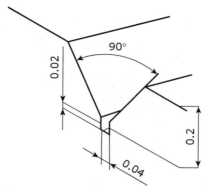

Figure 17.2 Geometry of the groove.
Dimensions in mm.

second step involved the realization of a microgrooving tool (described in detail below) with the negative profile of the grooves and dimensions according to Figure 17.2. Subsequently the reference surface and grooves were generated by mechanical material removal using the manufactured tools.

In order to realize a clean and efficient cut without burrs and to minimize rubbing due to the springback of the green ceramic material, the design of the tools included adequate clearance angles. For the first tool a 2° bottom clearance angle was used, while the grooving tool included a side clearance angle of 1° and a bottom clearance angle of 1.5°. Furthermore the frontal rake angle was 90° for both tools. The tools are shown in Figures 17.3 and 17.4.

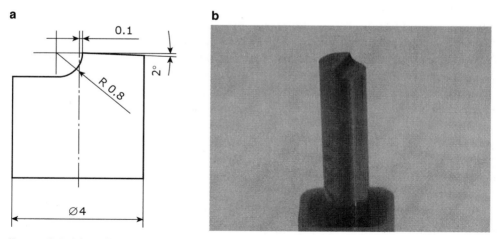

Figure 17.3 (a) single edge tool design; (b) manufactured facing tool.

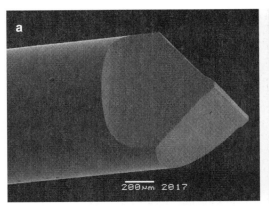

Figure 17.4 (a) microgrooving tool; (b) tool tip with the miniature square feature.

The tools were realized in micrograin tungsten carbide by microelectrical discharge milling, on the shaft of microend mills used for ordinary micromilling. This process allows the realization of complex 3D microgeometry in conductive materials by means of a circular rod electrode with the desired diameter. In micro-EDM milling, machining is performed by a sequence of electrical discharges occurring in an electrically insulated gap between a tool electrode and a workpiece. During the discharge pulses, a high temperature plasma channel is formed in the gap, causing local melting and evaporation of workpiece and electrode material. In this way a controlled material removal is realized, allowing a machining accuracy of less than $2\,\mu m$ [3]. Figure 17.4 shows the details of the microgrooving tool. In order to limit the machining time, a high value of the discharge energy was chosen. Thus the surface appears rough, and the accuracy of the features is limited. In particular, the square feature shown in Figure 17.4 (right), was measured by means of 2D SEM to be $46\,\mu m$ wide and $10\,\mu m$ high, while the nominal values were $40\,\mu m$ and $20\,\mu m$ respectively. The radius of the frontal cutting edges was estimated to be $1-3\,\mu m$.

The grooves were realized by mounting the manufactured tools on the same micro EDM milling machine. In this case the machine was only used to provide the necessary linear displacements, taking advantage of its high positioning accuracy. In fact, although conventional EDM machines are not suitable for mechanical loadings, owing to the small chip cross-section and the low workpiece material hardness, the cutting forces were negligible, thus no distortions of the machining system were expected. On the other hand, the positioning repeatability of the micro-EDM milling machine used was less than $0.6\,\mu m$ along three orthogonal axes at 95% confidence level, as verified by means of machine tool testing using a laser interferometer. Such a high positioning accuracy was considered necessary in order to match the prescribed machining tolerances. A flat surface was thus generated on the green ceramic by means of the 4 mm tool. The grooves were then generated using the manufactured microtool. In order to prevent the material from bulging aside, the grooves were generated by means of eight tool passes at progressive depths. The maximum incremental depth was $50\,\mu m$. Figure 17.5 shows the grooving process, while Table 17.1 summarizes the parameters used during machining. Since the tool is nonrotating, the feed

Figure 17.5 Grooving on green LTCC tape.

Table 17.1 Summary of process parameters used during planing and grooving.

Process parameter	4 mm single edge tool	Microgrooving tool
Number of passes	2	8
Max incremental depth	--	50 µm
Cutting speed	0.5 m/min	0.5 m/min

rate, which is comparable with what used in micromilling operations coincides with the cutting speed. This is a limitation of grooving processes in general, since optimal cutting speeds, which could be as high as 200 m/min cannot be realized.

17.2 Verification of grooves

The geometry and topography of the manufactured grooves were verified by means of a stylus contact profilometer. One of the machined grooves is shown in Figure 17.6 The average depth, over a length of 44 mm, was 203.5 µm with a standard deviation of 1 µm. The

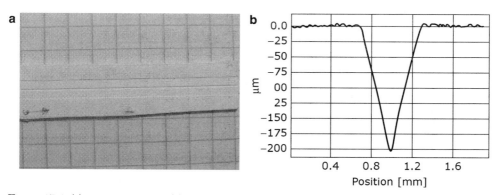

Figure 17.6 (a) groove on green LTCC tape; (b) measured profile.

roughness of the sides of the groove was Sa 0.53 µm and Sz 5.1 µm. As for the measurement of the bottom of the groove, which consisted of only few profiles, the evaluation based on 3D roughness parameters would be incorrect. Thus 2D roughness parameters were calculated on the middle profile among those representing the groove's bottom. The profile roughness of the groove's bottom was Ra 0.48 µm and Rz 3.28 µm. These rather high roughness values are probably the consequence of the very low cutting speed (0.5 m/min) allowed by the setup used.

Conclusions

A method for the generation of shaped microgrooves in ceramic materials has been presented. It involves the realization of shaped grooving tools in hard materials such as tool steel or tungsten carbides by means of other 3D structuring processes and the use of such tools as grooving tools. For the generation of the tool profile, appropriate processes must be selected. Micro-EDM milling is a suitable process, however, in order to achieve good tool accuracy and fine surface quality, finishing cycles and/or postprocessing must be implemented. The high hardness of the tools allows using substrates not only in the green state but also in the presintered state. The roughness of the sidewalls strongly depends on the roughness and profile of the grooving tool. Depth accuracy and roughness of the grooves can potentially be improved by reducing the roughness of the edges of the microgrooving tool and by increasing the cutting speed. The method is a suitable one when high accuracy of the groove's depth is required over a long distance. Furthermore, when a higher productivity is required, the method can be improved implementing parallel multigrooving using a tool with multiple profiles.

Acknowledgments

The authors are grateful for the funding of this research within the framework of the EC Network of Excellence Multi-Material Micro Manufacture: Technologies and Applications (4M). All the ceramic microstructures displayed in this paper were machined by the members of the group.

References

[1]   Andrijasevic D., Bissacco G., I. Giouroudi, Smetana W. and Brenner W. 2006. High precision micro bonding process using a focused stream of hot air. *Proceedings of the 6th Euspen International Conference*, 28 May–1 June 2006, Baden bei Wien, Austria, 2, 228–231.

[2]   Bissacco G., Andrijasevic D. and Hansen H.N. 2006. Realization and characterization of high precision microgrooves on green ceramic. *Proceedings of the 6th Euspen International Conference*, 28 May–1 June 2006, Baden bei Wien, Austria, **2**, 192–195.

[3]   Bissacco G., Valentincic J., Wiwe B.D. and Hansen H.N. 2007. Characterization of pulses in micro EDM milling based on wear and material removal. *Proceedings of the Third International Conference on Multi Material Micro Manufacture (4M–2007)*, 2–5 October 2007, Borovets, Bulgaria, 297–300.

18

Patterning of ceramics and glass surfaces

W. Pfleging and M. Rohde

At Forschungszentrum Karlsruhe research in material processing for micro- and nanotechnology is focused on laser-assisted surface functionalization, laser patterning and bonding technologies such as laser microwelding, laser brazing and laser transmission welding. This application-oriented research is related to applications in the life sciences, microreaction technology, biological interfaces and microoptics. In micro- and nanosystems different kinds of materials and thin films such as polymers, metals and ceramics are of interest. Laser technologies are powerful tools which can be used to enable a rapid fabrication of functional prototypes. Furthermore, new technical approaches including the combination of different material types such as polymer–ceramic or metal–ceramic can be realized by advanced laser technologies. This chapter discusses new approaches in laser technology for the patterning and functionalization of ceramic surfaces.

18.1 Patterning of quartz by CO_2-laser radiation

Glass has excellent chemical, optical and thermal properties suitable for applications in micro- and nanotechnology. State-of-the-art laser micropatterning of glass uses photolithography and wet-chemical etching. A flexible patterning of glass via direct-writing processes, which makes an appropriate use of laser methods for ablation, cutting and drilling is investigated. To date, this technical approach has only been successfully developed for polymer-based materials. For glass materials, such as quartz, the main challenges are caused by the high brittleness and mechanical–thermal properties of glass which, in general, leads to crack formation during laser ablation or melting and resolidification. Therefore the use of "cold ablation" processes enabled by short-pulsed UV-laser radiation or ultrashort laser pulses has been investigated [4–7]. However, "cold ablation" always means small ablation rates and therefore a significant increase in material processing time. A significant decrease in processing time can be achieved by an appropriate use of CO_2-laser radiation.

CO_2-laser radiation only enables a thermally driven ablation process. For a crack-free ablation the glass substrate temperature should be increased up to 200°C during laser patterning [8]. In our recent approach the crack-free ablation of quartz was also realized at room temperature. The ablation was performed with a CO_2-laser system which operated in

Gaussian mode (Synrad, Firestar v40). Argon or nitrogen were used as the laser processing gas. The average laser power used was less than 20 W and scanning velocities during ablation and cutting were in the range 1–100 mm/s.

The surface quality is obviously strongly influenced by the deposited laser energy. A simple model for the energy balance ("absorbed laser energy" = "energy for temperature rise" + "energy for phase transitions" + "energy loss by heat conduction") during the laser process delivers an expression for the ablation depth d as a function of laser power P_L and feed rate v of the laser beam [9]:

$$d \propto \frac{P_L}{v}$$

Depending on the laser power and the feed rate the energy input and therefore the ablation depth varies. P_L/v is the so-called line energy and is a very important parameter in continuous wave ablation. With decreasing line energy the dimension of the heat-affected zone, the ablation depth and the channel width decrease (see Figure 18.1). In order to obtain different aspect ratios, or if larger channel depths at fixed channel widths are desired, one has to repeat the ablation scan process several times. Under optimized conditions it is possible to generate channel structures with any channel width within the 40–150 µm interval. The use of nitrogen as the processing gas enables higher ablation rates and an improved surface quality than when oxygen or air are used as the processing gas.

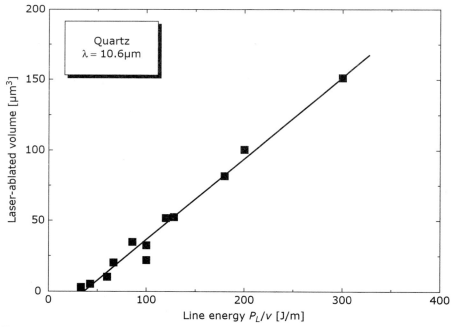

Figure 18.1 Laser-ablated volume of quartz measured for a channel with a length of 1 mm as function of line energy (laser power 3–9 W, laser scan velocity 10–90 mm/s, processing gas: N_2).

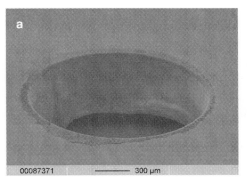
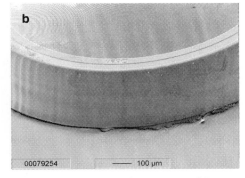

Figure 18.2 SEM of laser patterned quartz plates with a thickness of 700 µm (a) and 300 µm (b).

With an appropriate laser scanning process strategy, complex channel structures as well as large area patterning cannot be realized. The hole shown in Figure 18.2a was obtained by laser trepanning at a laser scan velocity of 1 mm/s and a pulse repetition rate of 4 kHz. Crack formation was completely suppressed. Nevertheless, the formation of debris could only be avoided for thin material thicknesses smaller than 700 µm as shown in Figure 18.2b. The CO_2-laser process for patterning of glass surfaces can be used to fabricate adaptive membrane lens systems [10].

18.2 Patterning of ceramics

With Q-switch Nd:YAG laser radiation (wavelength 1064 nm, pulse length 100–200 ns) one obtains relatively rough ceramic surfaces even with optimized laser and process parameters ($R_a = 0.8$ µm and $R_z = 11.2$ µm) [11]. This fact was used to join the polymers and ceramic parts. On the one hand polymers are well suited to laser-assisted micropatterning [12, 13] and on the other hand LTTC is an established ceramic for the production of sensors and actuators [14]. A strong bonding of transparent and microstructured polymers with LTTC could be used to realize an established ceramic sensor system with simultaneous optical monitoring. This type of sensors would be of great interest to biological and medical research.

The basic idea of joining of transparent polymers (thermoplastics) with LTTC is to heat up the interface of polymer and ceramic by the use of high power diode laser radiation (wavelength 940 nm). In general, thermoplastics are transparent to this laser beam. More experimental details have been described elsewhere [2]. The laser induces a temperature rise slightly above the glass temperature of the polymer. The "molten" polymer will flow into the pores of the ceramic surface. Laser ablation with Q-switch Nd:YAG laser radiation is performed in order to increase the pore density at the LTCC surface. Figure 18.3 shows a laser-produced hole in LTCC. The width and depth of the hole is about 100 µm. Additionally, small pores with diameters in the µm-range were generated at the sidewall of the crater.

The mechanical strength of the bonding depends on the density of laser-generated craters on the LTCC surface. Mechanical strength measurements of PMMA welded

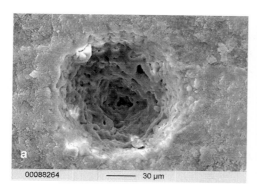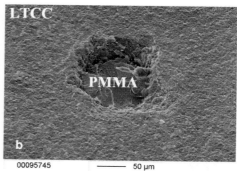

Figure 18.3 SEM images from laser-ablated craters in LTTC before welding (a) and after welding and tensile strength measurement (b) (Q-switch Nd:YAG, wavelength 1064 nm, pulse length 200 ns).

on LTCC show that the bonding strength at the interface could exceed the mechanical stability of the LTCC. For a density of 34 × 34 holes per square centimeter or higher the bonding failure is observed in the LTTC at a tensile strength of (4.9±0.7) MPa. Therefore, the bonding was studied for a hole density of 17 × 17 holes/cm². It was observed that the tensile strength in this case is about (1.8±0.2) MPa. PMMA material fills the craters and after tensile strength measurements the mechanical anchoring between polymer and the ceramic persists. The bonding failure in this case was observed in the PMMA, directly at the top of the craters in LTCC (Figure 18.3b). For a hole density of 17 × 17 holes/cm² the ratio between laser-treated area and the untreated area is about 1:44. That means that the effective tensile strength of PMMA is measured to 79 MPa. The bonding of microfluidic polymer chips on to LTCC sensor patterns is currently under investigation.

Laser ablation of ceramics with UV-laser radiation leads to a significant improvement in surface quality [15]. This is caused by two aspects: first, the improved laser beam absorbance at short wavelengths; and secondly, the high laser power densities obtained by using short laser pulses, which leads to a more effective sublimation and a significant decrease in melt formation. For this purpose the ablation of ceramics (Al_2O_3 and ZrO_2) was investigated with excimer-laser radiation (193 nm, 248 nm, pulse length 20–25 ns) and UV-Nd:YAG laser radiation (355 nm, pulse length 500 ps).

Figure 18.4 shows the laser ablation of ceramics for different types of laser radiation sources. The ablation rate increases linearly with laser fluence. For excimer-laser radiation and for laser fluences smaller than 5 J/cm² the ablation rate is smaller than 80 nm per pulse. Nevertheless, for large area patterning or for the fabrication of curved shapes the use of excimer-laser radiation is preferred. For low laser fluences <2 J/cm² the surface roughness at the bottom of the laser patterned area is similar to the roughness of the ceramic substrate (R_a = 1–2.5 μm). For laser fluences >2 J/cm² the average surface roughness is in the range R_a = 180–200 nm (R_z = 1.0–1.2 μm). Excimer-laser radiation enables a smoothing of the ceramic surface.

For a 355 nm laser wavelength the beam penetration depth into the material is larger than for 248 nm or 193 nm. Furthermore, for 355 nm the laser pulse length is significantly

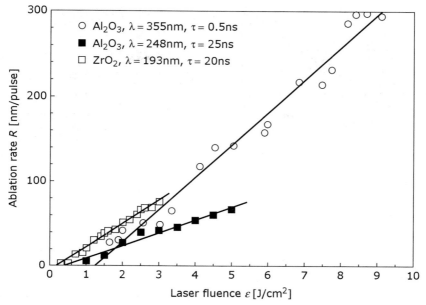

Figure 18.4 Laser etch rate of Al_2O_3 and ZrO_2 as function of laser fluence for different laser wavelengths λ and laser pulse widths τ.

smaller and the maximum laser fluence ε is higher compared to 193 nm and 248 nm (see Figure 18.4). This means, that the maximum laser pulse power intensity is increased by a factor of about 100 in comparison to the excimer-laser radiation. An increased absorption length and a significantly higher laser power intensity lead to an increase in the ablation rate (Figure 18.4). Additionally, it is expected that for short laser pulses and a high laser fluence the formation of melt could be suppressed. In fact, Figure 18.5e and f shows, that for a laser pulse length of 500 ps @ 355 nm no melt or debris formation on ceramic surfaces could be observed. For UV laser radiation with 20 ns @ 193 nm and 25 ns @ 248 nm a small melt seam is observed at the contour of the generated patterns. After removing these seams by cleaning in an ultrasonic bath small defects remain at the contours (Figure 18.5, b and c).

It has been established that the microstructuring of ceramic surfaces significantly improves their tribological properties [16].

18.3 Laser-induced surface modification of ceramics

Due to their poor thermophysical and mechanical properties, the use of single-phase commercial ceramics in technical applications is restricted. The introduction of a second phase, which can be selected in order to optimize these properties, can lead to a reinforcement of the mechanical strength and also to an enhancement of the thermal conductivity. Different thermal processing techniques can be applied to achieve this property modification

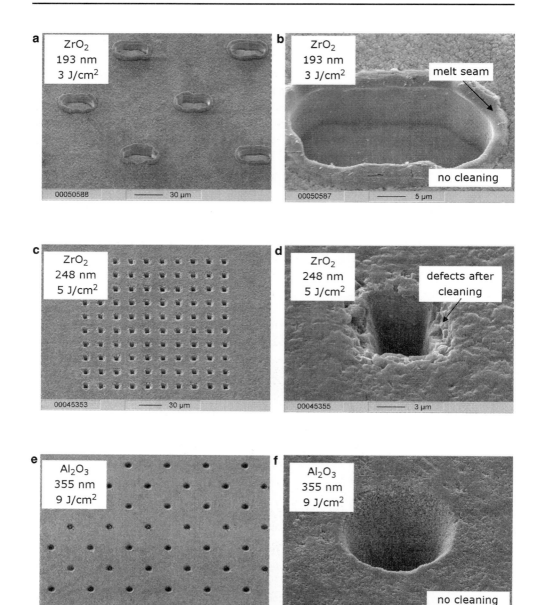

Figure 18.5 SEM images of laser patterned ceramics Al_2O_3 and ZrO_2 for different laser wavelengths λ and laser fluences ε.

by producing ceramic–metal composites with metal particles embedded in a continuous ceramic matrix. Among these methods, laser-supported modification techniques have the advantage that the mechanical and tribological properties [17, 18] can be improved and the thermal and electrical conductances can be adjusted [19, 20] while the property modifications are restricted to a localized surface area leaving the bulk of the ceramic in its original state.

The ceramic substrate can be modified by scanning the beam of a high power CO_2-laser across the surface of the sample with a laser power adjusted in order to melt the substrate locally. Metal particles were introduced into the melt pool using an injection technique or a preplaced powder coating. After the solidification, a ceramic–metal composite was developed with properties depending on the metal powder used in the process. For the laser processing a rectangular beam profile of $6 \times 1\,mm^2$ can be used with a typical laser power of 240 W and a scanning velocity of 250 mm/min. Laser beam profiles with a Gaussian intensity distribution can be applied in order to achieve modifications at a smaller spatial scale with a minimum line width down to 200 μm. During the process the substrate is heated to a temperature of 1500°C to avoid thermally induced development of cracks within the ceramic matrix. The substrate can be modified to a depth of about 500 μm with a metal volume fraction as high as 50%. Typical cross-sectional microstructures of alumina substrates modified with TiN and WC, respectively, (Figure 18.6) show that the hard metal particles, which appear as white areas in the microscope image, are embedded in the grey ceramic matrix.

The incorporation of the hard metal phase in the ceramic matrix induces changes in the mechanical as well as the thermophysical properties. The fracture toughness, which

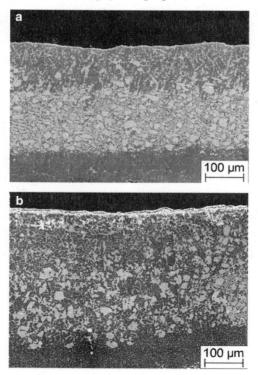

Figure 18.6 Microstructures of laser-modified ceramics: Al_2O_3 substrate modified with TiN (a) and WC (b), respectively.

characterizes the ability of the material to withstand crack propagation, increases significantly and also the tribological properties of the modified ceramic are also improved [17]. The wear of the modified ceramic surface is strongly reduced compared to the original ceramic and the variation in the friction coefficient as a function of the sliding velocity is reduced. The effective thermal conductivity increases due to the fact that hard metal phases with a relatively high thermal conductivity are embedded within the ceramic matrix. In order to improve this particular thermophysical property it is important that the thermal contact resistance between the matrix and the hard metal particle is as low as possible. Obviously, this occurs as a result of the laser process which leads to a close mechanical and thermal contact between particle and matrix.

The measured effective thermal conductivity for the laser-modified ceramic is shown in Figure 18.7. The total thickness of the substrate was about 1.5 mm and the depth of the modification was approximately 500 μm. Due to the laser modification process the thermal conductivity is significantly raised over a wide temperature range up to 400°C. At temperatures higher than 500°C the thermal conductivity of the modified ceramics falls to the same level as the original ceramic substrate. Among the laser-modified ceramics, the substrates, which were modified with WC particles, show higher thermal conductivity values than the ceramics dispersed with TiN particles for all measured temperatures although the volume fraction of WC is about 30% compared to approximately 50% for TiN particles. Qualitatively, this can be explained by the higher thermal conductivity of WC, which is in the range 80–129 W/mK [21] at room temperature for bulk tungsten carbide, compared to 22–28 WmK for bulk titanium nitride [22].

The laser dispersion also allows for direct writing of electrically conducting lines into a ceramic substrate, thus modifying the electrical transport properties. In contrast to metallic

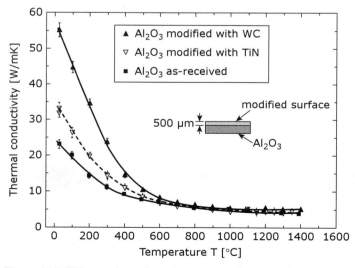

Figure 18.7 Effective thermal conductivity as a function of temperature for modified alumina compared to as-received ceramic.

coatings, this process is faster and the substrate is modified on a local scale without lithographic processing. Since the laser-modified tracks are incorporated into the surface rather than placed on top of it, these conducting lines are mechanically and thermally very stable. Therefore, they exhibit a relatively high current carrying capacity.

Typical cross-sectional microstructures of the laser-dispersed ceramics are shown in Figure 18.8. Cu particles were injected through a powder nozzle into the surface of the alumina ceramic (Figure 18.8a) during the laser-induced melting process. The eutectic microstructure of a WO_3/Al_2O_3 composite has been generated (Figure 18.8b) by dispersing W particles into an alumina substrate followed by a laser remelting step in air which leads to a complete oxidation of the metal phase. The typical powder particle size of the Cu and W powders used in the process varied in the range 5–15 µm. Before and during the laser process the whole substrate was heated in order to avoid thermally induced cracking. The heating temperature was 1500°C for the Al_2O_3 substrate.

Different microstructures could be generated within the laser-modified tracks by changing the additives or the process parameters. Introducing metal particles with a higher melting point than the ceramic leads to a metal–ceramic composite with the metal particles embedded within the ceramic matrix. This is the case for the W particles dispersed into the ceramic melt pool. During the laser-induced remelting the W particles are transported within the liquid ceramic process zone due to Marangoni convection and sedimentation.

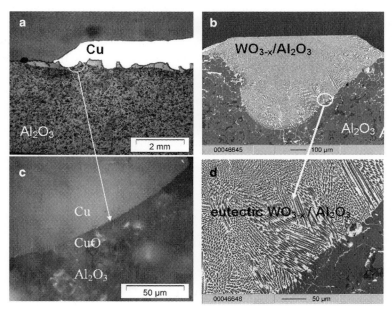

Figure 18.8 Cross-sectional micrographs of electrically conducting lines manufactured by laser modification of alumina: Cu-line on alumina substrate (a) and WO_{3-x}/Al_2O_3 composite produced by laser dispersing (b) of W particles and subsequent laser remelting in air. Details of microstructure are shown in lower figures.

After resolidification of the ceramic an electrical conducting path has been built up due to the percolating network of the metallic particles within the insulating ceramic matrix. A completely different behaviour can be observed if Cu powder is used as the additive. Since the melting point of Cu is much lower than that of the ceramic, the Cu particles melt and agglomerate to larger drops (see micrograph on the right in Figure 18.8). The wetting and adherence between the metal and ceramic phase is enhanced by a thin CuO-film at the interface between the Cu drop and the alumina matrix.

The combination of laser dispersing of W particles with a following thermal treatment with the laser generates a WO_3/Al_2O_3-composite with the different phases arranged in a eutectic microstructure with a narrow spacing of the WO_3 and Al_2O_3 lamella, as shown in the micrographs on the right-side of Figure 18.8.

The application of different additive materials in the laser process not only leads to the different levels of the electrical resistance at room temperature it also results in conducting tracks with different temperature coefficients of resistivity. This effect is shown in Figure 18.9 where the measured electrical resistivity is plotted as a function of the temperature. The ceramic modified with Cu shows an increasing resistivity with increasing temperature. This positive temperature coefficient is typical for metallic conductors. Qualitatively the same behaviour can be observed in the Al_2O_3 ceramic modified with W but with a slightly higher temperature coefficient. The electrical characteristic of the WO_3 alumina system is completely different. The resistivity decreases with increasing temperature. The negative temperature coefficient is typical for a thermally activated conduction mechanism which can be observed in semiconductors.

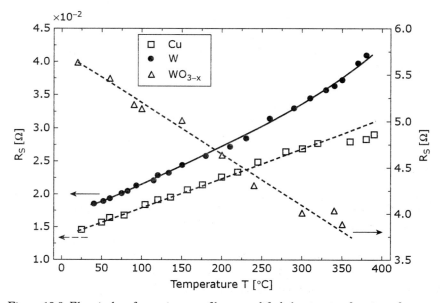

Figure 18.9 Electrical surface resistance of laser-modified alumina as a function of temperature.

Conclusions

Laser material processing is an appropriate tool for the patterning and modification of ceramic surfaces which has been demonstrated in detail for quartz, LTTC, Al_2O_3 and ZrO_2. For feature sizes down to $40\,\mu m$ glass surfaces can be structured and cut by CO_2-laser radiation with high surface quality without crack formation and with high processing velocities. For this purpose process strategies were developed using a high quality laser beam ($M^2 \approx 1.1$). Ceramics such as LTTC, Al_2O_3 and ZrO_2 can easily be structured by using near infrared-laser radiation with a pulse length of 100–200 ns. A further improvement of surface quality and reduction of feature sizes are obtained by using short pulsed UV-laser radiation. High laser fluences are necessary in order to obtain significant ablation rates larger than 100 nm/pulse. For serial material processing the use of frequency-tripled Nd:YAG laser radiation with short pulses (500 ps) enables high ablation rates while for large area patterning the use of high repetition and short pulse excimer-laser radiation is most efficient.

Surface properties of ceramic substrates like the thermal and electrical conductivity, respectively, can be modified by laser dispersing and alloying using a CO_2-laser. Additives like metals or semiconducting oxides can be introduced into the ceramic matrix during localized melt processing with the laser. After resolidification of the laser-induced melt pool a metal–ceramic composite appears with modified properties compared to the original ceramic. The thermal conductivity within the modified region increases significantly. The electrical conductivity and its temperature coefficient can be adjusted by an appropriate selection of the additive.

Acknowledgements

We are grateful to our colleagues M. Beiser and M. Hoffmann for their technical assistance in SEM/EDX. We also thank M. Torge for her support in laser processing and surface characterization and H. Besser for helpful discussions. We gratefully acknowledge the financial support by the program NANOMIKRO of the Helmholtz Association. Finally, the EU gave support within the Sixth Framework Programme, "Network of Excellence in Multi-Material Micro Manufacture" and "Network of Excellence in Microoptic". Furthermore, the authors wish to thank K. Poser and S. Schreck for the preparation of the laser-modified ceramic samples. The studies on laser modification and structuring of ceramics were supported by the Deutsche Forschungsgemeinschaft (DFG) in the context of the Sonderforschungsbereich 483 "High performance sliding and friction systems based on advanced ceramics".

References

[1] Pfleging W., Przybylski M. and Brückner H.J. 2006. Excimer laser material processing–state of the art and new approaches in microsystem technology. *Proceedings of the SPIE*, **6107**, 61070G-1–61070G-15.

[2] Pfleging W. and Baldus O. 2006. Laser patterning and welding of transparent polymers for microfluidic device fabrication. *Proceedings of the SPIE*, **6107**, 61075-1–61075-12.

[3] Pfleging W. and Adamietz R., Brückner H.J., Bruns M. and Welle A. 2007. Laser-assisted modification of polymers for microfluidic, microoptics and cell culture applications. *Proceedings of the SPIE*, **6459**, 645911-1–645911-9.

[4] Niino H., Ding X., Kurosaki R., Nazaraki A., Sato T. and Kawaguchi Y. 2003. Surface microstructuring of transparent materials by laser-induced backside wet etching using excimer laser. *Proceedings of the SPIE*, **5063**, 193–201.

[5] Niino H., Kawaguchi Y., Sato T., Narazaki A., Ding X. and Kurosaki R. 2004. Surface microfabrication of fused silica glass by UV laser irradiation. *Proceedings of the SPIE*, **5339**, 112–117.

[6] Nikumb S., Chen Q., Li C., Reshef H., Zheng H.Y. and Qiu H.D. 2005. Low precision glass machining, drilling and profile cutting by short pulse lasers. *Thin Solid Films*, **477**, 216–221.

[7] Karnakis D.M., Knowles M.R:, Alty K.T., Schlaf M. and Snelling H.V. 2005. Comparison of glass processing using high-repetition femtosecond (800 nm) and UV (255 nm) nanosecond pulsed lasers. *Proceedings in the SPIE*, **5718**, 216–227.

[8] Meng-Hua Y., Ji-Yen C., Cheng-Wey W., Yung-Chuan C. and Tai-Horng Y. 2006. Rapid cell-pattenring and microfluidic chip fabrication by crack-free CO2 laser ablation of glass. *Journal of Micromechanic Mircoengineering*, **16**, 1143–1153.

[9] Jensen M.F., Noerholm M., Christensen L.H. and Geschke O. 2003. Microstrucutre fabrication with a CO2 laser system: characterization and fabrication of cavities produced by raster scanning of the laser beam. *Lab on a Chip*, **3**, 302–307.

[10] Schneider F., Müller C., Wallrabe U., Ulmer U., Eberhard D., Besser H. and Pfleging W. 2007. Adaptive Silikonmembranlinsen mit integriertem Piezo-Aktor. *Proceedings of MikroSystemTechnik 2007*, 15–17 October 2007, Dresden, Germany, VDE–Verlag, Berlin–Offenbach.

[11] Hanemann T., Ruprecht R., Haußelt J., Zum Gahr K.H. and Pfleging W. 2000. Rapid fabrication of microcomponents. *SPIE Conference on Microfabrication, Integration and Packaging*, 9–11 May 2000, Paris, France, **4019**, 436–443.

[12] Pfleging W., Schierjott P. and Khan Malek C. 2007. Rapid fabrication of functional PMMA microfluidic devices by CO_2-laser patterning and HPD-laser transmission welding. *Laser Assisted Net Shape Engineering 5*, Vol. 2, Geiger M., Otto A., Schmidt M.(Eds), Meisenbach-Verlag, Bamberg, Germany, 1207–1220.

[13] Pfleging W., Lu Y., Washio K., Bachmann F.G. and Hoving W. [Eds]. 2007. Laser-based micro-and nanopackaging and assembly. *Proceedings of Photonics West: Lasers and Applications in Science and Technology*, San Jose, California, USA., 20–25 January 2007, Bellingham, Washington, USA, SPIE Proceedings Series 6459.

[14] Thelemann T., Thust H. and Hintz M. 2002. Using LTCC for Microsystems. *Microelectronics International*, **19**(3), 19–23.

[15] Pfleging W., Hanemann T., Torge M. and Bernauer W. 2003. Rapid fabrication and replication of metal, ceramic and plastic mold inserts for application in microsystem technologies. *Proceedings of the Institution of Mechanical Engineers, Part C. Journal of Mechanical Engineering Science*, **217**, 53–63.

[16] Schreck S. and Zum Gahr K.H. 2005. Laser-assisted structuring of ceramic and steel surfaces for improving tribological properties. *Applied Surface Science*, **247**, 616–622.

[17] Zum Gahr K.H. and Schneider J. 2000. Surface modification of ceramics for improved tribological properties. *Ceramic International*, **26**, 363–370.

[18] Poser K., Rohde M., Schneider J. and Zum Gahr K.H. 2005. TiN-particle reinforced alumina for unlubricated tribological applications mated with metallic counterbodies. *Materialwissenschaft und Werkstofftechnik*, **3–4**, 122–128.

[19] Duitsch U., Schreck S. and Rohde M. 2003. Experimental and numerical investigations of heat and mass transport in laser induced modification of ceramic surfaces. *International Journal of Thermophysics*, **24**, 731–740.

[20] Baldus O., Schreck S. and Rohde M. 2004. Writing conducting lines into alumina ceramics by a laser dispersing process. *Journal of the European Ceramic Society*, **24**, 3759–3767.

[21] Friedrich C., Berg G., Broszeit E. and Berger C. 1997. Datensammlung zu Hartstoffeigenschaften. *Materialwissenschaft und Werkstofftechnik*, **28**, 59–76.

[22] Taylor R.E. and Morreale J. 1964. Thermal conductivity of TiC, ZrC and TiN at high temperatures. *Journal of the American Ceramic Society*, **47**, 69–73.

19

Free form microprocessing of ceramic materials by electrodischarge machining

E. Ferraris, K. Liu, J. Peirs, B. Lauwers and D. Reynaerts

This chapter reviews EDM capabilities in the microprocessing of ceramics. It has three main parts: the first part addresses specific issues concerning micro-EDM, including novel architectures for pulse generators, high precision equipment, tool wear and process configuration. In the second part, EDM machining characteristics for a selected group of advanced ceramics including Si_3N_4-, Al_2O_3-, ZrO_2- based composites and silicon/boron carbides (SiC, B_4C) are reviewed and compared with respect to material removal rate, tool electrode wear and surface integrity. The type of material removal mechanism is also discussed. Most attention is paid to Si_3N_4–TiN composite, as the most intensively investigated material at both macro- and microscale. The final part presents the successful machining of typical ceramic microapplications, including spraying micronozzles and turbomachinery components.

19.1 Introduction

Among nonconventional methods commonly used for the free form shaping of ceramic microcomponents, EDM is a very promising and attractive solution [1–23].

EDM is a versatile process, which uses the erosive action of electrical discharges generated between a tool electrode and a workpiece to remove material mainly by melting and evaporation. No mechanical contact between the tool and workpiece is involved during the process, and very complex shapes and fragile components can be easily machined with high accuracy, independently from the hardness of the material. In addition, the recent advances in high frequency spark generators, faster servosystems and milling process strategies further extended EDM capabilities to micro scale applications [24], and smaller features with closer tolerances and improved surface quality can be obtained. As far as ceramics are concerned, however, the electrothermal nature of the process must be taken into account. According to Koenig and Dauw [3], all materials having electrical resistivity lower than $100\,\mu$cm can be machined by EDM. Therefore, spark erosion becomes a feasible technology for ceramic processing once a sufficient electrical conductivity is provided to these materials. Typically, electroconductive secondary phases, such as borides, nitrides or carbides of transition metals (TiB_2, TiN and TiC) are added into Si_3N_4, zirconia or alumina in order to form an electroconductive ceramic composite [4, 5]. This addition may further result in improved toughness, strength, and hardness with respect to the matrix itself. Nevertheless, most of these secondary phases easily oxidize and property degradation may occur because of the different high-temperature thermal stability of the compounds added. In order to evaluate the application limits of ceramic composites in long-term high-temperature applications, several studies have been done on the effect of thermal treatment and environment on the oxidation behaviour, fracture toughness, flexural strength and Young's modulus of the materials developed [25–33]. Certain ceramics also display sufficient electrical conductivity to be machined efficiently by EDM without the addition of conductive phases, as for instance pure silicon carbide and boron carbide [22].

19.2 Micro electrodischarge machining

EDM is a thermoelectric process that erodes material from the work piece by a series of discrete sparks between the workpiece and the tool electrode, both submerged in a dielectric fluid. The sparks, occurring at very high frequency, effectively and continuously remove the workpiece material by melting and evaporation. Then, the dielectric flow allows the ejection of the material debris resolidified within the gap while assuring optimal conditions for subsequent spark generation. This process has found tremendous applications in the last few decades, such as mould and die manufacturing, and small and burr-free hole drilling. Furthermore, EDM has acquired more and more attention in the domain of high precision manufacturing and complex 3D microstructuring [24, 34, 35]. In order to meet the demands in micromanufacturing, novel architectures of pulse generators, advanced CNC systems and special processing strategies have also been developed [36].

Pulse generator

In the case of micro-EDM, the suitable pulse energy is typically in the order of few µJ. High frequency and density of energy is also demanded in order to achieve higher machining speed and improved surface quality. Unlike conventional EDM, where rectangular pulses (namely iso-energetic static pulses) are obtained by switching the DC power on and off, relaxation type pulse generators are typically used in micro-EDM (see Figure 19.1) [37]. In these systems, the energy released for each pulse and the discharge duration are directly controlled by the capacitance of the RC circuit. Thus, very fine pulses with high density and low energy input are feasible.

Equipment

In order to reduce dimensional errors and to improve machining accuracy, the implementation of accurate motion drives and fast reaction servocontrol assuring precise movement of the feeding system and stable control of the sparking gap is also necessary. In addition, high precision measurement systems and special tool holders with very low eccentricity are integrated. Micro-EDM centres with submicrometre resolutions (0.1 µm) and positioning accuracies up to ±1 µm are typically available.

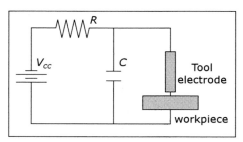

Figure 19.1 Schematic of a RC (resistance–capacitance) pulse generator.

Tool polarity

With the exception of wire-EDM configurations, tool wear is a major issue in micro-EDM. In the time of a short pulse, which is generally the case for microprocessing, the anode melts first, resulting in a larger molten area as compared to the cathode. Therefore, in micro-EDM the polarity of the tool is generally set negative to reduce tool electrode wear. Among the several compensation strategies, tubular electrodes and layer-by-layer strategies (i.e. the methods typically employed in milling EDM) are preferably used instead of complex shaped electrodes (as for die sinking EDM) [38]. Furthermore, besides copper and brass, tungsten carbide microtool electrodes are often used thanks to their easier manufacturing and higher erosion resistance [39].

Process configuration

Micro-EDM is categorized into similar manufacturing configurations as macro-EDM, with the main difference being the scale of the machined geometry. Similarly, microwire EDM (micro W-EDM), die-sinking micro-EDM (micro-S-EDM), micro-EDM milling (micro-M-EDM) and micro-EDM drilling are the generally recognized machining features. Microwire-EDM grinding (WEDG), first introduced by Masuzawa *et al.* [40] (see Figure 19.2 (a)), is widely used for forming very thin rods with very high-aspect ratio as the tool electrode for micro-EDM

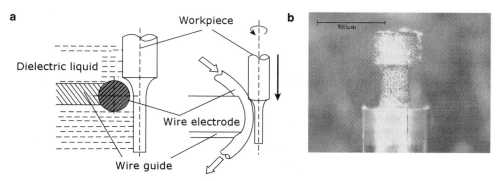

Figure 19.2 (a) Schematic of WEDG concept; (b) example of an axi-symmetric WEDG-ed tool electrode in WC.

Table 19.1 Overview of traditional micro-EDM process capabilities.

Configuration	Geometric complexity	Min. feature size	Max. aspect ratio	Surface quality R_a (µm)
WEDM	2 1/2D	3 µm	~100	0.1--0.2
Die-sinking	3D	~20 µm	~15	0.05–0.3
Milling	3D	~20 µm	10	0.05–1
Drilling	2D	5 µm	~25	0.05–0.3
WEDG	Axi.-sym	3 µm	30	0.8

drilling. It also allows shaping more complex tools (Figure 19.2 (b)) which can be further applied for fabricating special 3D structures and undercut features by using other configurations such as micro-M-EDM and S-EDM [38], [41]. By integrating a WEDG unit into a micro-EDM machine, run-out problems of the microtool electrode from the rotation axis of the spindle can be avoided by dressing the electrode after clamping. Therefore, WEDG is also regarded as an important micro-EDM configuration. An overview of the capabilities of these machining variants is presented in Table 19.1.

19.3 Micromachining performance of erodable ceramic materials

The first work on EDM of advanced ceramics was carried out on SiC at the beginning of the 1980s. After that, a certain number of papers concerning a wider range of ceramic materials were published [1–23].

These works mainly came from the research community and had the objective to determine the influence of process variables (e.g. pulse duration, discharge current, tool polarity, pulse shape etc.) on material removal rate, tool wear, and surface finishing. Owing to the need of achieving enough electrical conductivity, the influence of the chemical composition, namely types and amount of the doping compounds also aroused a great interest among researchers [4, 5, 8, 19]. Among the studies, design of experiments (DOE) techniques (e.g. Taguchi method and factorial designs) were often used as investigation methodologies, since they allow simultaneously analyzing several process variables over the responses of interest by means of a limited number of experimental runs [42]. In addition, investigations mostly focused on the finishing regime owing to the importance that surface integrity and presence of defects, such as cracks, have, especially in ceramics, on strength and wear resistance during service. Typically, evaluations of the surface/subsurface quality and amount of damage were conducted by measurements of the surface roughness and observations of topography and cross-sectional view of the EDM layer. The latter together with analyses of EDAX spectra and debris also provided a deep insight into the material removal mechanisms. Besides traditional melting and evaporation, spalling, chemical decomposition and oxidation have been identified as new leading material removal mechanisms during EDM on ceramics. A change in the material removal mechanism was first supposed by Koenig and Dauw [3], who observed that no direct correlation between cutting rate/surface roughness and electrical properties of the material exists at low conductivity. More recent works [6, 8, 14] state that spalling, i.e. brittle detachment of portions of material from the base, is characteristic of composites (e.g. Al_2O_3–SiCw–TiC, SiC–TiB_2 and Sialon–TiN), in which the thermal expansion mismatch between constituents promotes the formation of large subsurface cracks due to the thermal shock induced by high-energy electrical discharges. On the contrary, decomposition and decomposition/oxidation typically occur during machining of Si_3N_4–TiN composites in oil and de-ionized water as dielectric media, respectively [14, 18–20, 23].

Despite the available information, the literature is still very fragmented and far from providing a comprehensive knowledge, especially concerning micro-EDM characteristics. In addition, literature reports are difficult to compare with respect to the setting parameters, owing to the variety of the equipment and generators used, whose characteristics specifically depend on, and are proprietary knowledge of, the individual manufacturer. Whereas Si_3N_4–TiN is the most intensively studied ceramic composite at both macro- and microscale, very few papers deal with micro-EDM processing of carbide and oxide-based ceramics. In the following, the most relevant aspects and recent progress in micro-EDM processing of ceramics will be reviewed. In case of missing micro-EDM data, discussion will include the most promising results obtained on traditional equipment. The following abbreviations will be used: open voltage (u), discharge current (i_e), discharge duration (t_{on}), pulse interval (t_{off}), duty cycle (i.e. the percentage of discharge duration relative to total cycle time) (η), and dielectric pressure (P).

19.3.1 Silicon nitride based ceramics

Silicon nitride is one of the best ceramic materials for structural applications, mainly because of its special combination of outstanding mechanical and thermal properties, such as high strength, high toughness, high thermal stability, as well as thermal shock and oxidation resistance at high temperature and in aggressive environments. Though preliminary experiments have supported the probability to spark erode insulating Si_3N_4 ceramics by the assisting electrode method [12, 13], typically TiN particles in the percentage of 30–40 vol.% are incorporated into the Si_3N_4 matrix in order to form an erodable ceramic composite (see Table 19.2). Besides the increase in electrical conductivity, this addition has further benefits. The toughness of Si_3Ni_4–TiN has been reported to be enhanced by more than 50% compared to pure Si_3N_4. According to Gucciardi et al. [28], the room temperature strength may improve up to 1000 MPa.

Thanks to the enhanced properties, EDM on Si_3N_4–TiN has drawn considerable attention over the past decades. Liu [10] has shown the electrical conductivity of Si_3N_4–TiN composites as a function of the TiN content. By conducting S-EDM experiments using brass and copper tools, he also observed that the material removal rate (MRR) increases with increasing spark current, and concluded that surface integrity becomes worse with increasing pulse energy due to the formation of larger craters and micro-cracks. He also proved a decrease of the flexural strength after EDM processing and demonstrated the effectiveness of EDM in ceramic microprocessing by successfully machining microdemonstrators,

Table 19.2 Comparison of physical and mechanical properties of commercially available erodable ceramics.

Material Properties	Si_3N_4–TiN Kersit® Saint Gobain	SiC Ekasic D® Waker Ceramic	B_4C TETRABOR® Waker Ceramic	Al_2O_3– SICw– TiC Crystaloy© Ind. Cer. Tech., INC	ZrO_2–TiN KSG20 NTK
Composition (Vol %)	64%–36%	–	–	NA	66%–34%
Density (g/cm³)	3.95	3.1	2.5	2.5	5.83
Vickers Hardness	1600 (HV 1)	2600 (HV 0.5)	3200 (HV 0.5)	2400	1326 (HV 10)
Fracture Toughness (MPa m$^{1/2}$)	7	3.2	3–4	9.6	8.8
Young's Module (GPa)	370	410	450	406	279
Flexural Strength (MPa)	900 (3pt@ RT) 500 (3pt@800°C)	350 (4pt@RT)	400 (4pt@RT)	690 (4pt@RT)	665 (4pt@RT)
Electrical resistivity (Ωcm) @20°C	1.6×10^{-3}	1–5	0.1–10	9×10^{-3}	0.19
Thermal conductivity (W/mK) @RT	28	110	40	63	7.71
Thermal expansion coefficient (10^{-6}/K)	5.7 20°C–1200°C	4–5.8 20°C– 1000°C	4.5–5.6 20°C– 1000°C	7 20°C–1000°C	NA

including microholes (ϕ 75 μm) and square cross-trenches of 200 μm in width [11]. Martin et al. reported that Si_3N_4–TiN composites are more easily machined than tungsten carbide and tool consumption is lower [4]. Lauwers et al. [22] found that S-EDM on Si_3N_4–TiN proceeds faster than M-EDM, but surface quality is worse. High discharge current combined with short pulse duration should be applied for achieving a stable roughing regime. They also investigated the Si_3N_4–TiN material removal mechanisms with variable machining conditions, and proposed some chemical reactions to explain the observed phenomena [14]. Liu et al. [19] has provided optimum TiN content leading to maximum MRR and observed higher cutting rate and worse surface finishing in de-ionized water as compared to oil. In addition, they studied EDM of Si_3N_4–TiN with incorporation of alternative secondary phases, including WC and TiCN, which displayed worse performance in terms of MRR, but better surface finish. Furthermore, Liu *et al.* [18, 20, 23] systematically investigated micro-M-EDM characteristics of Si_3N_4–TiN composites in de-ionized water and oil dielectric using tungsten carbide (WC) tool electrodes and dedicated micro-EDM equipments (i.e. a four-axis AGIE Compact I micro-EDM machine and a SARIX SX-100H-HPM micro-EDM centre specifically developed for micro-M-EDM operations).

As general consideration, Si_3N_4–TiN displays very high machining rate and the tool consumption can be very low. However, the surface integrity remains poor, even after several finishing pulse regimes.

The same trend was also observed when using dedicated micro-EDM equipments. As reported by Liu *et al.*, MRR increases with increasing the peak discharge current. In addition, a high open voltage is necessary owing to the large voltage drop within the material [19]. This latter trend also helps maintaining low relative tool wear (RTW), thus creating favourable roughing machining conditions. Setting the following process parameters u–180 V, i_e 3.2A in de-ionized water and u -100 V, i_e 10A in oil dielectric (relaxation type pulses), MRR increased up to 0.28 mm³/min and 0.305 mm³/min respectively, and RTW remained very low (0.58% vs. 0.05%). In this context, it is interesting to note that the same parameters applied to steel would lead to MRR three times lower and to much higher RTW. On the other hand, a larger sparking gap was observed in de-ionized water (~27 µm). Furthermore, the surface finish was poor in both media (i.e. 2.45 µm Ra in de-ionized water and 2.37 µm Ra in oil dielectric). Lower pulse energy was necessary to achieve better surface quality. As visualized in Figure 19.3, gradually reducing the energy input by varying mainly the open circuit voltage and the discharge current, the surface roughness (SR) was enhanced to a large extent and the best results achieved in de-ionized water and oil-dielectric were comparable (i.e. 0.77 µm and 0.74 µm Ra, respectively). However, no further improvements could be achieved when applying even lower energy amount. Furthermore, the MRR drastically decreased (0.01 mm³/min) and the RTW increased by a factor of 10 (~5%).

For the reasons of high MRR, low RTW and limited surface quality comparing to the machining of steel, an additional contributing material removal mechanism should occur besides typical melting and evaporation. As proposed by Lauwers et al. [14], these aspects are related to oxidation and/or decomposition of Si_3-N_4–TiN induced by the thermal energy of the EDM process. Widely recognized chemical reactions during decomposition and oxidation are as follows:

$$Si_3N_4 \rightarrow 3Si(l) + 2N_2\uparrow$$
$$Si_3N_4 + 3O_2 \rightarrow 3SiO_2 + 2N_2\uparrow$$
$$2TiN + 2O_2 \rightarrow 2TiO_2 + N_2\uparrow$$

The decomposition reaction occurs when the temperature reaches a value of more than 1700°, which is certainly satisfied during EDM machining. In addition, as in principle oxidation reactions may not occur in oil-dielectric, decomposition is characteristic of this

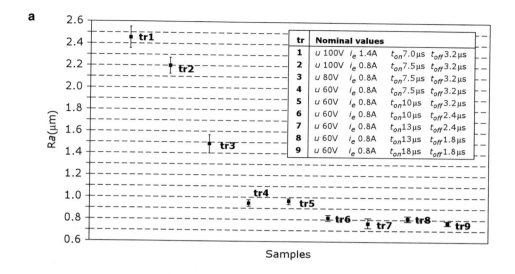

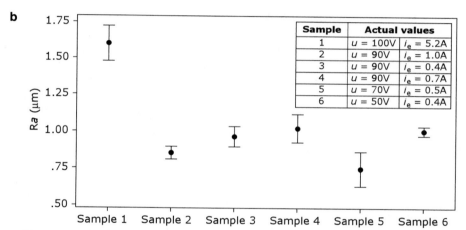

Figure 19.3 Surface quality optimization of Si_3N_4–TiN samples micromachined in (a) de-ionized water, (b) oil dielectric.

latter medium, while decomposition and oxidation simultaneously occur during EDM machining of Si_3N_4–TiN composites in de-ionized water, contributing to a higher MRR response. Accordingly, the analyses of the EDAX spectra of the microprocessed surfaces, compared to the bulk material, revealed a lower content of nitride and silicon on samples machined in oil-dielectric [23]. The content of Ni, Si and Ti further decreased (these latter two elements most probably because of decomposition of the oxide components and escape of the related gases [19]) on samples processed in water-based media, while the content of oxygen increased drastically.

The aforementioned reactions are also responsible for the limited achievable surface finish. The generation of nitrogen gas bubbles prohibits the formation of regular craters and

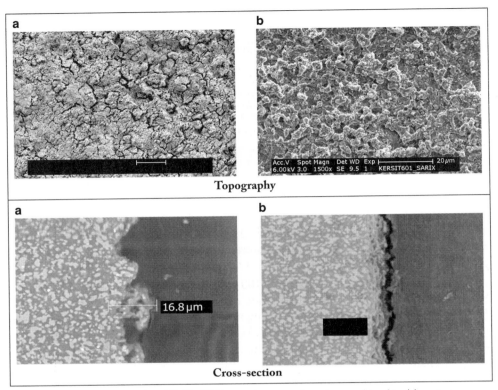

Figure 19.4 SEM topography of Si_3N_4–TiN ceramic composite micromachined in (a) water (Ra: 1.49 µm); (b) in hydrocarbon oil (Ra: 1.60 µm). Cross-section thickness of porous layer of Si3N4–TiN ceramic composite micromachined in water (Ra: 0.77 µm) and in hydrocarbon oil (Ra: 0.74 µm).

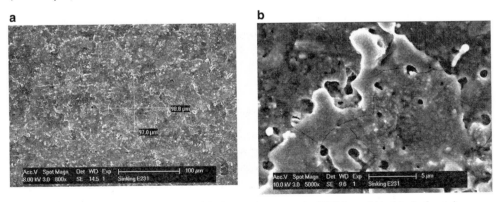

Figure 19.5 SEM topography of Si_3N_4–TiN ceramic composite EDM machined in hydrocarbon oil and with iso-energetic pulses, characterized by high peak of current and short discharge time. Machining conditions: u-80 V, ie 12 A, ton 1.6 µs and toff 12.8 µs. Tool electrodes negatively charged. SR equals to 0.9 µm. (a) characteristic crater dimension of the molten surface; (b) enlarged view of microcracks.

leads to the creation of voids, resulting in a foamy and porous layer loosely attached to the bulk material, as observed on the SEM surface topographies and cross-sectional views of the micro machined samples (Figure 19.4). The thickness of the damaged EDM layer was comparably equal to 10 times Ra. As a consequence of the EDM process, samples having SR ~ 1.45 μm displayed a flexural strength loss of ~15% [20].

On the contrary, the machining of Si_3N_4–TiN composites at fully melting/evaporation regime was recently demonstrated by the authors by applying iso-energetic static pulses, characterized by high peak of current (i_e) and short discharge duration (t_{on}). Experiments were conducted on a die-sinking machine, Roboform 350γ-Charmilles Technology, in hydrocarbon oil as the dielectric. Though this machine is not delicately designed for micromanufacturing purpose, the advanced generator can generate various pulses from iso-energetic static to relaxation type, and it still has modules and technology settings specialized for micro machining. Negatively charged copper infiltrated graphite (Poco EDM-C3) tools were applied as the tool electrodes. As shown in Figure 19.5, the samples machined reveal a characteristic EDM molten surface (machining conditions: u-80 V, i_e 12A, t_{on} 1.6 μs and t_{off} 12.8 μs). No foamy or porous layer is present and the formation of regular craters of ~100 μm in diameter is observed. The achieved SR was ~0.9 μm (Ra), while the thickness of the damaged EDM layer was 7.6 μm. However, the generation of microcracks due to the residual thermal stress induced by the process is significant (see Figure 19.5(b)), and a reduction of the flexural strength of ~20% was measured by conducting 3pt bending tests, which is relative high compared to the strength loss measured on the samples displaying the foamy and porous layer and having even higher SR. The resulting MRR was rather high (1.8 mm³/min), while the RTW increased up to 22%.

In this context, it is interesting to note that comparable results cannot be achieved when using dedicated micro-EDM equipments, namely allowing relaxation type pulses only. In addition, based on the knowledge of the present authors, no similar results are available throughout the literature. In order to deeply investigate the observed phenomena and further characterise the relationship among pulse shape, discharge energy, and material removal mechanism during machining of Si_3–N_4–TiN composites, tailored experiments are currently in progress.

19.3.2 Silicon and boron carbide-based ceramics

SiC and B_4C are among the hardest ceramics (Table 19.2). In addition, they combine a very low density with good mechanical properties even at high temperature. Thanks to the high thermal conductivity and low thermal expansion coefficient, SiC is widely used in high-temperature applications, such as heat exchangers and gas turbines. On the other hand, B_4C is extensively applied as protective arm material and for the development of high-speed bearings, spray- and burning-nozzles due to its outstanding wear resistance.

Although the electrical resistivity is very high, SiC and B_4C have just enough conductivity to allow EDM machining. Nevertheless, their processing can be very difficult and the range of process parameters promoting stable machining conditions is quite limited.

Lauwers *et al.* [22] identified stability regions for SiC and B_4C as a function of i_e, t_{on}, and t_{off} varying within an overall low-energy working space ($i_e < 10A$, $t_{on} < 25\,\mu s$ and $t_{off} < 50\,\mu s$). For the two materials, they also compared M-EDM and S-EDM machining characteristics leading to the development of a combined machining strategy for the efficient processing of 3D shapes (see Section 19.4). All experiments were conducted on a Charmilles Technology M-EDM centre (RoboMill 200®) in oil-dielectric and with positively charged copper tool electrodes. In case of SiC, the results showed a narrow transition zone between stable and unstable machining; a slight change in one of the parameters (e.g. t_{off} from $25\,\mu s$ to $6.4\,\mu s$) is sufficient to promote machining conditions characterized by violent arcing behaviour, low MRR (~$0.8\,mm^3/min$) and bad surface quality (Ra ~$11\,\mu m$). On the contrary, the material showed very good machinability within the region of stability. At low-energy conditions, SR reduced to ~$0.6\,\mu m$ (Ra) for both process configurations, while at higher pulse energy, MRR increased up to $0.2\,mm^3/min$ for S-EDM and $1.1\,mm^3/min$ for M-EDM owing to the better flushing condition (P 1bar) and high rotation speed of the tool electrode (6000 rpm); SR still maintained reasonable (Ra ~$1.2\,\mu m$).

In case of B_4C, more combinations of parameters were possible, and the larger process window allowed designing M-EDM experiments by means of a two-level Taguchi method. i_e, t_{on}, and t_{off} displayed major effects and influence trends very similar to those of steel. The optimized setting for the roughing regime led to a MRR of $5.3\,mm^3/min$ and a SR of $3.8\,\mu m$ (Ra). On the other hand, low-energy characteristics could not be improved so far when applying iso-energetic pulses (SR ~$2\,\mu m$ Ra). Similar to SiC, S-EDM showed worse performance, especially in terms of achievable MRR ($<< 1\,mm^3/min$). For the same material, comparable S-EDM results were obtained by Puertas *et al.* [15] within a slightly different working space. Adjusting further the values of t_{on} and t_{off} (i.e. t_{on} $10\,\mu s$ and η 60%), SR reduced to $1.2\,\mu m$ (Ra) at low-energy conditions ($i_e = 5A$). Nevertheless, MRR remained modest ($0.36\,mm^3/min$) and RTW was ~3.4%. On the contrary, surface finishing drastically improves when dealing with traditional W-EDM relaxation type pulses. As reported by Sanchez *et al.* [9], a SR of ~$0.5\,\mu m$ (Ra) was achievable. Apparently, typical melting and evaporation occur during machining of B_4C.

Besides pure SiC, silicon infiltrated silicon carbide (SiSiC) materials were often investigated. Special attention was paid to the commercially available REFEL F® [1-3, 9, 15-17], which displays exceptional tribological properties. In this context, a systematic study was reported by Puertas and co-workers [15-17], who conducted die-sinking experiments on a ONA-DATIC EDM machine, properly adapted with a stage of capacitors, and at working conditions similar to those generally applied in case of μ-EDM applications, e.g. negative tool polarity and overall low-energy conditions. The material showed a good machinability and characteristics similar to those obtained in case of pure SiC. Within the process window analyzed (i.e. $3\,A < i_e < 5\,A$, $-120\,V < u < -200V$, $20\,\mu s < t_{on} < 70\,\mu s$ and $0.4 < \eta < 0.6$), they observed that MRR rises up to $0.4\,mm^3/min$ when increasing i_e and u to the highest value, while RTW lies within the range of 5-15% and lowers when applying increasing t_{on}. Furthermore, SR lower than $1\,\mu m$ (Ra) was easily obtained by decreasing i_e and t_{on} and, unexpectedly, by increasing u. To achieve high surface quality, the grain size of the sintered materials is very crucial, owing to the nature of the material removal mechanism, which involves separation of individual SiC grains [3, 9].

A further interesting work concerning SiC-based ceramics was reported by Truemans *et al.* [8], who studied material removal by spalling during machining of different ceramics. In particular, they found that SiC–TiB$_2$ flakes may separate from the base under arc-related discharge conditions without inducing catastrophic fracture of the workpiece and limiting the depth of the damage induced, thus providing efficient and reliable roughing conditions.

No mechanical investigations concerning strength loss of silicon carbide or boron carbide ceramics due to EDM machining are available in the literature.

19.3.3 Alumina- and zirconia-based ceramics

Electro erodable Al$_2$O$_3$- and ZrO$_2$-based ceramics have been recently introduced into the market displaying very competitive physical and mechanical characteristics (see Table 19.2). However, no μ-EDM results have been reported so far and investigations are still very fragmented and preliminary [4, 5, 14, 21].

In order to form electroconductive alumina-based composites, TiC compounds are typically added to the ceramic matrix. A pioneering work in the development of this kind of composites was conducted by Martins *et al.* [4], who investigated the EDM feasibility of zirconia-toughened Al$_2$O$_3$–SiCw by adding TiC particles (up to 30 wt.%) having a mean grain size of 1.5 μm. Despite the presence of the whisker phase, the electrical conductivity of the original composite was in fact too low to perform EDM. Meanwhile, higher whisker contents led to difficulties in sintering and the incorporation of an additional phase (i.e. TiC compounds) was necessary. W-EDM experiments showed that EDM machining was possible for a TiC content higher than ~27.5 wt.% (1 μcm of electrical resistvity). However, the obtained composite revealed a decrease in the flexural strength by ~10% with respect to the zirconia-toughened Al$_2$O$_3$–SiCw matrix. More recent studies on EDM of Al$_2$O$_3$–SiCw–TiC and Al$_2$O$_3$–TiC mixed ceramics can be found [5, 14, 21]. Observations revealed that both typical and nontraditional material removal mechanisms occur during machining of these composites, depending on the applied energy conditions.

As shown in Figure 19.6 (a, b), droplets of molten material are clearly visible on EDM surfaces machined with low or medium energy input, independently from the applied process strategies (W-EDM relaxation type pulses in de-ionized water versus S-EDM iso-energetic pulses in oil-dielectric). As a consequence of the prevailing melting and evaporation material removal mechanism, SR easily reduced below 1 μm (Ra) [14]. However, (see Figure 19.6(c)), as very high spark energy conditions are applied, spalling mostly occurs due to the susceptibility of the material to the formation of large cracks, advancing parallel (up to a few hundred μm) and perpendicularly (up to a few μm) to the top surface. Thus, at these energy conditions a degradation of the mechanical properties and a premature catastrophical fracture of the material during service are expected induced by the damaged surface/subsurface layer. As observed by Lee and Deng [43], flexural strength and Weibull's modulus reduce up to 40% and 66%, respectively, in Al$_2$O$_3$–TiC EDM–samples having SR ~2.26 μm (Ra).

topography   cross section

(a) W-EDM (Charmilles ROBOFIL 2000) in de-ionized water: u -160V, i_e <10A, t_{on} 2.8 µs, t_{off} 3 µs

topography   cross section

(b) S-EDM (Agie AGIETRON INNOVATION 2) in oil dielectric: u 250V, i_e 10A, t_{on} 4.2 µs, t_{off} 18 µs

topography   cross section

(c) S-EDM (Agie AGIETRON INNOVATION 2) in oil dielectric: u 250V, i_e 72A, t_{on} 7.5 µs, t_{off} 18 µs

Figure 19.6 Surface topography and cross sections of W-EDM and S-EDM-ed Al_2O_3-SiCw-TiC composite samples machined in de-ionized and oil dielectric at: low (a); medium (b); (c) high energy conditions, respectively.

In order to form electroconductive zirconia-based ceramics, the addition of NbC, TiC and TiN is typical. A pioneering work in this direction was conducted by Matsuo and Oshima [5], who investigated W-EDM characteristics of ZrO_2 doped with NbC, TiC and Cr_3C_2. They observed that there exists an optimum carbide content, which optimizes the MRR. This optimum was about 28 vol% for NbC and 30 vol% for TiC. In addition, regarding ZrO_2–NbC, the maximum

Table 19.3 Overview of EDM capabilities in microprocessing of advanced ceramics.

Ceramic composite	Regime	MMR mm³/min	RTW (%)	SR (Ra µm)	material removal mechanism	Process configuration	Additional information
Si₃N₄–Tin	R*	0.28 vs 0.3	0.58 vs 0.05	2.45 vs 2.37	decomposition/oxidation	µ-M-EDM in water vs oil WC electrode−	relaxation pulse SG ~27 µm
	semi-F*	0.173	0.92	1.26	decomposition	µ-MEDM in oil WC electrode−	relaxation pulse FSR ~15%
	F*	0.01 vs -	5 vs -	0.77 vs 0.74	decomposition/oxidation	µ-M-EDM in water vs oil WC electrode−	relaxation pulse SG ~4 µm
	F	1.8	22	0.9	melting	S-EDM in oil Cu infiltrated graphite−	iso-pulse FSR ~20% SG ~25 µm
SiC	R*	1.1 vs 0.2	-	1.1 vs 1.2	-	M vs S-EDM in oil Copper electrode+	iso-pulse
	F	0.4 vs 0.1	-	0.6-0.5			
SiSiC	R	0.378	8.04	1.86	grain detachment	S-EDM, Copper electrode−	relaxation pulse
	F	0.024	8.92	0.75			
B4C	R*	5.3	-	3.8	melting	M-EDM in oil Copper electrode+	iso-pulse
	F	0.36	3.4	1.2	melting	S-EDM Copper electrode+	W-EDM SR~0.5 µm
Al2O3-SICw-TiC	R	-	-	-	spalling	S-EDM in oil	iso-pulse FSR 40%
	semi-F/F	-	-	<1	melting	W/S-EDM in water/oil	relaxation/iso-pulse
ZrO2-TiN	R	30.3 mm²/min	-	2.71	melting	W-EDM in water	relaxation pulse
	F	low	medium-high	0.6	melting	W-EDM in water	FSR 15%

* optimized values, FSR = flexural strength reduction, R = roughing, F = finishing, SG = sparking gap, +/− = tool polarity

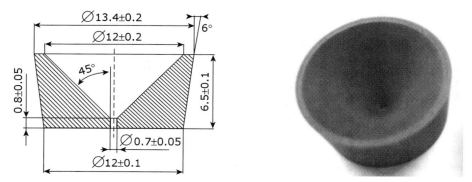

Figure 19.7 EDM example of a powder spraying nozzle produced in B_4C (dimensions in mm).

MRR was nearly the same (i.e. 40 mm²/mm) when varying the t_{on} and t_{off} parameter, while maintaining constant their ratio. On the other hand, for the optimized composite surface finish was poor and brittle fractured surfaces were mostly observed. EDM samples with increasing carbide content (up to 50 vol%) tended to show more characteristic molten surfaces, and SR roughness easily reduced below 1 μm. Studies on W-EDM of ZrO_2–TiN have been reported by Lauwers *et al.* [14]. Investigations mostly focused on the understanding of the prevailing material removal mechanism. From the experiments, there was no evidence of removal processes different from melting and evaporation. Based on the latest measurements conducted by the authors, SR of ~0.6 μm is achievable, while the flexural strength reduces by ~15%. Table 19.3 reviews the EDM process capabilities on ceramic composites.

19.4 Applications

Thanks to their intrinsic characteristics, ceramic composites are generally applied where high strength, hardness, wear and/or high temperature resistance are required. Typical applications include dies, cutting tools, bearing parts, nozzles and turbine blades.

Among them, nozzles for powder spraying are examples of components, which are subject to wear. Whereas traditional nozzles are made from tungsten carbide, Figure 19.7 shows a sample made from boron carbide, which displays exceptional wear resistance properties due to the very high Vickers hardness of 3200 (HV0.5) (see Table 19.2).

The component was produced by means of a combined milling–die sinking EDM strategy [22], which allowed to halve the total machining effort compared to pure S-EDM (see also Section 19.3.2). Specifically, the bulk of the material was removed by M-EDM using 2 mm diameter tubular copper electrodes and a layer-by-layer strategy. Finishing was done by three S-EDM steps using two copper and one tungsten–copper electrodes. Finally, relaxation type pulses were applied to achieve high dimensional accuracy and optimal surface quality. The total process took 330 minutes, 170 of which for M-EDM. The nozzles were all made within the required tolerance and the lifetime during application (powder spraying) was three times longer compared to traditional nozzles.

a    b

Figure 19.8 Si$_3$N$_4$–TiN ceramic composite impeller for an ultra-miniature gasturbine (diameter 20 mm) (a), tool approach during die sinking (b).

Ceramic materials are also interesting for turbomachinery, especially for hot components, such as combustors and turbine blades. Turbine impellers and blades are the most critical components since they are subjected to both high temperatures and centrifugal loads. Figure 19.8(a) shows a 20 mm diameter impeller made from a ceramic composite, intended for the development of a miniature gas turbine generating 1 kW of electrical power [44]. Due to the high rotational speed (500,000 rpm), the component is expected to endure a centrifugal stress of 580 MPa at full speed (max. principal stress) and an inlet temperature up to 1200 K. These thermomechanical working conditions make the material choice very crucial. The Kersit 601 Si$_3$N$_4$–TiN ceramic composite was here selected thanks to the favourable strength to density ratio (see Table 19.2) as well as to its low electrical resistivity, which allowed the material processing by EDM. Whereas turbines of regular size are assembled from separate blades and a carrier disc, this method is not feasible for ultraminiature gas turbines with impellers of only a few centimetres in diameter. Such small impellers have to be made as a single piece (monolithic), but as aerodynamics requires 3D blade geometries, the shape becomes very complex. Furthermore, the extreme operational conditions put strict requirements on shape accuracy and surface integrity of the final component, putting great challenge also to the application of near-net shapes technologies and green machining.

The blades were produced by S-EDM on a four-axis Roboform 350 γ die sinking EDM centre, using copper-infiltrated graphite electrodes (Poco EDM-C3 with grain size <5 μm). Each electrode was machined with the negative shape of the cavity between two adjacent blades, with subtraction of an undersize determined by the sparking gap and material allowance. Specifically, the electrodes were produced by micromilling (conventional cutting) on a five-axis Kern MMP 2522 micromilling centre, with milling tools down to 0.5 mm in diameter. Depending on the applied strategy, 10–16 electrodes were sunk into the Kersit blank to achieve the desired blade shapes (see Figure 19.8(b)). Total machining and handling time took around 20 and 13 hours for milling and S-EDM, respectively, even though they can be done in parallel. Test samples machined with the same S-EDM parameters showed a SR of about 0.9 μm (Ra) and a three-point flexural strength of 729 ± 61 MPa at room temperature (see also Section 19.3.1), which meets the requirements of the envisaged application. In addition, the dimensional control of the manufactured turbine was conducted on a Mitutoyo FN 905 CMM system (coordinate measurements

Figure 19.9 W-EDM-ed Al_2O_3–TiN ceramic gear 1 mm in diameter.

machine), revealing a geometric accuracy within $\pm 20\,\mu m$.

As further application, Figure 19.9 shows a 1 mm diameter gear machined by W-EDM in an Al_2O_3–TiN ceramic composite (equal volume percentages). A broken edge on the left-most tooth illustrates a well-known clamping problem for W-EDM: at the end of a closed external contour, the part breaks off in an uncontrolled way leaving a burr or broken edge. For larger steel parts this problem is often solved by fixing the semi-finished part with magnets; the same concept is inapplicable on ceramics. On the other hand, fixing microcomponents with glue might be problematic as the glue tends to flow into already cut grooves, thus forming a nonconductive obstruction for the EDM wire.

Conclusions

This chapter has systematically reviewed EDM capabilities in microprocessing of advanced ceramics. The machining characteristics of a selected group of ceramic composites, including Si_3N_4-, Al_2O_3-, ZrO2- based ceramics and silicon-, boron carbides are discussed and compared with respect to MRR, RTW, surface integrity and material removal mechanism. Particular attention has been paid to the Si_3N_4–TiN composite, as being the most intensively investigated material at both macro- and microscale owing to the enhanced mechanical behaviour resulted from the addition of the secondary electroconductive phase. In addition, discussion mostly focused on the finishing regime owing to the importance that surface integrity and presence of defects, such as cracks, have, especially in ceramics, on strength and wear resistance during service.

Si_3N_4–TiN displays very high machining rate and low tool wear. However, the surface integrity remains poor, even after several finishing pulse regimes. The observed machining characteristics are mainly due to the leading material removal mechanism, i.e. the decomposition of Si_3N_4 at high temperatures (1700°C). The generation of nitrogen gas bubbles contributes to faster machining, but it is also responsible for the formation of a foamy and porous layer, which results in a limited surface quality (Ra ~0.8 μm). This phenomenon is characteristic of oil type dielectrics when applying relaxation type pulses, and it is further enhanced during machining of Si_3N_4–TiN composites in de-ionized water due to the concurrent occurrence of decomposition and oxidation of the material compounds. On the other side, no foamy structures are observed and the formation of regular craters is obtained in hydrocarbon oil-dielectrics when applying iso-energetic pulses with high density of energy. However, the achievable surface integrity still remains crucial. Due to the thermal impact of the processes, a large amount of microcracks forms at the surface considerably reducing the attained final strength.

SiC/SiSiC, B_4C and ZrO_2–TiN composites display better surface finish conditions as compared to Si_3N_4–TiN. Similar to steel, SR lower than 1 μm (Ra) can be easily obtained by decreasing the pulse energy. Apparently, typical melting and evaporation occur during machining B_4C and ZrO_2–TiN composites, while the grain size of the sintered SiC/SiSiC materials seems to be very crucial in order to achieve high surface quality, owing to the nature of the material removal mechanism involving the separation of individual SiC grains. Besides the achievable MRR remains modest and it further decreases when applying relaxation type pulses (order of magnitude 10^{-1} and 10^{-2} mm^3/min with iso-energetic and relaxation type pulses at finishing regime, respectively).

Despite the narrow region of stability, the efficient machining of pure SiC and B_4C is confirmed at both roughing and finishing regimes. As far as B_4C is concerned, high competitive machining speed (~5 mm^3/min) can be obtained within the optimized roughing regime. Its successful processing into industrial spraying micronozzles based on a combination of milling and die-sinking EDM is also demonstrated.

The machining characteristics of Al_2O_3-based ceramics still need to be further investigated. Based on the preliminary observations, material removal mechanisms ranging from spalling to melting and evaporation may occur during processing of these composites, depending on the applied energy conditions. The attained final strength strongly depends on the obtained surface integrity and may drastically reduce (up to 40%).

References

[1]   Noble C.F., Aimal A.J. and Green A.J. 1983. Electro-discharge machining of silicon carbide. *International Symposium on Electro-Machining, ISEM VII*, 305.

[2]   Iwanek H., Grathwohl G., Hamminger R. and Brugger N. 1986. Machining of ceramics by different methods. *International Symposium Ceramic Materials and Components for Engineering*, **417**.

[3]   Koenig W. and Dauw D.F. 1988. EDM-Future steps towards the machining of ceramics. *Annals of the CIRP*, **37**(2), 623.

[4]   Martins C., Calles B., Vivier P. and Mathieu P. 1989. Electrical discharge machinable ceramic composites. *Materials Science and Engineering*, **A109**, 351.

[5]   Matsuo T. and Oshima E. 1992. Investigation on the optimum carbide content and machining condition for wire-EDM of zirconia ceramics. *Annals of the CIRP*, **41**(1), 231.

[6]   Gadalla A.M. 1992. Thermal spalling during electro-discharge machining of advanced ceramics and ceramic-ceramic composites. *Proceedings of Machining of Composite Materials Symposium*, 151.

[7]   Yan B.H., Huang F.Y., Chow H.M. and Tsai J.Y. 1999. Micro-hole machining of carbide by electrical discharge machining. *Journal of Materials Processing Technology*, **87**, 139.

[8] Trueman C.S. and Huddleston J. 2000. Material removal by spalling during EDM of ceramics. *Journal of the European Ceramic Society*, **20**, 1629.

[9] Sanchez J.A., Cabanes I., Lopez L.N. and Lamikiz A. 2001. Development of optimum electrodischarge machining technology for advanced ceramics. *Journal of Advanced Manufacturing Technology*, **18**, 897.

[10] Liu C.C. 2003. Microstructure and tool electrode erosion in EDMed of TiN-Si$_3$N$_4$ composites. *Materials Science and Engineering*, **A363**, 221.

[11] Liu C.C. 2003. Effect of the electrical discharge machining on strength and reliability of TiN-Si$_3$N$_4$ composites. *Cer Int.*, **29**, 679.

[12] Muttamara A., Fukuzawa Y., Mohri N. and Tani T. 2003. Probability of precision micro-machining of insulating Si$_3$-N$_4$ ceramics by EDM. *Journal of Materials Processing Technology*, **140**, 243.

[13] Tani T., Fukuzawaq Y., Mohri N., Saito N. and Okada M. 2004. Machining phenomena in WEDM of insulating ceramics. *Journal of Materials Processing Technology*, **149**, 124.

[14] Lauwers B., Kruth J.P., Liu W., Eeraerts W., Schacht B. and Bleys P. 2004. Investigation of the material removal mechanisms in EDM of composite ceramic materials. *Journal of Materials Processing Technology*, **149**, 347.

[15] Puertas I. and Luis C.J. 2004. A study on the electrical discharge machining of conductive ceramics. *Journal of Materials Processing Technology*, **153–154**, 1033.

[16] Luis C.J., Puertas I. and Villa G. 2005. Material removal rate and electrode wear study on the EDM of silicon carbide. *Journal of Materials Processing Technology*, **164–165**, 889.

[17] Puertas I., Luis C.J., Villa G. 2005. Spacing roughness parameters study on the EDM of silicon carbide. *Journal of Materials Processing Technology*, **164–165**, 1590.

[18] Liu K., Ferraris E., Peirs J., Lauwers B. and Reynaerts D. 2007. Process investigation of precision micro-machining of Si3N4-TiN ceramic composites by Electrical Discharge Machining (EDM). *International Symposium on Electro-Machining, ISEM XV*, **221**.

[19] Liu W., Brans K., Kruth J.P. and Lauwers B., EDM of Si$_3$-N$_4$-based electrical conductive ceramics. *International Symposium on Electro-Machining, ISEM XV*, **11**.

[20] Liu K., Ferraris E., Peirs J., Lauwers B. and Reynaerts D. 2007. Process optimization and surface quality improvement of micro EDM-milling of Si$_3$N$_4$-TiN ceramic composites. *18th Micro-Mechanics Workshop Europe*, 16–18 September 2007, Guimaraes, Portugal, **321**.

[21] Chiang K.T. 2007. Modelling and analysis of the effects of machining parameters on the performance characteristics in the EDM process of Al$_2$O$_3$+TiC mixed ceramic. *International Journal of Advanced Manufacturing Technology*.

[22] Lauwers B., Kruth J.P. and Brans K. 2007. Development of technology and strategies for the machining of ceramic components by sinking and milling EDM. *Annals of the CIRP*, **56**(1), 225.

[23] Liu K., Ferraris E., Peirs J., Lauwers B. and Reynaerts D. 2008. Micro-EDM process investigation of Si3N4–TiN ceramic composites for the development of micro fuel-based power units. *International Journal of Manufacturing Research*, **3**(1), 27.

[24] Pham D.T., Dimov S.S., Bigot S., Ivanov A. and Popov K. 2004. Micro-EDM recent development and research issues. *Journal of Materials Processing Technology*, **149**, 50.

[25] Beaume F.D., Cutard T., Frety N. and Levaillant C. 2002. Oxidation of a silicon nitride-titanium nitride composite: Microstructural investigations and phenomenological modelling. *Journal of the American Ceramic Society*, **85**, 1860.

[26] Bracisiewicz M., Medri V. and Bellosi A. 2002. Factors inducing degradation of properties after long term oxidation of Si3N4–TiN electroconductive composites. *Applied Surface Science*, **202**, 139.

[27] Klein R., Medri V., Desmaison-Brut M., Bellosi A. and Desmaison J. 2003. Influence of additives content on the high temperature oxidation of silicon nitride based composite. 2003. *Journal of the European Ceramic Society*, **23**, 603.

[28] Gucciardi S., Melandri C., Medri V. and Bollosi A. 2003. Effects of testing temperature and thermal treatments on some mechanical properties of a Si3N4_/TiN composite. *Materials Science and Engineering*, **A360**, 35.

[29] Jian D., Vanmeensel K., Vleugels J. and Van der Biest O. 2004. Si_3N_4-based composites with micron and nano-sized $TiC_{0.5}N_{0.5}$ particles. *Sil. Ind. Special Issues*, **69**(7–8), 267.

[30] Medri V., Bracisiewicz M., Krnel K., Winterhalter F. and Bellosi A. 2005. Degradation of mechanical and electrical properties after long-term oxidation and corrosion of non-oxide structural ceramic composites. *Journal of the European Ceramic Society*, **25**, 1723.

[31] Mazerolles L., Feldhoff A., Trichet M.F. and Ricoult M.B. 2005. Oxidation behaviour of Si3N4–TiN ceramics under dry and humid air at high temperature. *Journal of the European Ceramic Society*, **25**, 1743.

[32] Brach M., Sciti D., Balbo A. and Bellosi A. 2005. A short-term oxidation of a ternary composite in the system AlN-SiC-ZrB_2. *Journal of the European Ceramic Society*, **25**, 1771.

[33] Lavrenko V.A., Desmaison J., Panasyuk A.D. and Desmainson-Brut M., Fenard E. 2005. High-temperature oxidation of AlN–SiC–TiB_2 ceramics in air. *Journal of the European Ceramic Society*, **25**, 1781.

[34] Reynaerts D., Van Brussel H., Meeusen W., Driesen W. and Dierickx V. 2001. Micro-Electro Discharge machining: review and applications. *Keynote Paper Presented at Third Chemnitz colloquium on production technology*, **8**.

[35] Fleischer J., Masuzawa T., Schmidt J. and Knoll M. 2004. New applications for Micro-EDM. *Journal of Materials Processing Technology*, **149**, 246.

[36] Masuzawa T. 2001. Micro-EDM. *Proceedings of the 13th International Symposium for Electromachining ISEM XII*, **3**.

[37] Liu K, Ferraris E., Peirs J., Lauwers B. and Reynaerts D. 2007. Process capabilities of Micro-EDM and its applications. *Proceedings of the Third International Conference on Multi-Material Micro Manufacture*, **267**.

[38] Meeusen W. 2003. In *Micro-electro-discharge machining: Technology, computer-aided design and manufacturing and applications*. PhD thesis, Katholieke Universiteit Leuven, Belgium.

[39] Bertholds A., Heeren P.H., Larsson O., Van Brussel H., Beuret C. and Reynaerts D. 1996. Microstructuring of silicon by electro-discharge machining (EDM) part I: Theory. *Proceedings of Eurosensors X*, **251**.

[40] Masuzawa T., Fujino M. and Kobayashi K. 1985. Wire Electro Discharge Grinding for micromachining. *Annals of the Cirp*, 34(1).

[41] Ferraris E. 2006. In *Development of complex micro-devices: design, fabrication technology, reliability study*. PhD thesis, Univerisity of Tor Vergata Rome, Italy.

[42] Montgomery D.C. 2001. In *Design and Analysis of Experiments*, 5th edn., Wiley, NewYork, New York, USA.

[43] Lee T. and Deng J. 2002. Mechanical surface treatments of electro-discharge machined (EDMed) ceramic composite for improved strength and reliability. *Journal of the European Ceramic Society*, **22**, 545.

[44] Ferraris E., Liu K., Peirs J., Bleys B. and Reynaerts D. 2007. Production of a miniature Si_3Ni_4-TiN ceramic turbine impeller by die-sinking EDM. *7th International Workshop on Micro and Nanotechnology for Power Generation and Energy Conversion Applications*, **229**.

20

Layer manufacturing of ceramic microcomponents

P. Johander and K. Lindqvist

This chapter deals with layer manufacturing, a topic that has been extensively studied during the last 20 years. However, development of layer manufacturing using ceramic materials for producing microcomponents has not, so far, been specifically studied. Layer manufacturing using stereolithography is in principle the only technique that has been used for dedicated development of micromanufacturing of ceramic components. Therefore, this chapter focuses on the possibilities and limitations of existing layer manufacturing techniques which possess the potential to be used for manufacturing of ceramic microcomponents.

20.1 Introduction

During the last 20 years tremendous progress has been made in the development of user-friendly CAD programs. The technique to build objects by solid freeform fabrication (SFF) has been dependent on this development and also on the operational availability of computer controlled devices enabling the generation of 3D structures. The first SFF machine to be used in Europe was a stereolithographic equipment for manufacturing polymer products, which was installed at Electrolux R&D in 1989. Since then, the technique has been developed to include many methods and materials.

Common to all SFF methods is that the object to be formed must be defined in a CAD file. The geometry of the object is "sliced" by a special software. The pixel image of each layer will control the buildup sequence, which will be repeated layer by layer. The component is fabricated into the 3D shape without the use of a mould. There are several advantages of rapid prototyping and SFF. Visual and functional prototypes can readily be made without moulds and tooling and, by modifying the CAD file, revisions can easily be done. For example, during the development phase of an aero engine, casting cores for precision investment casting of turbine vanes can be made by SFF instead of by injection moulding. Modelling and performance can hereby be tested and verified at a lower cost compared to conventional forming methods. Structural features which are impossible to realize by conventional techniques can be made and unique (individual) components, e.g. dental components, can easily be manufactured using an SFF technique. The possibility of developing and designing new materials, functionally graded materials, composites, etc. increases dramatically with the use of SFF.

In general, the process for producing layer manufactured products can be divided into three process steps: layer deposition, solidification and rinsing. A general overview of the steps will be given and a more detailed description of the methods will follow:

- ink-jet printing;
- stereolithography (STL);
- selective laser sintering (SLS); and
- a few other techniques.

20.2 Basic principles

The results from the SFF process are very much dependent on the data delivered by the CAD system. The information from the CAD system is transferred to the SFF equipment software as a STL file. In this file the CAD model is broken down into triangles and vectors, see Figure 20.1, thus simplifying the description of the geometry. The operator working with the CAD information transfer, must determine the accuracy of the geometrical approximation performed by the STL file. The decision is a balancing act between high geometric accuracy, resulting in large STL files that take a long time to process, and less accuracy, which means fewer facets in the object geometry and shorter processing times. The accuracy of the layer manufacturing machine also has to be considered. Thus, the final STL file will be an approximation of the CAD model [1].

The slicing of the model is performed by a special software, which controls the SFF equipment, and the slice thickness is one of the many parameters that must be set before running the process. Varying the slice thickness along the geometry would be a method of increasing the surface finish but suitable software for handling such a procedure is not yet available.

With layer manufacturing it is possible to manufacture very complex 3D structures. It is easy to manufacture hollow structures in one piece, which would be impossible using other methods. An example is the cylinder top shown in Figure 20.2. Thus, it is very easy to design and manufacture very complex 3D structures for micromechanical system integration in various applications.

The structural information in each layer is transferred to the actual physical layer by means of light, laser or ink-jet printing. The physical layer is normally a metal or a ceramic

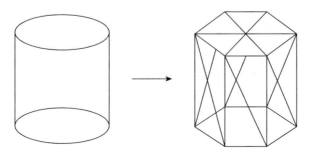

Figure 20.1 CAD model broken down into triangles and vectors.

Figure 20.2 Optical image of a downscaled model of cylinder top of a Volvo 850 manufactured in zirconia at Fcubic.

powder that is spread out with a powder dispenser but it could also be deposited in the form of a ceramic suspension by spraying or doctor blading.

The layer manufacturing process could be divided into three main steps. In the first step the ceramic layer is deposited and in the second step the structural information is printed in the layer by light, laser or ink-jet printing. The last step is the lowering of the production platform and then the process is repeated (see Figure 20.3). If the ceramic is deposited as a suspension, an additional drying step is required.

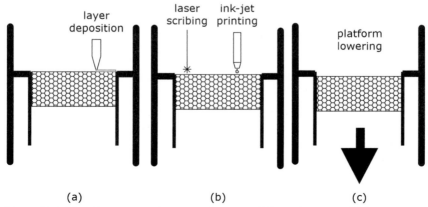

Figure 20.3 Three basic steps in layer manufacturing: (a) layer deposition; (b) laser scribing or ink-jet printing; (c) platform lowering.

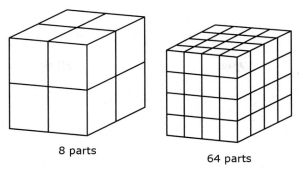

8 parts

64 parts

Figure 20.4 Scale effect is favourable for direct manufacturing of microparts.

20.3 Direct manufacturing

Layer manufacturing is an SFF technique mostly used for prototype manufacturing in polymers. In recent years it has also been applied as a means for direct (rapid) production of parts, especially for production of small parts. The scaling effect is favourable to the production of small parts as the number of parts that could be produced in one batch increases by the cube when the part size is halved (see Figure 20.4).

One *definition of direct manufacturing is:* the full geometric and functional information of the product is contained in a digital model, which could instantly be manufactured in the production equipment.

- The advantages of direct manufacturing are:
- production close to customer;
- production on demand;
- no storage of spare parts;
- easy upgrading;
- less material consumption;
- lower energy consumption;
- more functions integrated;
- more complex geometries; and
- favourable to production of microcomponents.

20.4 Manufacturing of microparts

So far, layer manufacturing has mainly been used to produce macro/mesoscopic parts. However, several papers on the production of ceramic microparts have now been published [2–5]. Deposition method and solidification process decide how small a feature is possible to manufacture by layer manufacturing. The highest resolution is obtained by STL, followed by SLS and then ink-jet printing (see Tables 20.1, 20.2 and 20.4 later in this chapter). The layer thickness is one important parameter that determines the resolution. In traditional layer manufacturing the layer thickness is a trade-off between the resolution and the manufacturing speed. In the solidification process the entire layer must be solidified in order

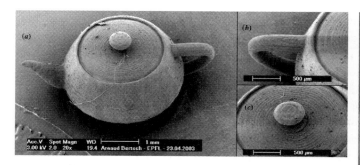
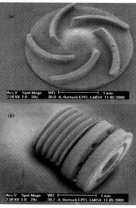

Figure 20.5 Example of parts manufactured with μ-STL.

to avoid a weak interface between the layers [6, 7]. The resolution in each layer is determined by the spot size of the laser or the drop size of the ink-jet printer. The materials also limit the smallest feature size that could be manufactured. The particle size limits the resolution when dry powder is used. Light diffraction in ceramic suspensions affects the feature size in STL of ceramic parts. Figure 20.5 shows some examples of parts manufactured by means of μ-STL.

20.5 Layer deposition

Dry powder-based layer manufacturing will result in porous parts with a porosity in the range 40–60 %. Higher density and thus lower porosity could be obtained by reducing the powder diameter. However, if the powder diameter is reduced the attractive forces will increase between the ceramic particles (relative to other forces acting on the particles) and it will be more difficult to rake the ceramic powder into a uniform layer. The density can be improved or the porosity can be decreased by a postsintering process or by infiltration.

In ceramics manufacturing the most common way to disperse the ceramic particles in a liquid is to add a surfactant. Submicron and nanoparticles can thus be densely packed without agglomerates forming. The viscosity of the ceramic suspension is critical in order to obtain a uniform layer.

This approach has been used for deposition of slurry layers in layer manufacturing by various deposition techniques such as dispensing [8], spraying, [9] and tape casting [10]. In the STL process the platform is lowered into the ceramic slurry, which will flow over the platform. Layer deposition of suspensions requires a drying step in the manufacturing cycle.

20.6 Solidification process

The transfer of the structural information to the top layer includes some sort of solidification process. After the transfer of the structural information the layer consists of the actual solidified part and the nonsolidified material that acts as support. The two main methods for transferring the structural information are to use light (including laser) or ink-jet printing, but there are various methods for solidifying the parts. In traditional STL a liquid monomer is photopolymerized by an ultraviolet light source [2]. Similar principles could be used to solidify ceramic slurries that contain a monomer and a photo-initiator [11]. The curing could also be done in the subsequent step, in which the part is developed. This technique is used in the fcubic process developed at Swerea IVF [12].

The regions of the powder bed, which are not solidified by the ink, act as temporary support material. Final fixation of the part is done during a postprint heat treatment, where also a sintering process takes place. The fcubic process differs from other freeform fabrication (FFF) processes as the part is actually created in this postprocess. In the 3D printing (3DP) process a polymeric binder is printed that glues the particles together [13].

20.7 Rinsing process

As the support material has to be removed in the post processing it is important that this can easily be accomplished. Rinsing dry powder is very easy as the powder is free flowing and very little energy is required to remove the excess powder. Removing dried dispersed support material is considerably more complicated. The excess ceramic material must be redispersed and this demands a lot more from the binder system. The formation of strong chemical bonds between the particles should be avoided by adjusting the slurry pH to a level where the particles are stable in the slurry. The solidified binder should be stable during the redispersion while the dried slurry bed should be easy to disperse in a liquid, preferably water [14]. Stiction problems are related to the manufacturing of 3D microparts by μ-STL. As the parts are wet when lifted out from the μ-STL machine, too narrow structures will be drawn towards each other due to capillary forces, which could completely ruin a microstructure. Drying by sublimation instead of by evaporation will diminish this problem considerably [15].

20.8 Layer manufacturing by ink-jet printing

The SFF process 3DP was developed at MIT and many papers on the development of this process have been published [13–23]. The development at MIT was halted around 2002 and not much has been published since. The 3DP process is based on ink-jet printing of a binder onto a thin layer of powder material. The process was developed both for free-flowing dry powders and slurry-based layer materials using drop-on-demand (DoD) ink-jet printers or continuous ink-jet printers.

Another SFF technique under development is direct ceramic ink-jet printing (DCIJP) [16]. Ink-jet printers with nozzle diameters in the range 20–75 μm are used to print diluted ceramic suspensions. This technique has limitations regarding 3D complexity if not a sacrificial support material is printed at the same time as the structural layer.

Structural ceramic components were initially fabricated by means of 3DP using spray-dried 50 μm granules [17]. Spray-dried powders are easy to spread into uniform layers but printed components will have packing densities of less than 35% [18]. The slurry-based

3DP process was therefore developed to overcome the difficulties of spreading fine submicron powders and thus increase the density (~ 60%), and also make it possible to deposit layers as thin as 10 μm [9].

In the 3DP process the part generation starts by spraying a well-dispersed suspension containing water, dispersing agent, 30–35 vol% ceramic powder and polyethylene glycol (PEG) through a 127 μm nozzle to form the powder bed. The wet layer is dried using a lamp. The binder, a 10% water-based solution of a styrene acrylic copolymer, can then be printed with ink jet. The finished but still embedded parts are placed in an oven to thermally cross-link the binder, which is cured at 150°C for 1 h in an inert environment to prevent degradation of PEG. The unprinted region of the powder bed can be redispersed in water to retrieve the part.

The selection of binder is a critical issue for the 3DP process. The binder must penetrate the powder bed and have sufficient strength to survive the part retrieval process. It also has to be insoluble in water to survive the redispersion of the unprinted powder. Important properties are surface tension, viscosity, conductivity, pH and chemical stability. Figure 20.6 shows a schematic picture of the penetration behaviour of three different binder systems on a dried slurry bed. A dry powder bed, formed from spray-dried granules, may easily be penetrated by all binder systems due to the high porosity. The pores are much smaller in

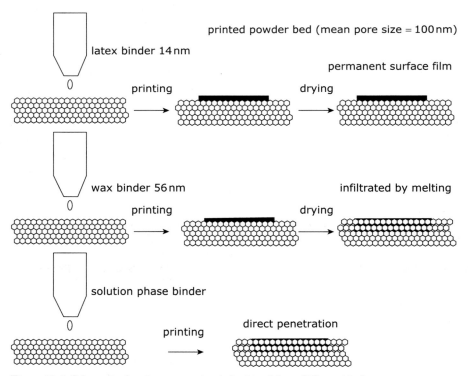

Figure 20.6 Schematic showing penetration behavior of three different binder systems.

a dried slurry bed and the latex and the wax form a surface film when printed, while the solution phase binder (styrene-acrylic copolymer) easily penetrates the bed formed from a suspension [19]. The solution phase binder has to be cured at an elevated temperature to strengthen the part and to make the binder insoluble in water. The spraying (casting) of the slurry onto the powder bed is another critical issue. The areas printed with binder are more hydrophobic compared to the unprinted areas. Due to difference in wetting behaviour the liquid slurry tends to migrate towards the unprinted region. The wetting behaviour could be improved by decreasing the binder hydrophobicity or spraying the powder bed with a wetting agent before deposition of the next layer [20]. The strength of the particle (particle contact in the unprinted powder bed) is a critical property for controlling the redispersion behaviour during the part retrieval. The low molecular weight of PEG reduces the strength of the particle–particle contacts. The PEG will be left at the necks of the particles as the slurry dries. PEG is soluble in water and the chains will extend as liquid (water) enters the powder bed during the redispersion process, acting as an additional force pushing the particles apart. The process is a result of the osmotic pressure caused by the PEG.

20.9 Resolution limiting factors

The resolution of the ink-jet printed structures depends on several material properties such as porosity, surface energy of the powder bed and ink properties, such as viscosity and surface tension. Table 20.1 compares some of the ink-jet systems on the market. The process parameter that has the largest influence on the resolution is the drop size.

A single droplet was printed with a HP 51626A head with a 140 pl drop volume. The obtained diameter of the print was 200 μm [19]. A modern ink-jet printer could today print down to 1–3 pl drops so the potential for improvements of the resolution is large.

The line width is directly related to the viscosity and surface tension of the ink. An increase of the viscosity from 1.2 to 4.2 mPas reduces the line width from 800 to 400 μm and in the same way, if the surface tension increases from 25 to 50 mN/m the line width decreases from 800 to 400 μm [19]. The line width is also influenced by the mean pore size.

Table 20.1 Typical performance data for an ink-jet printing system working with transparent polymers.

Research group company	Ink-jet head	Drop volume Pl	Resolution	Layer thickness μm	Smallest feature size	Speed	Reference
Sachs	HP DOD	140	180 μm	50			19
fcubic	HP DOD	30	40 μm	40	300 μm	10 sec/ part	21
Z-Corp			40 μm	100		25–50 mm per hour	22

A powder bed with a mean pore size of 0.049 μm resulted in a line width of 650 μm while a powder bed with a pore size of 0.221 μm resulted in a line width of 270 μm. The printed binder is drawn into the powder bed by capillary forces. Concerning a powder bed with larger pores the binder, after application, immediately penetrates the pores and does not spread sidewise into the powder bed [19].

Parts of various materials have been formed by the 3DP slurry-based method: alumina and silicon nitride parts were formed by Grau *et al.* [17], dielectric ceramic components [20] and zirconia-toughened alumina [23]. As mentioned above, very thin layers (10 μm) can be deposited with 3DP slurry-based processing and a very good surface finish can be achieved. However, one single line had, at least during the studies of binder selection, a width of 150 μm. This value must be reduced to increase the accuracy of the process when small components are shaped. The 3DP process has the ability to produce parts with locally controlled composition by printing with different inks.

20.10 Stereolithography

STL was one of the FFF technologies. The technique is based on the polymerization of liquid monomers when exposed to UV radiation from a laser [24]. Figure 20.7 shows a schematic drawing of the STL process. As for other SFF methods, a 3D computer image is sliced into cross-sectional layers of a given thickness (150–200 μm). Using the information of each slice, laser radiation is scanned on the surface of the liquid monomer to cure the layer. The liquid monomer flows across the first polymerized layer, and to ensure that the second layer of monomer is of the desired thickness, a recoat blade moves across the surface. The laser scans this new surface, thereby polymerizing the second layer. This process is repeated many times until the part is finished.

To be able to use STL for direct buildup of ceramic prototypes, many requirements have to be met. The ceramic suspension must have a high solid loading (50–65 vol%) and low viscosity (in most cases), preferably without a pronounced yield point. The ceramic powder must be effectively dispersed in the UV (or visible light)-curable monomer, which requires

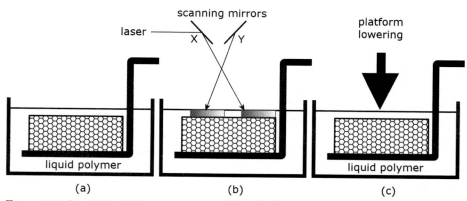

Figure 20.7 Principle of STL process adapted from [25].

suitable dispersants. The suspension must be sufficiently transparent to (UV) radiation to permit an acceptable depth of cure. The light penetration in the suspension is limited by light absorption in the particles and the curable monomer but also by light scattering in highly concentrated suspensions. The theoretical expression of the cured depth (C_d) is derived from the Beer–Lambert law and can be written as [26]:

$$C_d = D_p \ln\left(\frac{E}{Ec}\right) \qquad (20.1)$$

where D_p is the depth of penetration, E (J/m²) the exposure and E_c (J/m²) the critical exposure or the minimal exposure to provide polymerization of the monomer. For a loaded monomer, D_p is a function of the volume fraction Φ of ceramic powder, of the mean ceramic particle diameter d and of the efficiency factor for the extinction coefficient Q of the ceramic–resin system:

$$D_p = \frac{d}{Q \cdot \Phi} \qquad (20.2)$$

Q describes the material scattering ability and depends on the refractive index difference between the ceramic powder and the UV-curable solution ($\Delta n = n_{powder} - n_{resin}$) and the particle diameter and wavelength. This could be used for light scattering of gases, for example air, but not for ceramic suspensions:

$$Q = \left(\frac{\Delta n}{n_{resin}}\right)^2 \left(\frac{d}{\lambda}\right)^2 \qquad (20.3)$$

Light scattering Q for highly concentrated suspensions has been described by an empirical equation [27]:

$$Q = \frac{S}{\lambda}(\Delta n)^2 \qquad (20.4)$$

where S is the interparticle distance. The distance between the particles will affect how the radiation penetrates the suspension and provides a better description of the scattering phenomenon than the particle diameter. Substituting Q in Equation (20.2), results in an expression with particle diameter d, solid loading Φ and the interparticle spacing S.

$$D_p = \frac{d\lambda}{S \Delta n^2 \Phi}$$

Chartier et al. [26] studied the effect on cure depth of some parameters such as the photo-initiator concentration, the refractive index and the concentration of the ceramic powder and the incident radiation energy. Alumina or zirconia particles (average particle size: ~ 0.6 µm) were dispersed with a phosphate ester in a mixture of methyl, ethyl, ketone, and ethanol. The solvent was evaporated and the powder was added to a mixture of UV-curable acrylate binder with photo-initiator and dispersant. UV curing was achieved by passing the suspension in a petri dish under a UV lamp with a peak intensity of 360 nm. The average UV power concentrated on the suspension was about 1.8 W/cm². Figure 20.8 shows the cured depth versus exposure for zirconia and alumina. The cure depth of the alumina suspension is three times greater than that of zirconia due to the lower refractive index of alumina ($n_{Al} = 1.70, n_{Zr} = 2.16$). The scattering phenomenon is then less pronounced for alumina. The effect of varying the photo-initiator concentration showed that a minimum amount is required to obtain a certain cured depth but that a higher concentration does not increase the cured depth. When the cure depth was studied as a function of the photo-initiator concentration

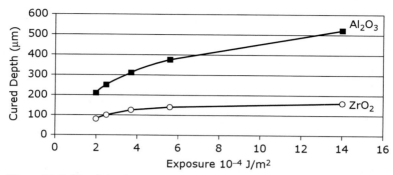

Figure 20.8 Cured depth versus exposure for zirconia and alumina suspensions (30 vol% of powder, 2 wt% of photo-initiator) adapted from [26].

and exposure conditions, it was found that the cure depth increased with increasing exposure but the optimum concentration of initiator was constant. The influence of volume fraction on cure depth was studied on suspensions containing 10–45 vol% of zirconia and 10–50 vol% of alumina. The cure depth significantly decreases when the volume fraction increases due to the scattering phenomenon. The linear behaviour is in agreement with Equations (20.1) and (20.2). It is therefore possible to predict the cured depth for any powder concentration.

Griffith and Halloran [27] discussed the scattering of UV radiation in turbid suspensions starting with Beer's law. They proposed a general scattering equation similar to the one proposed by Chartier:

$$D_c = \frac{2|d|}{3\tilde{Q}} \frac{n_0^2}{\Delta n^2} \ln\left(\frac{E_0}{E_{crit}}\right) \text{ where } \tilde{Q} \propto \frac{S}{\lambda} \text{ and } Q = \tilde{Q}\frac{\Delta n^2}{n_0^2} \quad (20.6)$$

D_c is the cure depth, d is the average particle size, S the interparticle spacing, λ is the wave length, n_0 is the refractive index of the medium, Δn the refractive index difference between the ceramic and the medium, E_0 is the energy density from the UV-laser and E_{crit} the minimum energy density required for photopolymerization of the resin.

Griffith and Halloran [27] obtained an interesting result for the cure depth dependence of the particle size. An alumina suspension with an average particle size of 0.46 µm and a solid loading of 50 vol% obtained a cure depth of 400 µm, whereas a suspension with the average particle size of 0.61 µm and equal solid loading obtained a cure depth of 300 µm. This is in accordance with the equation given above, which gives a larger cure depth with a decreasing interparticle distance. It is clear that the spacing between the particles affects the radiation penetration of the suspension.

20.11 Process parameters and binders

Hinczewski *et al.* [25] investigated ceramic suspensions suitable for STL. Their work showed that ceramic suspensions suitable for STL must have a viscosity of less than 5 Pa · s to ensure layer recoating. It must be UV curable with a useful depth and resolution and the green parts must be easily debound and sintered. Alumina (mean particle size: 0.5 µm) was deagglomerated and mixed into the UV-curable system consisting of photo-initiator

(2,2-dimethoxy-2-phenylactophenon or dmpa) dissolved in photo-polymerizable monomer (di-ethoxylated bisphenol A dimethacrylate or diacryl 101). A reactive diluent (acryloyloxy-ethyl n-butyl carbamate or acticryl) was also added to decrease the viscosity of the resin. Special attention was paid to the viscosity of the ceramic suspension. It was noticed that a temperature of 60°C decreased the viscosity with a factor of 6 (to 8 Pa s) for a suspension of 53 vol% alumina with 2.2 wt% dispersant in the diacryl resin. The diluent is a monofunctional acrylate that polymerises under UV radiation and leads to a flexible polymer. The diluent allowed reduction of the viscosity of the monomer without modification of the reactivity. The cure depth is reduced when the ceramic is added and the cure width is increased due to scattering in the ceramics. A cure depth above 200 μm was easily obtained for the alumina-loaded monomer but the cure width was in no case less than 450 μm.

20.12 Layer manufacturing by microstereolithography

Most of the layer micromanufacturing of ceramic micro components has been done by μSTL [28, 29]. The first equipment used for μSTL is in principle a conventional stereo-lithographic machine but with a higher dimensional resolution. The UV-laser beam can be focused to 1–2 μm and cure a layer thickness in the range of 1–10 μm. A submicron (0.5 μm) resolution of the x–y–z translation stages is also necessary. Aqueous and nonaqueous resins [24] were used together with alumina (average particle size: 0.2 μm). A line width as fine as 1.2 μm was obtained and microgears with diameters of 400 and 1000 μm were fabricated with a good definition [28]. Laser scanning is a sequential manufacturing technique and the process time is therefore quite long. Two other techniques have been developed for projection of the whole pattern in one exposure with an liquid crystal display (LCD) mask or a digital mirror [30–32]. The disadvantages of the LCD mask is that the UV absorption is high and therefore visible light and a nonstandard UV curing system must be used. The LCD mask also has quite large pixels 33 × 33 μm compared to digital mirror 13 × 16 μm and quite a long switching time (20 ms) compared to digital mirror (20 μs). In all aspects the digital mirror performs better for dynamic projection. Examples of 3D ceramic microparts fabricated with μSTL are shown in Figure 20.5.

A summary of the performance of mask projection μSTL is found in Table 20.2. The data in Table 20.1 is valid for transparent polymers. Due to the light scattering in the ceramic slurry the line width increases with approximately 10% [7].

With two-photon-polymerization it is possible to manufacture submicrometre 3D parts. The photo polymerization is done in the liquid by crossing two laser beams and is therefore restricted to transparent polymers. The light absorption and scattering in ceramic suspensions prevent this from being used for ceramic micromanufacturing [40].

A limitation of the traditional STL is that the ceramic slurry has to have a low viscosity and the solid content therefore has to be limited. Another technology that does not require a container for the ceramic suspension is based on doctor blading to spread a high viscos-ity suspension [41]. The high viscosity suspension prevents settling and allows long-term storage. Fully dense materials with isotropic properties comparable to classical forming techniques have been produced from alumina suspension with a solid loading of 68 vol%.

Table 20.2 Performance of mask projection μSTL from [39].

Research group	Light source	Mask	Resolution	Component size	Speed	Reference
Bertsch	Laser 515 nm	LCD	5 × 5 × 5 μm	6 × 8 × 15 mm²	Not reported	33
Chatwin	Laser 351 nm	SLS	5 μm lateral resolution	Not reported	60 s exposure time/layer 50 μm thick	34
Monneret	Broadband visible light	LCD	2 μm lateral resolution	Not reported	60 s /layer 10 μm layers	35
Bertsch	Lamp visible	DMD	5 × 5 × 5 μm	6 × 8 × 15 mm²	1 mm/h	36
Bertsch	Lamp UV	DMD	10 × 10 × 10 μm	10.25 × 7.68 × 20 mm²		37
Hadipoespito	Lamp UV	DMD	20 μm lateral resolution	Not reported	Not reported	38
Rosen	Lamp UV	DMD	6 μm lateral resolution	1.1 × 1.8 mm	90 s for one 100 μm layer	39

Ventura et al. [42] have presented a STL method called Direct Photo Shaping (DPS). The process is based on the photo-curing of polymerizable compositions curable by visible light. Each layer is selectively photo-imaged by digital light projection via a digital micromirror device (DMD) array that performs the function of an electronic maskless tool. The process allows a rapid build-up time because of the flood exposure, thus curing the entire profile at once unlike the case with a scanning laser. No laser is needed, thus, the process allows a low cost and a very high resolution. DMD arrays with resolution of 1280 × 1024 and fill factor of about 90% are available. Each pixel is a 16 μm square mirror. The DMD device is an array of aluminium mirrors, with precise control that could modulate light. The DMD array is interfaced with a suitable light source and optics, and each pixel mirror is electronically controlled to reflect the incident light in or out of the projection area. The DMD array can be used to project grey images, and the intensity of the grey can be modulated to control light transmission. Thus, the surface finish of a part was improved by applying a 4-pixel wide anti-aliasing filter on the wedge-shaped part boundary. The anti-aliasing filter turns the pixel-to-pixel black and white stair-steps into smooth grey-scale boundaries. This resulted in a power density gradient along the periphery of each slice that smoothed the surface over several layers.

20.13 Mechanical properties

The mechanical properties of silicon nitride formed by DPS were compared to properties obtained by slip casting the same powder. Tile specimens were formed and gas-pressure sintered to > 99% of theoretical density and machined into flexure bars. No delamination between layers was observed by SEM examination. Table 20.3 shows the mechanical properties obtained from each test. The material shaped by DPS showed similar properties to that formed by slip casting. The flexure strengths for the DPS material were not dependent on the build orientation.

Corbel *et al.* [43], evaluated the mechanical properties of alumina specimens produced by STL of a suspension with 51 vol% solid loading and hexanediol diacrylate HDDA as a photo-curable binder. The specimens had an average relative density of 96% and were evaluated by three-point flexural test. The specimens were tested in order to study the influence of: (1) the hatch spacing (distance between parallel scanning lines: 0.2, 0.3 or 0.4 mm); (2) the layer thickness: 50 μm or 100 μm; and (3) the orientation of the test sample. The results showed that a flexural stress of 366 MPa was obtained for a hatch spacing of 0.4 mm (layer thickness 100 μm) whereas the stress was 131 MPa for a hatch spacing of 0.1 mm. This result confirms the creation of internal or residual stresses during curing and debinding. The internal stresses are greater when there is a large overlapping of the lines, i.e. low hatch spacing. Concerning the influence of the layer thickness, the flexural stress for specimens made with 0.4 mm space hatching increased from 366 to 428 MPa when the layer thickness was divided by 2 (from 100 μm to 50 μm). These results showed that the shrinkage due to polymerization and thus the generated stresses increase as the layer thickness increases. Contrary to the hatch spacing, which determines the distribution of the polymerization ratio, the layer thickness does not modify the existing stresses in a significant way. Flexural stresses of 435 and 459 MPa were

Table 20.3 Mechanical properties of layer manufacture (DPS) compared to slip cast silicon nitride.

Test/conditions	DPS silicon nitride results	Standard slip-cast silicon nitride
Flexural strength/machined surface/20°C	786 MPa	734 MPa
Flexural strength/as-sintered surface/20°C	797 MPa	588 MPa
Flexural strength/machined surface/1370°C	558 MPa	580 MPa
Fracture toughness/20°C	9 MPa(m)$^{1/2}$	8.1 MPa(m)$^{1/2}$
Flexural stress–rupture/ 1316°C/310 MPa applied for 10 h	No failures	No failures

obtained when a load was applied perpendicular to the plane of the layers. When the load was parallel to the layers the flexural stress was 93 MPa. This low value was due to bad adhesion between the layers. Delamination was in fact observed as a problem at the end of the heat treatment.

20.14 Selective laser sintering

Two principles are used for SLS of ceramics. The first one is based on selective sintering of a ceramic powder coated with a polymer allowing a green-state body to form which is sintered in a postprocess. The other is based on "direct" sintering which implies that the ceramic is melted and the part could be obtained directly without any postprocessing [44]. The traditional SLS uses dry powder composed of polymer or metal and is relatively easy to melt and sinter. Layer manufacturing with dry powder ceramics is much more difficult to melt and sinter and the material obtained will have high porosity. A commercial process based on green-body laser sintering and postsintering in a furnace is available [45]. The industrial use is quite limited due to the poor mechanical properties. The normal approach to increasing the density and lowering the porosity is to use ceramic slurries. This method has also been implemented for direct laser sintering of ceramics [46, 47]. The slurry must be dried before sintering so the water content is below 3% [46]. The process window is quite narrow for direct laser sintering of ceramics. The scan velocity and laser energy must be optimised for the ceramic to melt and fuse. Direct laser sintering of ZrO slurries was evaluated and it was found that the surface roughness increases with increasing energy and lower scan speed. The open porosity obtained was between 24–33% and depends on the scan velocity, the lower the scan speed the lower the open porosity. At higher scan velocity the sintering is not complete and therefore the strength of the sintered layers is also lower [46]. The energy required to obtain stable parts is shown in Figure 20.9. The overlapping between the laser scans and the scan velocity was used to vary the irradiated energy. Delamination occurs if the energy is too low and the scan speed too high. If the energy is too high the ceramics evaporate while at low scan speed and low energy the thermal stress results in delamination and cracks [47].

Table 20.4 Performance of SLS.

Research group company	Laser	Material	Resolution μm	Layer thickness μm	Smallest feature size	Speed	Reference
Klocke				50			46
Exner	q-switched	metal	20	1	50		44
Exner	q-switched	ceramic	80	15			44
Phenix		ceramic	20 × 20 × 20		300		48
Phenix		metal	20 × 20 × 20		100		48

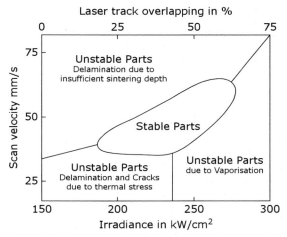

Figure 20.9 Process window for direct laser sintering of Al_2O_3–SiO_2 [47].

Conclusions

Layer manufacturing and SFF have been developed into an industrial business for prototype fabrication during the last 20 years. The SFF methods for manufacturing of ceramic components have not been entirely successful mainly due to high porosity and poor strength. The concept of direct manufacturing or rapid manufacturing is very promising for small parts as the number of parts that can be manufactured increases by the cube if the part size is halved. Another very important possibility with layer manufacturing is length-scale integration. In many applications, in which microparts are used, they must be integrated in a meso/macroscopic system. With layer manufacturing is it possible to manufacture very complex mesoscopic 3D parts and at the same time manufacture integrated microscopic parts in one single manufacturing operation. The most suitable method for manufacturing of microparts is µSTL. The demonstrated lateral resolution is about 5–10 µm and the smallest structure that could be manufactured is about 50–100 µm. The resolution for other techniques such as laser sintering and ink-jet printing is about 20–40 µm and the smallest feature size that could be manufactured is 200–300 µm. The resolution obtained with ink-jet printing could be further improved with new ink-jet heads that could print 3 pl instead of 30 pl. The micromanufacturing capability depends very much on the (materials) suspension processing and material development. New ceramic nanomaterials that are optimized for layer manufacturing could further improve the possibility to manufacture ceramic microcomponents.

References

[1] wikipedia.org (accessed 15 November 2008); www.ennex.com (accessed 15 November 2008); www.solidview.com (accessed 15 November 2008).

[2] Jacobs, P.F. 1992. *Rapid Prototyping and Manufacturing: Fundamentals of Stereo lithograph*, Society of Manufacturing Engineers, Dearborn, Missouri.

[3]   Provin C. and Monneret S. 2002. Complex ceramic-polymer composite microparts made by stereolithography. *IEEE Transations on Elecronics Packaging Manufacturing*, **25**(1), 59–63.

[4]   Regenfuss P., Streek A., Hartwig L., Klötzer S., Brabant Th., Horn M., Ebert R and Exner H. 2007. Principles of laser micro sintering. *Rapid Prototyping Journal*, **13**(4), 204–212.

[5]   Sun C., Fang N., Wu D.M. and Zhang X. 2005. Projection micro-stereolithography using digital micro-mirror dynamic mask. *Sensors and Actuators A*, **121**(1), 113–120.

[6]   Grau J.E., Uhland S.A., Moon J., Cima M. and Sachs E.M. 1999. Controlled cracking of multilayer ceramic bodies. *Journal of the American Ceramic Society*, **82**(8), 2080–2086.

[7]   Hinczewski, C., Corbel, S. and Chartier, T. 1998. Ceramic suspensions suitable for stereolithography. *Journal of the European Ceramic Society*, **18**(6), 583–590.

[8]   Greco A., Liccilli A. and Maffezzoli A. 2001. Stereolithograpy of ceramic suspensions. *Journal of Materials Science*, **36**(1), 99–105.

[9]   Klocke F., Derichs C., Ader C. and Demmer A 2007. Investigations on laser sintering of ceramic slurries. *Production Engineering Research and Development*, **1**(3), 279–284.

[10] Heinrich J.G., Gahler A., Gunster J., Schmucker M., Zhang J., iang D. and Ruan M. 2007. Microstructural evolution during direct laser sintering in Al_2O_3 system. *Journal of Materials Science*, **42**(14), 5307–5311.

[11] Bertsch A., Jiguet S. and Renaud P. 2004. Microfabrication of ceramic components by microstereolithography. *Journal of Micromechanics and Microengineering*, **14**(2), 197–203.

[12] Bauer W., Kaufmann U., Harrysson U. and Johander P, 2006. Free form fabrication of 3D ceramics parts with ink jet printing. Paper presented at *CIMTEC, Advances in Science and Technology*, 4–9 June 2006, Acireale, Italy.

[13] Yoo J., Cima M., Sachs E. and Suresh S. 1995. Fabrication and microstructural control of advanced ceramic components by three dimensional printing. *Ceramic Engineering and Science Proceedings*, **16**(5), 755–762.

[14] Moon J., Grau J.E. and Chima M.J. 2000. Slurry chemistry control to produce easily redispersible ceramic powder compacts. *Journal of the American Chemical Society*, **83**(10), 2401–2408.

[15] Wu D., Fang N., Sun C. and Zhang X. 2006. Stiction problems in releasing of 3D microstructures and its solution. *Sensors and Actuators A*, **128**(1), 109–115.

[16] Song, J.H., Edirisinghe, M.J. and Evans, J.R.G. 1999. Formulation and multilayer jet printing of ceramic inks. *Journal of the American Ceramic Society*, **82**(12), 3374–3380.

[17] Grau J., Moon J., Uhland S., Cima M. and Sachs E. 1997. High green density ceramic components fabricated by the slurry-based 3DP process. In *Proceedings of the Solid Freeform Fabrication Symposium*, August 1997, Austin, Texas, USA., 371–378.

[18] Kaufmann U., Harrysson U., Johander P. and Bauer W. 2006. Free form fabrication of 3D-ceramic parts with inkjet-printing. Paper presented at *CIMTEC 2006, Advances in Science and Technology*, 4–9 June 2006, Acireale, Italy.

[19] Moon J., Grau J.E, Knezevic V., Cima M. and Sachs E.M. 2002. Ink jet printing of binders for ceramics components. *Journal of the American Ceramic Society*, **85**(4), 755–762.

[20] Uhland S.A., Holman R.K., Cima M.J., Sachs E. and Enokido Y. 1999. New process and materials developments in 3-dimensional printing, 3DP. In *Solid Freeform and Additive Fabrication*, MRS **542**, Dimos D., Danforth S.C. and Cima M.J. (Eds.), Materials Research Society, Warrendale, Pennsylvania, USA, 153–158.

[21] http://www.fcubic.com/ (accessed 15 November 2008).

[22] Technology white paper Z Corporation 3D Printing Technology. http://www.zcorp.com/documents/108_3D%20Printing%20White%20Paper%20FINAL.pdf (accessed 15 November 2008).

[23] Yoo J., Cho K., Cima M. and Suresh S. 1998. Transformation-toughened ceramic multilayers with compositional gradients. *Journal of the American Ceramic Society*, **81**(1), 21–32.

[24] Griffith, M.L. and Halloran, J.W. 1996. Freeform fabrication of ceramics via stereo¬litho¬graphy. *Journal of the American Ceramic Society*, **79**(10), 2601–2608.

[25] Hinczewski, C., Corbel, S. and Chartier, T., 1998. Ceramic suspensions suitable for stereo¬litho¬graphy. *Journal of the European Ceramic Society*, **18**(6), 583–590.

[26] Chartier, T., Hinczewski, C. and Corbel, S. 1999. UV curable systems for tape casting. *Journal of the European Ceramic Society*, **19**(1), 67–74.

[27] Griffith, M.L. and Halloran, J.W. 1997. Scattering of ultraviolet radiation in turbid suspensions. *Journal of Applied Physics*, **81**(6), 2538–2546.

[28] Zhang, X., Jiang, X.N. and Sun, C. 1999.Micro-stereolithography of polymeric and ceramic microstructures. *Sensors and Actuators, A*, **A77**(2), 149–156.

[29] Chartier T., Chaput C., Doreau F. and Loiseau M. 2002. Stereolithography of structural complex ceramics parts. *Journal of Materials Science*, **37**(15), 3141–3147.

[30] Bertsch A., Bernhard A., Vogt P. and Renaud P. 2000. Rapid prototyping of small size objects. *Rapid Prototyping Journal*, **6**(4), 259–266.

[31] Bertsch A., Jiguert S. and Renaud P. 2004. Microfabrication of ceramic components by microstereolithography. *Journal of Micromechanics and Microengineering*, **14**(2), 197–203.

[32] Sun C., Fang N., Wu D.M. and Wang X. 2005. Projection micro-stereolithography using digital micro-mirror dynamic mask. *Sensors and Actuators, A*, **212**(1), 113–120.

[33] Bertsch. A, Zisszi. S, Jezequel. J, Corbel. S. and Andre. J. 1997. Microstereolithography using liquid crystal display as dynamic mask generator. *Microsystems Technologies*, **3**(2), 42–47.

[34] Farsari. M, Huang. S, Birch. P, Claret-Tournier. F, Young. R. and Richardson. J. 2000. A novel high-accuracy microstereolithography method employing an adaptive electro-optic mask. *Journal of Materials Processing Technology*, **107**(1), 167–172.

[35] Monneret. S, Provin. C. and Le Gall. H, 2001. Microscale rapid prototyping by stereolithography. *Proceedings of the 8th IEEE International Conference on Emerging Technologies and Factory automation (EFTA)*, Nice, France, **2**, 299–304.

[36] Beluze L., Bertsch A. and Renaued P. 1999. Microstereolithography a new process to build complex 3D objects. *SPIE Symposium on Design, Test and Microfabrication of MEMS/MOEMS*, **3680**(2), 808–817.

[37] Bertsch A., Lorenz H. and Renaud P. 2000. Rapid prototyping of small size objects. *Rapid Prototyping Journal*, **6**(4), 259–266.

[38] Hadipoespito G., Tang Y., Choi H., Ning G. and Li X. 2003. Digital micromirror device based microstereolithography for micro structures of transparent photopolymer and nanocomposites. In *Proceedings of the 14th Solid Freeform Fabrication Symposium*, 4–6 August 2003, Austin, Texas, 13–24.

[39] Limay A.S. and Rosen D.W. 2007. Process planning method for mask projection micro-stereolithography. *Rapid Prototyping Journal*, **13**(2), 76–84.

[40] Maruo S., Nakmura O. and Kawata S. 1997. Three dimensional microfabrication with two photon absorbed photopolymerisation. *Optics Letters*, **22**(2), 132–134.

[41] Chaput C., Doreau F. and Chartier T. 1999. Stereolithography for ceramic part manufacturing. In *Proceedings of the 8th European Conference on Rapid Prototyping and Manufacturing*, 5–7 July 1999, Nottingham, United Kingdom, Campbell R.I.(Ed.), 291–297.

[42] Ventura, S., Narang S., Guerit P., Liu S., Twait D., Khandelwal P., Cohen E. and Fish R. 2000. Freeform fabrication of silicon nitride components by direct photo shaping. In *Proceedings of the Conference on Solid Freeform and Additive Fabrication*, 24–28 April 2000, San Francisco, California, USA., **625**, Materials Research Society, Warrendale, Pennsylvania, USA. 81–89.

[43] Corbel, S., Hinczewski, C. and Chartier T. 1999. Mechanical properties of ceramic parts made by stereolithography and sintering process. In *Proceedings of the 8th European Conference on Rapid Prototyping and Manufacturing*, 5–7 July 1999, Nottingham, United Kingdom, Campbell R.I (Ed.).

[44] Regenfuss P., Streek A., Hartwig L., Klötzer S., Brabant Th., Horn M., Ebert R. and Exner H. 2007. Principles of laser micro sintering. *Rapid Prototyping Journal*, **13**(4), 204–212.

[45] http://www.phenix-systems.com/ (accessed 15 November 2008).

[46] Klocke F., Derichs C., Ader C. and Demmer A. 2007. Investigation on laser sintering of ceramic slurries. *Production Engineering Research Development*, **1**(3), 279–284.

[47] Heinrich J., Gahler A, Gunster J., Schmucker M., Zhang J., Jiang D. and Meiling Ruan M. 2007. Microstructural evolution during direct laser sintering in the Al_2O_3-SiO_2. *Journal of Materials Science*, 42(14), 5307–5311.

[48] http://www.phenix-systems.com/produits_en/pm_250.htm (accessed 15 November 2008).

21

Micro direct writing of ceramics

F. Bortolani, D. Wang and R.A. Dorey

MEMS incorporating integrated ceramic functional materials have applications in a wide range of sectors where the functional properties lead to microscale devices that can act as sensors, actuators, transducers, light emitters and energy scavengers. Conventionally such MEMS are produced using a subtractive approach leading to material wastage, the use of hazardous etchants and limited flexibility in prototyping and small volume production. Similar limitations have been encountered, and overcome, in the graphics industry through the use of digital printing technologies. In analogy to this work, explorations of digital direct writing technologies for the creation of MEMS and microsystems is underway.

Direct writing presents several advantages over the traditional thick-film deposition techniques including: direct printability of the pattern on the substrate; high printing speed; low wastage of material; high resolution; ability to produce patterns on a greater range of substrates; excellent control over the composition and surface chemistry of a system and fabrication of high chemical/functional complexity systems without severe cross-contamination. Direct writing is distinct from the more general SFF or rapid prototyping techniques in that material is directly deposited without using powder beds (e.g. SLS, or 3D printing), layering of green sheets (e.g. laminated object manufacture), or lithographic stamps (e.g. microprinting).

This chapter will examine some of the droplet-based direct writing technologies that have arisen in recent years for production of microscale ceramic architectures.

21.1 Introduction

MEMS are small-scale active devices consisting of integrated functional materials, electrical connections and mechanical elements on a range of substrates [1,2] including silicon, polymers or metals. Of the functional materials, active ceramics such as piezoelectrics, pyroelectrics, ferroelectrics, and ion conductors are of great interest due to their high properties and wide range of applications. The conventional approach to producing microscale features for MEMS is through the use of multiple layering techniques preformed in conjunction with selective protection and subtractive etching [3, 4]. Several techniques are available to deposit continuous ceramic films onto substrates including spin coating [5–7], electrophoretic deposition [8], and spray coating [9]. Screen printing has been used to directly pattern features on different substrates [10–12]. In screen printing the pattern is obtained from the ink deposition after its transition through a fine mesh with a defined feature. Resolution is limited by the size of the mesh: this method is not suitable for obtaining small features.

While these routes have been proven for producing a range of MEMS and microsystems there are disadvantages associated with them including: (1) material wastage; (2) the use of hazardous etchants such as hydrofluoric acid and nitric acid; and (3) the requirement for an expensive new mask set for each alteration in the design [13]. A similar situation has already been encountered in the graphics industry where lithography printing has successfully been superseded by digital printing technologies in a large number of fields. In analogy to the digital printing process, work on exploring digital direct writing technologies for the creation of MEMS and microsystems is underway. This chapter will examine some of the technologies that have arisen in recent years.

Direct writing [1, 14–19] techniques comprise a group of technologies, including micropen, robocasting, fused deposition and ink jet, where the material of interest is deposited where it is required without the need for moulds, etching or machining. These techniques allow the creation of 3D structures under computer control through a noncontact or contact layer deposition [14–16, 20]. Direct writing presents several advantages over the traditional thick film deposition techniques including: direct printability of the pattern on the substrate; high printing speed; low wastage of material; high resolution; ability to pattern a material that is incompatible with the resist, image development, or etching process; excellent control over the through-thickness composition and surface chemistry of a system; and allows for the fabrication of systems with high chemical/functional complexity without severe cross-contamination [21]. Direct writing is distinct from the more general SFF [22] or rapid prototyping techniques in that material is directly deposited without using powder beds (e.g. SLS or 3D printing), layering of green sheets (e.g. laminated object manufacture), or lithographic stamps (e.g. microprinting).

21.2 Technologies

21.2.1 Ink-jet printing

Ink-jet printing is a well-known technique, first developed in the 1960s [23], and established in many applications particularly the digital reproduction sector where it is used for production of graphic products without the need to first produce printing plates. Within the field of materials processing, ink-jet printing is still under development, particularly for microscale systems, and in recent years has extended its applications because of its versatility. In fact with this technique small fluid drops are repeatedly formed and directed to the substrate with high accuracy [13]. Moreover it allows printing on a wide range of materials, such as silicon wafers [24], glass or plastics [13]. All of these properties make direct ink-jet writing suitable for several applications such as: biomedical [25], optic and electronic [15, 20, 26]. In the field of biomedicine, Sumerel *et al.* [25] have demonstrated the use of ink-jet printer deposition of protein and adhesives on silicon substrates and for processing of hybrid composites.

The key parameter of ink-jet printing is its high resolution. Usually it is expressed in dots per inch and it is equivalent to the number of individual dots printed in a linear inch and relates to the ability to control the deposition and the size of the drops produced. The final resolution of a printed pattern is not only dependent on the printer resolution but is determined by several properties of the ink (e.g. viscosity, surface tension) and the drop size [25]. Therefore, ink development is a crucial point in the drive for higher resolution capability. The chemical behaviour, in particular the viscosity and the surface tension of the ink determine its printability with an ink-jet printer. These aspects will be considered in greater detail below.

Depending on the ink delivery system, ink-jet printing can be divided in two categories: continuous and drop-on-demand (DoD) [1, 13, 15, 16, 18, 27–30]. A continuous ink-jet printer forms a stream of droplets even when there is no print demand. The droplets are then charged and deflected to their destination in an electric field and those not required are collected or recirculated [31]. Conversely a DoD printer only ejects an ink droplet when it is required.

21.2.1.1 Continuous deposition

In continuous ink-jet printing ink is forced to flow through an orifice with a 30–200 µm diameter nozzle [23]. Passing through the orifice the ink flow breaks up into single drops which can be deflected by charged plates before reaching the substrate [16]. For this reason the ink must be conductive and therefore this technique is not suitable for many ceramic suspensions [29]. Since it is a noncontact process, the unprinted droplets can be recovered and recirculated through an ink recovery system [13, 16]. This technique produces uniform droplets [27] of approximately twice the size of the orifice diameter, typically in the range 20–150 µm [13], offers high speed process and allows a large area to be printed [27, 29].

The voltage of the electrode is limited to a few hundred volts, above which secondary atomization occurs due to the surface tension being overcome [32]. Once produced the droplets pass between deflection plates at a constant potential in the range 3–18 kV [31]. Charged droplets are either deflected by a set amount (constant charge: binary system) or a variable amount (variable charge: multiple system) [33]. In both systems, if the particle is uncharged it is collected and recycled.

21.2.1.2 Drop-on-demand deposition

Most of the activity in ink-jet printing is focused on DoD [34] systems. DoD refers to the ink droplets being ejected only when required. DoD was developed to overcome the limitations (start-up and shut down requirements, complex charging and the need for ink recirculation) of continuous jet printing systems [31].

The equalized and spaced droplets are produced by a pressure pulse acting on the fluid chamber [16, 18]. The pressure pulse can be created by exciting a piezoelectric actuator at a determined frequency or by heating the fluid reservoir to create a vapour bubble [32, 33]. When the pressure pulses reach the orifice, a droplet or a group of droplets is ejected [13]. The ink positioning is not achieved by droplet deflection, as in the continuous mode, but

Table 21.1 Typical ink and process parameters for continuous and DoD ink-jet printer.

	Continuous	DoD
Nozzle diameter (μm)	10–220	20–100
Drop diameter	~2 × nozzle diameter	
Operating frequency (kHz)	17–1000	3–25
Charging voltage (V)	100–300	n/a
Deflection voltage (V)	1000–18000	n/a
Drop velocity (m s^{-1})	5–50	3–15
Ink viscosity (mPa s)	1–20	10–12
Ink surface tension (mN m^{-1})	25–70	28–33
Ink conductivity (S m^{-1})	> 0.2	n/a
Particle size (μm)	< 1	< 1/100 nozzle diameter
Resolution (dpi)	170–19,000	

by the movement of the cartridge relative to the substrate [1]. Of the two, the piezoelectric actuation is more commonly used approach [18]. DoD techniques produce droplets from a few microns to several hundreds microns (typically 20–100 μm) [13, 18, 28] with no need for droplet recovery. Typical ink and process parameters for DoD and continuous ink-jet systems are summarised in Table 21.1.

21.2.1.3 Drop-on-demand printer inks

An accurate understanding of the behaviour of the printing ink is important in the creation of ceramic 3D structures using ink-jet printing. Principally the fluid dynamics involved in the drop formation and the ink behaviour on the substrate is required to accurately control the printing process.

The fluid dynamics of the ink control the ability to produce a droplet. Provided that the dimensionless 'jettability' parameter Z is in the range 1–10 a droplet can be ejected. Z is given by Equation (21.1) [35]:

$$Z = \frac{Re}{We} = \frac{\sqrt{\gamma \cdot \rho \cdot r}}{\eta} \qquad (21.1)$$

where Re and *We* are the Reynolds and Weber number, respectively, γ, η, ρ, (represent the surface tension, density and viscosity of the ink, respectively, and r is the nozzle radius. If Z is too small (<1) the viscous forces dominate and a high pressure is required to eject the droplet. If Z is too high (>10) a liquid column forms from the nozzle before the drop is developed [36] leading to the creation of satellite drops. Equation (21.1) highlights the importance of surface tension forces and viscous forces on an ink flow and has been used to develop several suspensions suitable for ink-jet printing [18, 19, 30, 36].

One of the key parameters affecting the fluid dynamics of the ink is its solids loading. From a manufacturing point of view a high solids loading is preferable as this maximizes

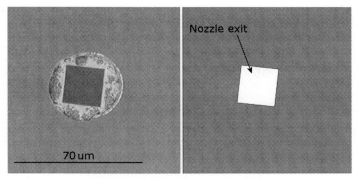

Figure 21.1 Optical micrograph and schematic of the external view of a clogged nozzle.

the deposition speed and minimizes shrinkage and associated stresses. This is often in direct conflict with the requirement of control Z which is most often described by dye or pigment-based inks used in the graphics industry [18]. For this reason the typical solid content is usually low (~5 vol%) [1, 15, 20] but higher concentrations of particle, up to 35 vol%, have been used [18] through careful control of the particle sizes, carrier fluid and dispersants.

In order to minimize nozzle clogging a dispersant can be added to the ink formulation to help the ceramic particles to remain in suspension. If the suspension is unstable its rheology will deviate from the ideal $1<Z<10$ reducing print efficiency and potentially leading to internal clogging of the nozzle. The amount of dispersant is a function of the effective surface area of the solid to be dispersed. The smaller the particle size, the higher the effective surface area, and hence the larger the area for the dispersant to cover and the more dispersant required. The level of dispersant added can vary in the range 0.1–6 wt% of the powder, but larger quantities are also be used. Additionally, a 10–20 vol% of humectants, such as ethylene glycol, can be added to reduce the drying rate and clogging [26] at the nozzle exit. Figure 21.1 shows a light micrograph of the external face of a clogged nozzle. It can be seen that the nozzle is not blocked on the outside. Instead the clogging process has taken place from the inside by particle sedimentation.

21.2.1.4 Printing defects

For some applications, such as sensors, actuators or transducers, a certain thickness of active material is required to obtain a system that is of commercial interest [1]. Such thicknesses are in the range 10–100 μm [1, 37]. However, a single droplet relic thickness depends on the rheology, density and composition of the active material and multiple depositions will be required to achieve the final thicknesses required. Such multiple deposition stages inevitably lead to macro- and microscopic defects [20] caused by substrate interactions, shrinkage, cracking and droplet positioning. Macroscopic defects can be defined as those that bring a loss of accuracy in size and shape of the feature. Conversely, microscopic defects relate to small-scale features such as internal voids or craters.

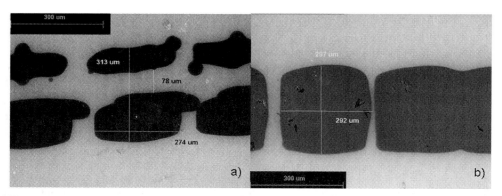

Figure 21.2 Pattern printed on photopaper with a cartridge printer height of: (a) 1 mm; (b) 0.25 mm.

Macroscopic defects arise from the formation of ink spray and satellite drops, incorrect drying times or inappropriate printing distances. It is possible to avoid or reduce them by regularly cleaning the nozzles, increasing the ink viscosity, controlling drying or reducing the distance between the cartridge and the substrate. Figure 21.2 shows the difference in the pattern printed with two different cartridge printer heights. It can be seen that with a lower cartridge print height (CPH) (b), the resultant printed area, is more homogeneous with smoother edges. The disadvantage of reducing the print height is that the likelihood of touching the printed material increases, especially when thick film structures are being produced.

The drying rate of the ink has a large impact on the macroscopic defects that arise. A high drying rate may lead to nozzle clogging which will affect both horizontal and vertical resolution [38]. Conversely, if the drying rate is too low the drop can spread on the substrate causing a loss of resolution [1, 15, 27].

The formation of microscopic defects also depends on the ink drying rate as this influences the way in which subsequent droplets interact with the deposited drops. The surface of the previously printed layer must be dried or hard enough to resist the impact of the subsequent drops. This can be achieved by eliminating the binder (not always possible), reducing the print speed (slower fabrication speed and increased risk of clogging) or by increasing the drying rate of the ink. In order to increase the drying rate it is possible to increase the temperature of the system or use more volatile carrier fluids. However, as stated above, higher drying rates can lead to the formation of macroscopic defects. This highlights how correcting one issue can lead to deleterious effects in another aspect of printing.

The nozzle diameter influences both the jetted droplet size and the line width and thickness. The latter two are also influenced by the properties of the fluid. Ibrahim et al. [38] have reported on the spreading and the drying time of a silver ink in order to obtain 3D structures. The spreading behaviour of the ink on a substrate is important [39–41] because it determines the precision of the final printed pattern and it depends on the ink contact angle on the substrate. Printing silver lines on three different substrates (stainless steel, glass and ink-jet film) they obtained a lower spread on the glass surface due to its greater contact angle.

For all of these reasons a ceramic ink suited for ink-jet printing should have appropriate viscosity and surface tension, in order to facilitate the formation of drops and at the same time to avoid nozzle clogging. Other factors that affect the final dimension and resolution of features include statistical variations of the flight direction of droplets [21], the atmospheric conditions [42, 43], temperature [44], the presence of surfactant [45] and the size of dispersed particles in the droplet [46]. A typical ink drop of 60 μm in diameter can produce, after contact with the substrate, a feature with a diameter up to 600 μm and a height of ~1 μm [1, 15].

21.2.1.5 *Application of ink-jet technology*

The advantages of direct ceramic ink-jet printing are: high speed, silent non-impact operation, electrical computer control, no postprinting treatment, multi-ink capability and the capacity for rapid change of image [31]. The possible applications of ink-jet printing are wide. Several studies have been conducted on drop formation [27, 28, 38] and deposition [20, 47, 48] in order to understand drop behaviour and obtain better printed areas. Important aspects are the ink development and the treatments that the printed pattern undergoes.

Wang and Derby [18] developed a PZT wax suspension suitable for ink-jet printing. The ink was composed of 35 vol% of PZT, a 60/40 paraffin oil/wax medium and a combination of dispersants (polyester and amine system). This suspension was successfully used to print a truncated pyramid, which was then heat treated and characterized. The resulting body presented very little porosity (less than 1%) and 25% linear shrinkage.

Fuller *et al.* [49] have used ink-jet printing to build an active microsystem using a single nozzle DoD. The active material was a colloidal ink composed of 10 wt% gold or silver nanoparticles (5–7 nm) dispersion. Heating the substrate at 300°C during the deposition left droplets of 80–100 μm in diameter and 100 nm thickness with placement repeatability of ±5 μm and gaps between lines of 10 μm. The printed structures were then sintered at 300°C for 10 minutes. They were able to build a resonant inductive coil, electrostatic drive motors and multiple-layered structures. For example, the drive motors were found to move insulating objects such as dielectric glass balls or pieces of tissue paper. Szczech *et al.* [50] made three different geometries of fine-line conductors, demonstrating the flexibility of ink-jet printing. To do this, nanoparticle fluid suspensions composed of 1–10 nm silver or gold particles in toluene-based solvent were used. Using this system they printed coplanar waveguide transmission lines to evaluate the performance of the silver suspension in the radio frequency regime of 0–40 GHz. The results showed an effective response at the given frequency, which has demonstrated the success of ink-jet printing.

21.2.2 Electrohydrodynamic atomization printing

Electrohydrodynamic atomization printing (EHDAP) is a computer-controlled method of depositing ceramic droplets subject to electrohydrodynamic atomization in cone-jet mode [51, 52]. EHDA occurs when a high electric field is applied between a needle containing a liquid and a ground electrode some distance below the metal capillary [53]. A balance between the electrical and physical forces caused the liquid to form a cone shape and atomise at its tip. This phenomenon can be explained by the Rayleigh limit [54], described by Equation (21.2).

$$Q_R = 2\pi(16\gamma\varepsilon_0 r^3)^{\frac{1}{2}} \qquad (21.2)$$

where Q_R is the Rayleigh limit charge on a drop, γ is the liquid surface tension, ε_0 is the permittivity of free space, and r is the drop radius. Highly charged drops are produced when Q_R is exceeded and the surface tension force is overcome. According to the geometry of the jet and droplet, the resulting atomization modes can be mainly classified as dripping, microdripping, spindle, multispindle, cone-jet and multijet. Among these the stable cone-jet mode is the most interesting functioning mode due to the production of small and uniform droplets [55]. The cone-jet mode produces near monodisperse droplets of a few micrometres in size [56] and can be used to process ceramic suspensions [57]. The use of a suspension with this technique offers two distinct advantages over ink-jet printing techniques: (a) fine deposition product; and (b) less risk of nozzle blockage during processing due to the larger orifice size (several hundred micrometres in size) [31]. By using a single point electrode below the metal capillary it is possible to focus the stream of atomised droplets and better control their deposition [58, 59]. In this manner it is possible to print architectures with a predetermined design.

21.2.3 Contact direct wiring

While ink-jet printing and EHDA are noncontact methods of directly depositing material where required, a number of techniques exist whereby the print head makes contact with the workpiece when depositing the material. Examples include micropen [60] and fused deposition moulding [61, 62] approaches [15]. Like ink-jet and EHDA printing, the micropen technique uses a computer controlled system to control the deposition of the working material. In the case of micropen, the working material is deposited in the form of a thin (>25 μm diameter) extrudate which contains 10–33 vol% of solids loading and takes the form of a viscous paste.

Fused deposition moulding, is related to the micropen approach in that the working material is extruded from a computer controlled tool. The primary difference is that the extrudate has a much higher solids loading (40–50 vol%) in fused deposition moulding allowing overhanging self-supporting structures to be fabricated. As fused deposition moulding operates by laying down filaments of material, pores can arise where multiple filaments meet. In addition such pores, and the microstructure in general, is not homogenous in all axes of the workpiece [61].

Conclusions

Direct writing – the controlled direct deposition and patterning of the working material – is being developed as a method of increasing processing flexibility during the production of ceramic 2D and 3D microcomponents for active applications. Unlike conventional photolithographic-based processing routes there is no need to create a mask set or moulds, conduct hazardous etching procedures or dispose of large quantities of waste.

Direct writing also brings the advantages and flexibility of digital imaging to a rapidly evolving industry.

Direct wiring techniques can be classified as noncontact and contact (or droplet and filament) deposition routes depending on whether individual droplets are formed and deposited by travelling across a gap between the print head and the workpiece (noncontact/droplet) or if a filament is extruded and laid onto the workpiece (contact/filament). Both methods have their relative merits, yet it is the noncontact, and primarily ink-jet printing, that has attracted the most interest due to the high speed, silence and ease of the operation.

References

[1] Dorey R. A. and Whatmore R.W. 2004. Electroceramic thick film fabrication for MEMS. *Journal of Electroceramics*, **12**(1–2), 19–32.

[2] Setter N. 2001. Electroceramics: looking ahead. *Journal of European Ceramamic Society*, **21**(10–11), 1279–1293.

[3] Roy S. 2007. Fabrication of micro- and nano-structured materials using mask-less processes. *Journal of Physics D: Applied Physics*, **40**(22), R413–R426.

[4] Baborowski J. 2004. Microfabrication of piezoelectric MEMS. *Journal of Electroceramics*, **12**(1–2), 33–51.

[5] Chen B., Huang C. and Wu L. 2003. Crack alleviation processing of lead zirconate titanate thin films deposited by sol-gel method. *Thin Solid Films*, **441**(1–2), 13–18.

[6] Dorey R. A., Stringfellow S. B. and Whatmore R.W. 2002. Effect of sintering aid and repeated sol infiltrations on the dielectric and piezoelectric properties of a PZT composite thick film. *Journal of European Ceramamic Society*, **22**(16), 2921–2926.

[7] Zhao M., Fu R., Lu D. and Zhang T. 2002. Critical thickness for cracking of $Pb(Zr_{0.53}Ti_{0.47})O_3$ thin films deposited on Pt/Ti/Si(100) substrates. *Acta Materialia*, **50**(17), 4241–4254.

[8] Chen Y. H., Li T. and Ma J. 2003. Investigation on the electrophoretic deposition of a FGM piezoelectric monomorph actuator. *Journal of Materials Science*, **38**(13), 2803–2807.

[9] Lu J., Chu J., Huang W. and Ping Z. 2003. Microstructure and electrical properties of Pb(Zr, Ti)O3 thick film prepared by electrostatic spray deposition. *Sensors and Actuators A*, **108**(1–3), 2–6.

[10] Cotton D. P. J., Chappell P. H., Cranny A. and White N. M. 2007. A new binderless thick-film piezoelectric paste. *Journal of Materials Science: Materials in Electronics*, **18**(10), 1037–1044.

[11] LeDren S., Simon L., Gonnard P., Troccaz M. and Nicolas A. 2000. Investigation of factors affecting the preparation of PZT thick films. *Materials Research Bulletin*, **35**(12), 2037–2045.

[12] Maas R., Koch M., Harris N. R., White N. M. and Evans A. G. R. 1997. Thick-film printing of PZT onto silicon. *Materials Letters*, **31**(1–2), 109–112.

[13] Wallace D., Hayes D., Ting C., Shah V., Radulescu D., Cooley P., Wachtler K. and Nallani A. 2007. Ink-jet as a MEMS manufacturing tool. *Proceedings of International Conference on Integration and Commercialization of Micro and Nanosystems*, 10–13 January 2007, Sanya, Hainan, Vol. B.

[14] Heule M., Vuillemin S. and Gauckler L. J. 2003. Powder-based ceramic meso- and microscale fabrication processes. *Advanced Materials*, **15**(15), 1237–1245.

[15] Lewis J. A. 2002. Direct-write assembly of ceramics from colloidal inks. *Current Opinion in Solid State and Materials Science*, **6**(3), 245–250.

[16] Lewis J. A., Smay J. E., Stuecker J. and Cesarano III J. Direct ink writing of three-dimensional ceramic structures., *Journal of American Ceramic Society*, **89**(12), 3599–3609.

[17] Smay J. E., Nadkarni S. S., Xu J. 2007. Direct writing of dielectric ceramics and base metal electrodes. *International Journal of Applied Ceramic Technology*, **4**(1), 47–52.

[18] Wang T. and Derby B. 2005. Ink-jet printing and sintering of PZT. *Journal of the American Ceramic Society*, **88**(8), 2053–2058.

[19] Zhao X., Evans J. R. G., Edirisinghe M. J. and Song J. H. 2003. Formulation of a ceramic ink for a wide-array drop-on-demand ink-jet printer. *Ceramics International*, **29**(8), 887–892.

[20] Song J. H., Nur H. M. 2004. Defects and prevention in ceramic components fabricated by inkjet printing. *Journal of Materials Processing Technology*, **155–156**(1–3), 1286–1292.

[21] Geissler M. and Xia Y. 2004. Principles and some new developments. *Advanced Materials*, **16**(15 special issue), 1249–1269.

[22] Pham D. T. and Gault R. S. 1998. A comparison of rapid prototyping technologies. *International Journal of Machine Tools and Manufacture*, **38**(10–11), 1257–1287.

[23] Heinzl J. and Hertz C. H. 1985. Ink-jet printing. *Advances in Imaging and Electron Physics*, **65**, 91–171.

[24] Okamura S., Takeuchi R. and Shiosaki T. 2002. Fabrication of ferroelectric Pb(Zr,Ti)O3 thin films with various Zr/Ti ratios by ink-jet printing. *Japanese Journal of Applied Physics, Part 1*, **41**(11B), 6714–6717.

[25] Sumerel J., Lewis J., Doraiswamy A., Deravi L., Sewell S. L., Gerdon A. E., Wright D. W. and Narayan R. J. 2006. Piezoelectric ink jet processing of materials for medical and biological applications. *Biotechnology Journal*, **1**(9), 976–987.

[26] Calvert P. 2001. Inkjet printing for materials and devices. *Chemistry of Materials*, **13**(10), 3299–3305.

[27] Blazdell P. F. and Evans J. R. G. 2000. Application of a continuous ink jet printer to solid freeforming of ceramics. *Journal of Materials Processing Technology*, **99**(1), 94–102.

[28] Dong H., Carr W. W. and Morris J. F. 2006. An experimental study of drop-on-demand drop formation. *Physics of Fluids*, **18**(7), 072102.

[29] Lee D. H., Derby B. 2004. Preparation of PZT suspensions for direct ink jet printing. *Journal of the European Ceramic Society*, **24**(6), 1069–1072.

[30] Windle J. and Derby B. 1999. Ink jet printing of PZT aqueous ceramic suspensions. Journal of Materials Science Letters, **18**(2), 87–90.

[31] Tay B. Y., Evans J. R. G. and Edirisinghe M. J. 2003. Solid freeform fabrication of ceramics. *International Materials Reviews*, **48**(6), 341–370.

[32] Lloyd W. J. and Taub H. H. 1988. Output Hardcopy Devices. Ink jet printing. *Output Hardcopy Devices*. Academic Press, Boston, Massachusetts, USA. 311–370.

[33] Le H. P. 1998. Progress and trends in ink-jet printing technology. *Journal of Imaging Science and Technology*, **42**(1), 49–62.

[34] Rogers J. A., Baldwin K., Bao Z. N., Dodabalapur A., Raju V. R., Ewing J. and Amundson K. 2001. Rubber stamped plastic circuits for electronic paper. Proceedings of *International Symposium on Advanced Packaging Materials: Processes, Properties and Interfaces*, 11–14 March 2001, Breselton, Georgia, USA, 98–103.

[35] Fromm J. E. 1984. Numerical calculation of the fluid dynamics of drop-on-demand jets. *IBM Journal of Research and Development*, **2**(3), 322–333.

[36] Seerden K. A. M., Reis N., Evans J. R. G., Grant P. S., Halloran J. W. and Derby B. 2001. Ink-Jet Printing of Wax-Based Alumina Suspensions. *Journal of the American Ceramic Society*, **84**(11), 2514–2520.

[37] Thiele E. S. and Setter N. 2000. Lead zirconate titanate particle dispersion in thick-film ink formulations. *Journal of the American Ceramic Society*, **83**(6), **2000**, 1407–1412.

[38] Ibrahim M., Otsubo T., Narahara H., Koresawa H., Suzuki H. 2006. Inkjet printing resolution study for multi-material rapid prototyping. *JSME International Journal, Series C: Mechanical Systems, Machine Elements and Manufacturing*, **49**(2), 353–360.

[39] Lau W. W. Y. and Burns C. M. 1973. Polystyrene melts on plane glass surfaces. *Journal of Colloid and Interface Science*, **45**(2), 295–302.

[40] Birdi K. S., Vu D. T. 1993. Wettability and the evaporation rates of fluids from solid surface. *Journal of Adhesion Science and Technology*, **7**(6), 485–493.

[41] Tay B. Y. and Edirisinghe M. J. 2002. Time–dependent geometrical changes in a ceramic ink droplet. *Proceedings of the Royal Society A.*, **458**(2025), 2039–2051.

[42] Ray A. K., Lee J. and Tilley H. L. 1988. Direct measurements of evaporation rates of single droplets at large Knudsen numbers. *Langmuir*, **4**(3), 631–637.

[43] Erbil H. Y. 1999. Determination of the peripheral contact angle of sessile drops on solids from the rate of evaporation. *Journal of Adhesion Science and Technology*, **13**(12), 1405–1413.

[44] Chandra S., DiMarzo M., Qiao Y. M. and Tartarini P. 1996. Effect of liquid-solid contact angle on droplet evaporation. *Fire Safety Journal*, **27**(2), 141–158.

[45] Ainsley C., Reis N. and Derby B. 2002. Rapid prototyping of ceramic casting cores for investment casting. *Key Engineering Materials*, **206**(2), 297–300.

[46] Maenosono S., Dushkin C. D., Saita S. and Yamaguchi Y. 1999. Growth of a Semiconductor Nanoparticle Ring during the Drying of a Suspension Droplet. *Langmuir*, **15**(4), 957–965.

[47] Jung H. C., Cho S., Joung J. W. and Oh Y. 2007. Studies on inkjet-printed conducting lines for electronic devices. *Journal of Electronic Materials*, **36**(9), 1211–1218.

[48] Tekin E., DeGans B., Schubert U. S. 2004. Ink-jet printing of polymers - From single dots to thin film libraries. *Journal of Materials Chemistry*, **14**(17), 2627–2632.

[49] Fuller S. B., Wilhelm E. J., Jacobson J. M. 2002. Ink-jet printed nanoparticle microelectromechanical systems. *Journal of Microelectromechanical Systems*, **11**(1), 54–60.

[50] Szczech J. B., Megaridis C. M., Gamota D. R., Zhang J. 2002. Fine-line conductor manufacturing using drop-on-demand PZT printing technology. *IEEE Transactions on Electronics Packaging Manufacturing*, **25**(1), 26–33.

[51] Zeleny B. J. 1914. The electrical discharge from liquid points, and a hydrostatic method of measuring the electric intensity at their surfaces. *Physical Review*, **3**(2), 69–91.

[52] Jaworek A. and Krupa A. 1999. Classification of the modes of EHD spraying, Journal of Aerosol Science. *Journal of Aerosol Science*, **30**(7), 873–893.

[53] Jayasinghe S. N. and Edirisinghe M. J. 2003. A novel process for simulataneous printing of multiple tracks from concentrated suspensions. *Materials Research Innovation*, **7**(2), 62–64.

[54] Rayleigh L. 1882. On the Equilibrium of Liquid Conducting Masses Charged with Electricity. *Philosophical Magazine*, **14**(5), 184–186.

[55] Loscertales I. G., Barrero A., Guerrero I., Cortijo R., Marquez M. and Ganan–Calvo A. M. 2002. Micro/nano encapsulation via electrified coaxial liquid jets. *Science*, **295**(5560), 1695–1698.

[56] Chen D. R., Pui D. Y. and Kaufman S. L. 1995. Electrospraying of conducting liquids for monodisperse aerosol generation in the 4 nm to 1.8 μm diameter range. *Journal of Aerosol Science*, **26**(6), 963–977.

[57] Teng W. D., Huneiti Z. A., Machowski W., Evans J. R. G., Edirisinghe M. J. and Balachandran W. 1997. Towards particle-by-particle deposition of ceramics using electrostatic atomization. *Journal of Materials Science Letters*, **16**(12), 1017–1019.

[58]   Rocks S. A., Wang D., Sun D., Jayasinghe S. N., Edirisinghe M. J. and Dorey R. A. 2007. Direct writing of lead zirconate titanate piezoelectric structures by electrohydrodynamic atomisation. *Journal of Electroceramics*, **19**(4), 289–293.

[59]   Jayasinghe S. N., Edirisinghe M. J. and Kippax P. G. 2004. Relic and droplet sizes produced by electrostatic atomisation of ceramic suspensions. *Applied Physics A.*, **78**(3), 343–347.

[60]   Tohver V., Morissete S. L. and Lewis J. A. 2002. Direct-write fabrication of zinc oxide varistors. *Journl of American Ceramic Society*, **85**(1), 123–128.

[61]   Leong K. F., Cheah C. M. and Chua C. K. 2003. Solid freeform fabrication of three-dimensional scaffolds for engineering replacement tissues and organs. *Biomaterials*, **24**(13), 2363–2378.

[62]   Miranda P., Saiz E., Gryn K. and Tomsia A. P. 2006. Sintering and robocasting of β-tricalcium phosphate scaffolds for orthopaedic applications. *Acta Biomaterialia*, **2**(4), 457–466.

22

Quality assurance and metrology

L. Mattsson, V. Schulze and J. Schneider

The cost-effective large-scale production of reliable microsystems for mechanical and tribological applications requires customized quality assurance. Equipment and procedures established for the characterization of materials and components on the macroscale very often cannot be successfully applied on the microscale. Therefore in recent years customized techniques have been developed for dimensional control, surface topography and roughness characterization as well as for microstructural analysis, mechanical and tribological testing. This chapter overviews the state-of-the-art of research and technological developments in these fields and thus responds on the necessary fundamentals of metrology, the constraints of the used equipment and technologies as well as the particular problems arising from characterization of microcomponents made of ceramic materials.

22.1 Introduction

The speedy characterization of finished ceramic parts is a key to success in safety- and time-critical microtechnological applications. By means of characterization, issues of quality, processing conditions and costs can be effectively addressed. Metrology is still a mitigating factor against achieving 100% quality for finished microparts.

Quality assurance requires that micromanufactured ceramic components and assembled parts have the right dimensions, a specified surface structure, appropriate mechanical properties and functions according to given specifications. Quality assurance is thus a task throughout the entire manufacturing chain, starting with the design and ending with a final functional test. It is very important that the design engineer or a design group of mixed competences have knowledge in materials properties as well as manufacturing processes and available verification tools. Geometric tolerances must be critically analyzed at the design stage with regard to the possibilities of measuring them by noncontacting or contacting probes. Far too many designs are made without considering the fundamental limits of the available measuring probes e.g. accessibility and measurement forces on fragile microstructures. An inspection plan which is prepared at the design stage is therefore strongly recommended. An example of its possible content is shown in Figure 22.1.

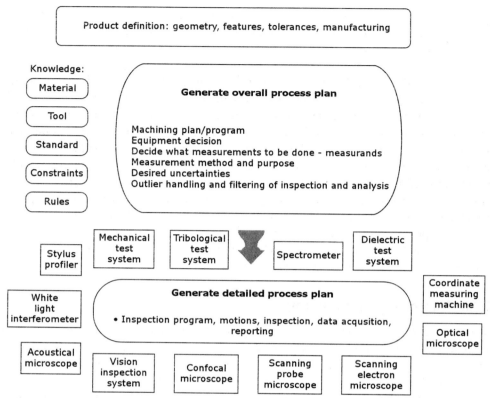

Figure 22.1 Inspection process definition model with possible microcharacterization tools for ceramic microcomponents.

After this general view of the inspection planning we concentrate on the measurement and characterization by reviewing metrology terminology, standards, constraints and equipment. The following section highlights dimensional metrology, roughness measurement and microstructural characterization of the processed ceramic. The last section is devoted to mechanical and tribological characterization.

22.2 Metrology

For microscale parts the measuring techniques must be carefully chosen to be free of significant errors caused by outside disturbances as well as the influence of the sample. The more accurate and finer one wishes to measure the dimensions the more the measurement can be affected by outside influences such as: vibrations, airflow, temperature, illumination, humidity, and measurement- as well as adhesion-forces. Therefore all equipment should preferably be calibrated to a suitable standard, which in turn must be traceable to some defined standard laid down by the International Standards Organization (ISO) [1, 2] and materialized and verified at a national standards institute.

Without these points of reference a measurement that has been taken cannot be supported. The unfortunate fact in micrometrology is that there is a lack of standards that cover the micro–nano range sufficiently well. A very recent example is the lack of high aspect ratio microstructure (HARMS) artifacts that can be used for validating 3D micrometrology systems. When such structures were finally made within the scope of the 4M Network of Excellence [3] and tested it was immediately revealed that several types of commercial optical profilers suffered from serious lateral measurement errors, up to 8 μm, because of the way they interpret the optical response of very deep and narrow features [4]. However, having had access to such a structure a company could optimize its white light interferometer to measure aspect ratios as high as 50 (400 μm deep and 8 μm wide) with lateral errors of, at most, a couple of microns. To keep microstructures within given dimensional tolerances is only one part of the quality assurance. Other very fundamental parts which need to be characterized include mechanical, tribological and dielectrical properties as well as application-dependent functional aspects. In this chapter we will briefly review the dimensional metrology aspects as well as characterization of surface topography and microstructure, mechanical and tribological properties.

22.3 Metrology terminology

This section discusses dimensional metrology as the influence of measurement errors become visible due to misalignments or poor fit. However, the terminology is recommended for any measurement or characterization process delivering numerical data, such as force, dielectric permitivity or refractive index measurements.

The difference between accuracy and resolution needs to be clearly understood. Accuracy may be defined as the correctness i.e. conformity of a measured quantity to its actual (true) value. As no one knows the exact (true) dimension of a component the measured and averaged result has to be supplied with accuracy information expressed as a $\pm$ uncertainty value with a statistical confidence related to it. Usually the uncertainty value is stated as a standard deviation σ (67 % confidence), 2σ (95%) or 3σ (99.7%). The smaller the uncertainty is in the measuring instrument the better the measurements will be and in a production line this means that for a given tolerance width more "tolerance" is given to the machining units and less needs to be offered for the measurement itself.

Resolution is often stated in many instrument specifications and it refers to the readable limits within which the measurement can be ascertained using a particular instrument. As an example a measuring instrument with a read out resolution of 0.1 μm may not be accurate to better than 10 μm because of flexing fixtures, image distortion in the optical system, or inappropriate software algorithms. In the newly released standards for metrology instruments the terminology for instrument specifications will be changed from uncertainty to maximum permissible error (MPE), giving a clearer indication of the performance of an instrument.

All instruments have inherent limits of accuracy and for a good reliability in measurement, it is, if possible, advisable to use instruments with MPE < 10% of the tolerance to be measured.

Precision in metrology has a well-defined meaning and differs from the use of the word in other areas e.g. precision machining, where it means machining to closer tolerances. Precision in metrology characterizes the degree of mutual agreement among a series of individual measurements, and can be very tight (within 0.5 µm) although the average of these measurements can be off from the correct (true) value by several microns perhaps because of a nonlinearity in the scale of the instrument. Thus, high precision in metrology does not necessarily mean high conformity to the true value. Repeatability is another way of expressing precision and it refers to the variation among a series of identical measurements performed under identical conditions, i.e. nothing but the measurement probe or sample stage is moved back and forth to the same place. Reproducibility is evaluated by letting different instrument operators perform the full measurement cycle, to reveal differences in fixture mountings, probe settings etc. as listed in a measurement plan. The uncertainties are therefore widened in reproducibility tests, but they are of considerable importance in practice as most errors and unforeseen instabilities will be incorporated in the measurement cycle. Round robin tests are typical for reproducibility evaluations of measurement systems and are the basis for the traceability chain in dimensional metrology.

The instrumentation for quality assessment must be reliable. For large production volumes statistical analysis should be employed to establish the reliability and allow confidence to be quantified. This is where the six sigma approach and performance measures such as defects per million opportunities and capability index $Cp = (\text{tolerance width})/(6\sigma)$ may be of interest for monitoring and improving production performance [5].

Verification of individual components by standard measuring techniques can assure that the component is within the tolerance limits but when those components are built into an assembly functionality tests can become more complicated, especially when the requirement is to check that the assembly is operating in the manner expected at real operating speeds. This sometimes calls for measuring equipment to be designed as an adjunct to the product design itself. All such metrology equipment has to be "workshop proof" which again imposes tough environmental use during production to ensure quality.

22.4 Shape and dimensional measurements

For a good overall review of the field of micro- and nanometrology the reader is referred to the CIRP keynote papers [6, 7]. Specific challenges for microcomponent metrology are discussed in [8]. Component dimensions need to be assigned tolerances, and in the ISO (Geometrical Product Specification) these are assigned as values of standard tolerance grades specified in ISO 286-1 [9]. For microcomponents it is obvious that the ISO separation of item size would need a finer scale, (see Figure 22.2) as all items < 3 mm are classified into one and the same category. Tolerances of form, orientation, location and run-out are treated in ISO 1101:2004 [10].

The most common equipment for dimensional measurements of today is the coordinate measurement machine (CMM). Typical measurement force is about 50 mN and standard ball probes can be obtained with diameters down to $\varnothing 500\,\mu m$. Ball material is typically ruby, i.e. alumina, but silicon nitride is also available and can be a better choice for measurement

Nominal size mm		Standard tolerance grades								
		IT01	IT0	IT1	IT2	IT3	IT4	IT5	IT6	IT7
Above	Up to and inclu- ding	Standard tolerance values μm								
—	3	0,3	0,5	0,8	1,2	2	3	4	6	10
3	6	0,4	0,6	1	1,5	2,5	4	5	8	12
6	10	0,4	0,6	1	1,5	2,5	4	6	9	15

Figure 22.2 Values of standard tolerance grades for items with nominal sizes < 10 mm according to ISO 286-1 [9].

of alumina parts and vice versa, as similar materials tend to attract them selves and increase adhesion forces. The comparatively large measurement force and size of the ball of conventional CMMs is a drawback for measurement of microparts. Dedicated micrometrology CMMs have been developed by Mitutoyo (UMAP) with a Ø30 μm probe and 1 μN touching force [11], Werth (Fiberprobe) with a Ø10 μm glass ball and the end of a flexible optical fibre [12, 13], and there are systems which focus on accuracy like IBS (ISARA) with a Ø300 μm ball and volumetric uncertainty of 30 nm [14], and Zeiss (F25) with a 250 nm uncertainty and Ø50 μm ball probe [15]. Very accurate positioning systems like the SIOS NMM-1 [16, 17] can be fitted with custom probes, and can thus be adapted to the particular parts being measured.

In order to measure microcomponents and devices in all three dimensions, access is required to all surfaces having their dimensional tolerances specified. This is where the ball probe instruments may have accessibility problems, for instance in HARMSs, with trench widths or hole diameters of the order of 20 μm or less. Although ball probes like the Fiber probe and the UMAP are small, they depend on optical imaging and if the received light cone is diminishingly small uncertainties will rapidly grow. In addition charging and capillary condensation can easily force the ball probe to attach to nearby surfaces in an unacceptable way.

Scanning probe microscopes like the atomic force microscope (AFM) and scanning probe microscope (SPM) are with a few exceptions [18] limited in lateral (x,y) dimensions to approximately 100 μm measurement length and in the vertical (z) direction to 6–10 μm. On the other hand they can be operated down to the subnanometer positioning and have extremely sharp tips so they are the obvious choice when features are approaching atomic sizes laterally and vertically. In contact mode there is a risk of damaging the surface, while the so-called tapping mode, an intermittent touching of the surface, is less prone to cause damage. The market for scanning probe microscopes has grown rapidly and a vast number of instruments are available today.

The alternative to contacting probe systems are optical coordinate metrology systems, or optical profilers based on imaging in the lateral x,y directions and height sectioning in the z-direction. Sectioning is then typically made by autofocusing, white light interferometry, chromatic aberration focusing, or confocal imaging. The performance of these instruments on high aspect ratio structures has been investigated [4] and was found to be useful at high aspect ratios of about 5 for the optical CMM at 35 µm groove width and an aspect ratio of about 14 for a white light interferometer at groove widths of 10 µm, which is better than mechanical probes. However, diffraction-induced problems cause unreliable measurements in the lateral directions, and they can be off by 5–10 µm. False height profiles caused by edge scattering are another problem to be addressed in some of these instruments. A most remarkable achievement has also been made by optimizing a white light interferometer, and data to be published of a 3D star structure reports correct measurements of 400 µm deep grooves at a groove width of 8 µm, i.e. at an aspect ratio of 50 [19].

The most common noncontacting tool, the SEM is often used for evaluating sizes of 3D microfeatures by the size bar attached to the images. But, without proper calibration of this bar under identical conditions regarding surface charging effects and magnification, one can not expect high quality dimensional measures. An ordinary SEM is therefore not considered to be a reliable metrology tool.

For optical measurement systems based on imaging of a light spot on a ceramic surface by triangulation, the high volume scattering of the underlying ceramic imposes a problem. This scattering will move the point of maximum intensity from the surface and into the subsurface, thus causing a systematic error. This is more pronounced for highly polished ceramic surfaces with low top surface scattering. The size of this error depends on the material. Contacting probe calibration is therefore recommended in these cases so this can be systematically compensated.

Vision systems, very common for inspection at high speed manufacturing, may suffer from the same problems and should be carefully calibrated before being used for dimensional measurements in the micron–submicron range. When approaching sub-micron accuracies the imaging and localization of an edge, or height step, is becoming really tricky. The optical image of an abrupt step is no longer a distinct change from black-to-white but rather a continuous intensity change. The question is thus, which grey-level corresponds to the position of the real edge?

Dimensions of embedded structures in high scatter ceramics are very difficult to deal with using conventional optical methods. The two methods that are recommended are: acoustical microscopy [20] and optical coherence tomography (OCT) [21, 22], which are also useful for subsurface defect localization. Ultrasonic techniques [23] and acoustic microscopy share two major shortcomings: they require immersion of the coated part in a liquid to ensure proper coupling of sound waves and they have a resolution limit of about 1 µm under optimal conditions. OCT has recently been demonstrated to detect subsurface defects in Si_3N_4 ceramic balls [24], and to measure the thickness up to 50 µm of thermal barrier coatings [25].

22.5 Surface roughness

Surface roughness is the measured quantity of the microtopography of a surface. Depending on the function of the surface, there are a large number of ways to calculate and assign parameters to the roughness. Some have little functional correlation – like the Ra parameter which is simply a mean value of the height variation of the topography, independent of if the topography is a flat base surface with a few ridges sticking up, or if it is the reverse, i.e. a flat top with a few grooves in it. Both give the same Ra value but behave completely differently when subject to a high load from another surface. Core roughness depth Rk, reduced peak height Rpk and reduced valley depth Rvk [26] are on the other hand well suited for specifying surfaces subject to high loads, wear and lubrication needs, respectively.

In the ISO terminology, surface roughness is the name for the subsection of surface structure containing the short spatial wavelength components, e.g. closely spaced small peaks and valleys. Waviness corresponds to the longer spatial wavelength undulations of the surface. In the ISO standard the split between these two sections is not fixed, but depends on the amplitude (height variation) of the surface structure [27, 28]. For typical engineered surfaces the split is at a spatial wavelength of 800 µm. Traditionally Ra, the arithmetic mean deviation of the assessed profile, has been used to characterize surface roughness in mechanical engineering and Rq in optical engineering. The latter is equivalent to the standard deviation of the surface height variations (often referred to as rms roughness) and has a direct correspondence to the light scattering properties of smooth surfaces [29]. The R-parameter standardized sampling length for surfaces having $Ra < 20$ nm is 80 µm, while surfaces in the $0.02 \, \mu m < Ra < 0.1 \, \mu m$ range require 250 µm and surfaces in the $0.1 \, \mu m < Ra < 2 \, \mu m$ range require a 800 µm sampling length. Generally, roughness should always be measured perpendicular to the direction of grinding scratches, i.e. one should look for the maximum roughness values when aligning the sample for measurement. However, obvious defects in the surface should not be included in the measurement. Good practice is therefore to investigate the sample carefully in a microscope before performing the measurements.

For roughness measurements of microcomponents the default R-parameters are of limited use, as they require a total measurement length of $(2 + 5)$ times the sampling length, which for a typical roughness of 1 µm is 0.8 mm according to the standard. With a demand of 5×0.8 mm long trace plus 2×0.8 mm for start and stop of a mechanical stylus profiler, we need an available surface length of at least 5.6 mm for performing the measurement. At the other end of the spatial wavelength spectrum, the possible resolution is set by the stylus radius for contacting probes or the optical or pixel resolution of imaging microscopes. The ISO standard for mechanical profilers has a sharpest nominal tip radius of 2 µm, and cone angle of 60°.

A much better practice for micro-components than using the default R-parameters is to make use of the P-parameters, e.g. Pa and Pq, as they refer to the primary profile and can be measured at any length along the surface as long as the sampling length (the measured lateral distance) is stated together with the value of the P-parameter. The only filtering of the primary profile should be the short wavelength cut off at a spatial wavelength of 2.5 µm for stylus profilers [30]. Thus, the surface roughness can be measured according to the ISO

4287 standard at a 50 μm wide ceramic micro-feature. In this case Pa or Pq should preferably be calculated for a sampling length of the central 40 μm across the feature to avoid any influence from rounded edges. When performing roughness measurements on tiny areas having height variations in the micron range, the statistical spread will be considerable from one profile to the other. Variations of 50% in Pq values are not at all uncommon. For good repeatability measures of surface roughness on tiny areas it might be necessary to polish them to keep the spread in Pa or Pq within the tolerance limit.

Standard equipment for surface roughness measurements is available from many vendors, but for tiny measurement areas special profilers are required. The most common are optical profilers based on microscope systems, using interferometry, confocal imaging, or autofocusing systems. Many of them have software updated to be in accordance with the ISO standards, but a skilled instrument operator who knows in detail what the instrument does is necessary. Otherwise, parameter and filter settings may easily cause erroneous results. Traditional workshop stylus profilers can hardly be used for microsurface roughness measurements, they are not precise enough in positioning at microsized features. But dedicated profilers for semiconductor wafer structure measurements can be used as well as a few dedicated step-height profilers, like the Talystep which has been used as a reference system for many years [31] and the Nanostep [32, 33].

If roughness profiles shorter than 100 μm and with height deviations < 6 μm need to measured AFM is the obvious choice. This is also the case if roughness information needs to be captured at submicron to nanometre spatial wavelengths, as there are no other techniques available there. Note that in this case it is off-ISO standard for mechanical profilers, which are limited to spatial wavelengths > 2.5 μm.

Some examples of AFM measurements on ceramics where the probe tip is in physical contact with the specimen surface is given in [34, 35]. Applications of the noncontacting confocal microscope are given in [36, 37].

22.6 Microstructural analysis

Characterization of ceramic microparts is a technically challenging task and may require impeding the integration in a production line. A number of studies [38–43] have reported on test methods for microparts. In this section the focus will be on test methods for vital structural properties in ceramic microparts namely: surface and bulk microstructure and the residual stress state. It is known [44–47] that these properties can significantly influence the mechanical and tribological behaviour and the overall service life of ceramics e.g. silica, silicon nitride, silicon carbide, zirconia or alumina.

The surface analysis can be grouped into chemical, related to the chemistry at the surface, and physical, related to the microtopography. For the chemical analysis common methods such as Auger electron spectroscopy (AES) and X-ray photoelectron spectroscopy (XPS) can easily be used [48–50]. The accuracy and reproducibility of the results from AES and XPS measurements depend both on the specimen and the instrument. Charging effects in nonconducting specimens such as ceramics and the inability of the instrument to resolve electron yield or weakly scattering phases are examples. Both methods are at best nondestructive when sputter

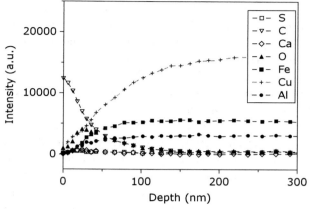

Figure 22.3 AES depth profile of a microcast aluminium bronze tensile test specimen.

depth profiles are not required. AES has a lateral resolution of about 10 nm while XPS yields to 15 μm. Figure 22.3 shows an AES depth profile obtained from a microcast aluminium bronze (5 wt% each Ni and Fe) specimen of cross-section $130 \times 260 \, \mu m^2$ where there is carbon contamination at the surface.

Physical analysis involves the methods discussed above in Section 22.5 such as AFM and optical microscopy utilizing height sectioning of the microtopography. Both methods provide 3D maps of the probed surface. As an example Figure 22.4 shows the surface topography of a zirconia microbeam test specimen of dimension $200 \times 200 \times 1200 \, \mu m^3$ obtained by a confocal microscope.

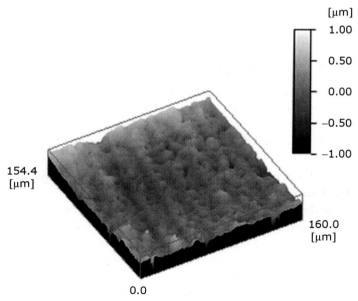

Figure 22.4 3D image of surface of an injection moulded zirconia microspecimen.

Imaging of materialographically prepared cross-sections using optical light microscope, SEM and focused ion beam imaging techniques are some common imaging methods. The specimen to be tested for the structure state must first be etched thermally or chemically depending on production conditions [51]. Although resolution can differ markedly amongst these imaging methods SEM appears to be the most viable option for ceramic materials because of the combined advantage of resolution 1–20 nm, easy data interpretation and acquisition time which takes only a few minutes. Figure 22.5 shows a SEM image of a 3 mol.% yttria stabilized zirconia (3YSZ). The grain size can be determined using the linear intercept method [52] while the porosity can be analysed by the binary method on a reduced grayscale spectrum [53]. A trick which can be used in order to rapidly image nonconducting specimens without elaborate preparation is to identify a cracked region on the specimen or intentionally create a cracked surface for imaging. Figure 22.6 shows the SEM image of a microgear part made by MicroPIM. Grains in the cracked surface are better resolved.

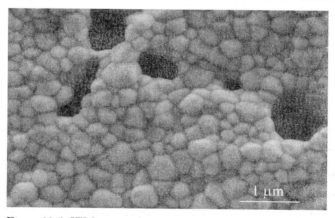

Figure 22.5 SEM image of a zirconia specimen showing porous regions.

Figure 22.6 SEM image of a crack region on a micro-gear wheel. Grains in the crack are better resolved.

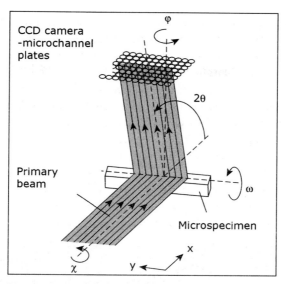

Figure 22.7 Sketch of MAXIM camera within a diffractometer setup.

For the residual stress analysis (RSA) of microparts made from ceramic materials a variety of approaches may be followed [42, 43, 54]. The XRD $\sin^2 \psi$ method is currently the most applied RSA method. However, there can be problems associated with the nature of the specimens or unsuitable instrumentation. For example, grain size effects may lead to poor signals. Also the multiphase nature of some microspecimens may make it impossible to determine the residual stresses effective in the material [55]. Studies carried out within the framework of the German Collaborative Research Centre 499 (SFB 499) have demonstrated that with special X-ray detectors such as the MAXIM [42], residual stresses in microspecimens can be analysed locally. A setup indicating the MAXIM output on a lateral resolution of ~13 μm is presented in Figure 22.7. Potentials for the determination of residual stresses in microparts have been shown [43] where the stress state in silicon nitride thin films are accurately deduced by monitoring the displacements created when focused ion beam tools are used to mill a micrometre size slot in the specimen.

22.7 Mechanical testing

For the mechanical characterization of ceramics and metals in microscale the mechanical tests are performed usually as three-point bending, tensile or fracture tests. Additionally test procedures based on microindentation like hardness measurement, compression tests or push-out tests are available. Only in a severely restricted number of cases were cyclic tests performed. Overviews to usual microtest techniques are given in [56–63]. Some additional methods are presented briefly below.

Microspecimens may be directly fabricated and tested without finishing, so that the as-fired surface effects are fully active. This can be important for MEMS parts. Hot moulding, microPIM or microcasting (of metals) are applicable in this context for the generation of tensile test specimens or microbending beams. Additionally specimens may be fabricated by etching processes [64], micro-STL based on thin films [65] or prepared from larger parts. A new and challenging approach in this context is the usage of the focused ion beam technique to prepare microcolumns for compression tests using nanoindentation [66]. This technique has been used up to now particularly for metals. At present the materials examined most strongly in the microdimension are single and polycrystalline silicon. Further materials in focus are silica, silicon nitride, silicon carbide, alumina and zirconia.

Microthree-point bending tests were performed in a large variety within the German Collaborative Research Centre 499 (SFB 499) using a self-designed microuniversal test device [67]. This consists of an electrodynamic actuator allowing for cross-head velocities of $2\,\mu m/s$. The lower support used for three-point bending shows a bending width of $800\,\mu m$ and radii of $50\,\mu m$ and was fabricated using EDM. The specimens which were tested were quasistatically or cyclically microcast Au and Cu alloys and hot-moulded ZrO_2 and Si_3N_4 [67–72]. They had dimensions of about $200 \times 200 \times 1{,}200\,\mu m^3$. The results for ZrO_2 presented in Figure 22.8 show the typical effect of Weibull statistics, leading to increases of the characteristic strength up to 3,181 MPa [70] compared to macrospecimens of the usual size $3 \times 4 \times 40\,mm^3$, which show strengths in the range of 1,200 MPa [73] for the same powder. Tensile tests on single SiC-whiskers with diameters in the range $4–7\,\mu m$ and lengths of about $10–25\,mm$ produced by a vapour–liquid–solid process have been carried out [74]. They also used a self-designed test device. Microindentation tests using spherical

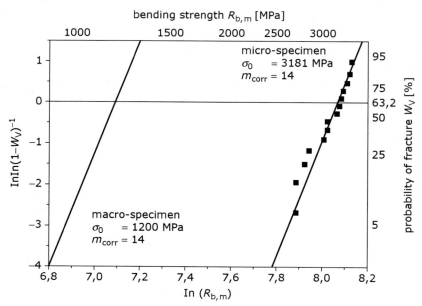

Figure 22.8 Weibull-characteristics of ZrO_2 in three-point bending, comparison of specimen sizes (micro: $200 \times 200 \times 1200\ \mu m^3$, macro: $3 \times 4 \times 40\,mm^3$) [35].

Figure 22.9 Test of root-strength of micro sun gear teeth made of ZrO_2 [107].

and Vickers indenters have been performed with a stepping motor driven test device [75, 76] and used a Berkovich indenter with loads in the range 0.2–15 mN for mechanical characterization of thin films made of Si and SiC. Last but not least, microcomponents (e.g. microgears) can be tested. The universal testing device described above [67] allows for single tooth fracture tests like that shown in Figure 22.9. Albers *et al.* [77] developed a microgear testing device which allows standardized tests like single and double flank gear rolling inspection, rotating backlash and efficiency test with torques of up to 50 mNm.

22.8 Microtribological testing

The performance and reliability of MEMS with movable components are strongly affected by the tribological behaviour, which depends on manufacturing processes as well as on alignment of the microcomponents, the operating conditions and the materials. High friction forces reduce the usable output torque (kinetic friction) or even prevent (static friction) the operation of microsystems such as turbines, motors or gears. Even very small amounts of wear resulting in loose wear debris can cause the loss of function of micromechanical devices. Hence, friction and wear have to be recognized as very important issues in microtechnology. Several studies on friction and wear of tribologically loaded microcomponents have clarified, that results from tribological tests on the macroscale cannot simply be transferred to the microscale. Because of the large surface area to volume ratio in micromechanical systems adhesion and/or capillary forces become critical and can even surpass externally applied forces. Therefore friction and wear behaviour of any material pair which is considered for utilization in microtribological systems have to be systematically characterized at the microscale [78–82].

Tribological tests with real micromechanical systems [78, 83–85] are very helpful for controlling functional behaviour and estimating service life, but often show a relatively large scatter of results due to the insufficient knowledge of the geometry (e.g. evenness, clearance), operating (e.g. contact pressure, humidity), surface (e.g. roughness or residual stresses) and/ or materials (e.g. hardness, Young's modulus, grain size) parameters. Therefore laboratory tests with simplified specimen geometry, strictly defined and controlled loading, surface, environmental conditions and materials properties are very important. Results from such tests allow a better understanding of relevant friction and wear mechanisms on microscale and are the essential basis for the successful design of tribologically loaded MEMS.

In recent years intensive research work has been done on tribological issues of silicon and thin-film based MEMS as well as on head materials for magnetic storage devices [78, 80, 86–91]. Most of the experimental studies were done using AFM and/or friction force microscopy (FFM) and focused on the characterization of adhesion and static friction phenomena, problems of lubrication (e.g. self-assembling monolayers) and last but not least dynamic friction. Characteristic loads within the nN range, low sliding speeds from typically a few µm/s up to about 10 mm/s and also the limited range of materials make it difficult to transfer results from these fundamental nanotribological tests to MEMS under high mechanical and tribological loadings.

The increasing demand for tribological tests suitable for the typical operating conditions of highly loaded microsystems, such as microgears, led to the development of microtribo testers covering the load range from µN to mN and sliding speeds up to a few hundred mm/s. Beside specialized laboratory equipment for abrasive and scratch tests various microtribometers were developed for tests in unidirectional and oscillating sliding or rolling contact, some of which are now commercially available (Tetra GmbH, Ilmenau, Germany; CSM Instruments, Peseux, Switzerland; CETR Inc., Campbell, CA, USA) [77, 92–98]. These tribometers allow an accurate tribological testing under well-defined and controlled conditions for loading and environment. Recently additional finite element simulations of microtribological contact problems have proved to be helpful by reducing the experimental effort and improving the understanding of critical operating conditions [99–101].

Results from microtribological tests with ceramic materials are shown below. Within the frame of Collaborative Research Centre 499 at the University of Karlsruhe tribological properties of Al_2O_3 and ZrO_2 produced by ceramic MicroPIM at the Institute of Materials Research III at the Forschungszentrum Karlsruhe as well as commercially available Si_3N_4 and ZrO_2 were extensively characterized both under sliding and rolling conditions [94,102–106]. Figure 22.10 summarizes results from sliding tests, which were run in ambient air with a relative humidity of 50% and in distilled water using a ball-on-disc configuration at a normal load of $F_N = 800$ mN and a sliding speed of $v = 400$ mm/s over a sliding distance of $s = 1,000$ m and clearly point out the significant influence of both the self-mated materials and environmental conditions. Rolling tests were run using a disc-on-disc

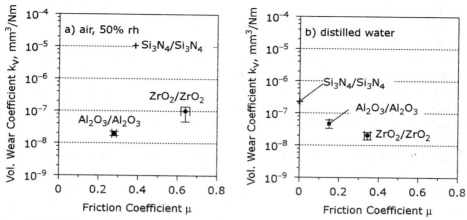

Figure 22.10 Volumetric wear coefficient versus friction coefficient of self-mated Si_3N_4, ZrO_2 and Al_2O_3 after 1,000 m in unidirectional sliding contact in: (a) air with a relative humidity of 50%; (b) distilled water (F_N = 800 mN, v = 400 mm/s); after [103].

configuration at a normal load of F_N = 250 mN and a rolling speed of v_1 = 800 mm/s for the driven disc over 10^6 revolutions at slip ratios of 1, 4 and 10%, respectively, (see Figure 22.11). In unlubricated contact, the friction coefficient increased with increasing slip ratio for both ceramic rolling pairs (Figure 22.11(a)). Under water lubrication, the influence of the slip ratio on the friction coefficient was less significant (Figure 22.11(b)). Whereas friction coefficient was extremely low and nearly independent of the slip ratio for Si_3N_4, friction coefficient of the self-mated ZrO_2 slightly decreased with increasing slip ratio under water lubrication.

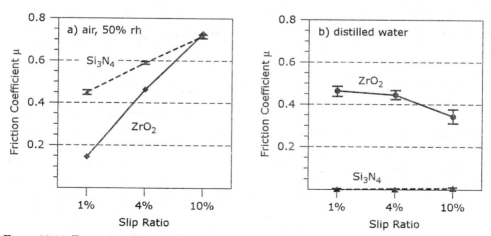

Figure 22.11 Friction coefficient of the self-mated Si_3N_4 and ZrO_2 after 10^6 revolutions in rolling contact in: (a) air with a relative humidity of 50%; (b) distilled water versus the slip ratio (F_N = 250 mN, v_1 = 800 mm/s), after [106].

Conclusions

The problem of quality assurance is one of the most essential keys for the effective fabrication of reliable microcomponents. Since many measurement techniques and instruments which are successfully used on the macroscale are not applicable in microtechnology specifically adapted equipment and procedures are necessary. This chapter has covered several aspects of quality assurance to be considered in design, manufacturing and verification of microceramic components. After a short introduction to the fundamentals of metrology a survey of recent developments in dimensional metrology and roughness measurement using both contacting (stylus) and noncontacting (optical) techniques was given. Furthermore the state-of-the-art in microstructural analysis, mechanical and tribological testing on the microscale was discussed and results of the latest research in these fields have been presented.

References

[1] ISO 5436–1:2000. Geometrical product specifications (GPS)–Surface texture: Profile method; Measurement standards–Part 1: Material measures, International Organization for Standardization, Geneva, Switzerland.
[2] ISO 3650:1998. Geometrical product specifications (GPS)–Length standards, International Organization for Standardization, Geneva, Switzerland.
[3] Dimov S.S., Bramley A.N., Eberhardt W., Engel U., Fillon B., Johander P., Jung E., Kirby P.B., Matthews C.W., Mattsson L., Richter M., Ritzhaupt–Kleissl H.J., Schoth A., Velten T. and Wenzel C. 2007. 4M network of excellence, progress report 2004–2006. In *Proceedings of 3rd International Conference on Multi–Material Micro Manufacture*, 3–5 October 2007, Borovets, Bulgaria, Whittles Publishing, Scotland, xvii.
[4] Mattsson L. 2007. Experiences and challenges in multi material micro metrology. *Proceedings of 2nd International Conference on Micro Manufacturing*, 10–13 September 2007,Clemson University, Clemson, South Carolina, USA. (on DVD).
[5] Magnusson K., Kroslid D. and Bergman B. 2003. *Six sigma–The pragmatic approach*. Student Literature, Lund, Sweden.
[6] Hansen H.N., Carneiro K., Haitjema H. and De Chiffre L. 2006. Dimensional micro and nano metrology. *Annals of the CIRP*, **55**(2) 721–743.
[7] Danzebrink H.U., Koenders L., Wilkening G., Yacoot A. and Kunzmann H. 2006. Advances in scanning force microscopy for dimensional metrology. *Annals of the CIRP*, **55**(2) 841–878.
[8] Mattsson L. 2006. Metrology of micro–components–a real challenge for the future. In *Proceedings of the 5th International Seminar on Intelligent Computation in Manufacturing Engineering (CIRP ISME '06)*, 25–28 July 2006, Ischia, Italy.
[9] ISO 286–1:1998 Geometrical product specifications (GPS)—ISO code system for tolerances of linear sizes—Part 1:Basis of tolerances, deviations and fits. *International Organization for Standardization*, Geneva, Switzerland.
[10] ISO 1101:2004. Geometrical product specifications (GPS)—Geometrical tolerancing— Tolerances of form, orientation, location and run–out. *International Organization for Standardization*, Geneva, Switzerland.
[11] UMAP, Vision System Hyper 302—hybrid vision/touchprobe measuring machine from Mitutoyo, Andover, United Kingdom.

[12] Brand U. and Kirchhoff J. 2005. A micro–CMM with metrology frame for low uncertainty measurements. *Measurement Science and Technology*, **16**, 2489–2497.

[13] Werth Messtechnik GmbH, Giessen, Germany.

[14] Isara, 3D–CMM from IBS Precision Engineering bv, Eindhoven, the Netherlands.

[15] F25, 3D–CMM from Carl Zeiss Industrielle Messtechnik GmbH, Oberkochen, Germany.

[16] Jager G., Manske E., Hausotte T. and Schott W. 2002. Operation and analysis of a nanopositioning and nanomeasuring machine, in *Proceedings of the ASPE Annual Meeting*, 20–25 October 2002, St. Louis, Missouri, USA, Carr J W(Ed). 299.

[17] NMM1, Nano Positioning and nano–measuring–machine from SIOS Messtechnik GmbH, Ilmenau, Germany.

[18] Dai G., Pohlenz F., Danzebrink H.U., Xu M., Hasche K. and Wilkening G. 2004. Metrological large range scanning probe microscope. *Review of Scientific Instruments*, **75**(19), 962–969.

[19] Mattsson L. 2007. Metrology of 3D microstructures–a demanding task with surprising results. Proceedings of *4th Erlangen Workshop on Microforming*, 27–28 November 2007, Engel U. and Geissdörfer S.(Ed).

[20] Briggs A. 1992. Acoustic microscopy–a summary. *Reports on Progress in Physics*, **55**, 851–909.

[21] Bashkansky, M., Duncan, M.D. and Reintjes, J. 1998. High–speed high–resolution subsurface defect detection in ceramics using optical gating techniques. *Review of Progress in Quantitative Nondestructive Evaluation*, **2**, 1785–1791.

[22] Veilleux J., Moreau C., Le´vesque D., Dufour M. and Boulos M. 2007. Optical coherence tomography for the inspection of plasma–sprayed ceramic coatings. *Journal of Thermal Spray Technology*, **16**, 435–443.

[23] Vincent A. and Moughil A. 1989. Ultrasonic characterization of zirconia–based thermal barriers. *NDT International*, **22**, 283–291.

[24] Bashkansky M., Lewis III D., Pujari V., Reintjes J. and Yu H.Y. 2001. Subsurface detection and characterization of hertzian cracks in Si3N4 balls using optical coherence tomography. *NDT&E International*, **34**, 547–555.

[25] Ellingson W.A., Visher R.J., Lipanovich R.S. and Deemer C.M. 2006. Optical NDT techniques for ceramic thermal barrier coatings. *Materials Evaluation*, **64**, 45–51

[26] ISO 13565–2:1997. Geometrical product specifications (GPS)—Surface texture: Profile method–Surfaces having stratified functional properties – Part 2: Height characterization using the linear material ratio curve. *International Organization for Standardization*, Geneva, Switzerland.

[27] ISO 4287:1997. Geometrical product specifications (GPS)—Surface texture: Profile method—Terms, definitions and surface texture parameters, *International Organization for Standardization*, Geneva, Switzerland.

[28] ISO 4288:1996. Geometrical product specifications (GPS)—Surface texture: Profile method–Rules and procedures for the assessment of surface texture. *International Organization for Standardization*, Geneva, Switzerland.

[29] Bennett J.M. and Mattsson L. 1999. *Introduction to Surface Roughness and Scattering*, 2nd revised edn. Optical Society of America, Washington D.C. USA.

[30] ISO 3274:1997. Geometrical product specifications (GPS)—Surface texture: Profile method–Nominal characteristics of contact (stylus) instruments. *International Organization for Standardization*, Geneva, Switzerland.

[31] Bennett J.M. and Dancy J.H. 1981. Stylus profiling instrument for measuring statistical properties of smooth optical surfaces. *Applied Optics*, **20**, 1785–1802.

[32] Lindsey K., Smith S.T. and Robbie C.J. 1988. Subnanometre surface texture and profile measurement with NANOSURF 2. *CIRP Annual*, **37**, 519–522.

[33] Bennett J.M., Elings V. and Kjoller K. 1993. Recent developments in profiling optical surfaces. *Applied Optics*, **32**, 3442–3447.

[34] Wang Z.F., Cao W., Shan X.C., Xu J.F., Lim S.P., Noell W. and de Rooij N.F. 2004. Development of 1 L' 4 MEMS–based optical switch. *Sensors and Actuators A: Physical*, **114**, 80–87.

[35] Auhorn M., Kasanická B., Beck T., Schulze V. and Löhe D. 2006. Mechanical strength and microstructure of Stabilor–G® and ZrO2 microspecimens. *Microsystem Technologies*, **12**(7), 713–716.

[36] Horsch Ch., Schulze V. and Löhe D. 2005. Vermessung und bewertung der oberflächentopographie von mikrobauteilen mittels konfokaler weißlichtmikroskopie. *Praktsche Metallographie*, **42**, 377–386.

[37] Prasad V., Semwogerere D. and Weeks E.R. 2007. Confocal microscopy of colloids. *Journal of Physical Condensed Matter*, **19**(11), 113102.

[38] Tay C.J., Quan C., Shang H.M., Wu T. and Wang S. 2003. New method for measuring dynamic response of small components by fringe projection. *Optical Engineering*, **42**, 1715–1720.

[39] Berek H. and Tiederle V. 1998. New method in accelerated reliability testing of electronic components for automotives under field conditions. In *Proceedings of the International Electronic Manufacturing Technology Symposium*, 27–29 April 1998.

[40] Wu D., Fang N., Sun C. and Zhang X. 2002. Adhesion force of polymeric three–dimensional microstructures fabricated by microstereolithography. *Applied Physics Letters*, **81**(21), 3963–3965.

[41] Auhorn M., Kasanická B., Beck T., Schulze V. and Löhe D. 2003. Microstructures, surface topographies and mechanical properties of slip cast and powder injection moulded microspecimens made of zirconia. *Zeitschrift für Metallkunde*, **94**, 599–606.

[42] Kasanická B., Wroblewski Th., Schulze V. and Löhe D. 2005. Analysis of residual stresses in micro powder injection moulded micro bending specimens made of zirconia. *Materials Science Forum*, **490–491**, 503–508

[43] Sabaté N., Vogel D., Gollhardt, Keller J., Michel B., Cané C., Gràcia I. and Morante J.R. 2006. Measurements of residual stresses in micromachined structures in a microregion. *Applied Physics Letters*, **88**, 071910-1–071910-3.

[44] Zum Gahr K.H. 1987. *Microstructure and Wear of Materials*, Elsevier, Amsterdam.

[45] Bonny K., De Baets P., Van der Biest O., Vleugels J. and Lauwers B. 2007. Edge effects in sliding wear behaviour of ZrO2–WC composites and WC–Co cemented carbides. *Materials Science Forum*, **561–565**, 503–506.

[46] Khan Z.A., Hadfield M., Tobe S. and Wang Y. 2006. Residual stress variation during rolling contact fatigue of refrigerant lubricated silicon nitride bearing elements. *Ceramics International*, **32**, 751–754

[47] Rögner J., Okolo B., Kurzenhäuser S., Müller M., Bauer W., Ritzhaupt–Kleissl H.J., Kerscher E., Beck T., Schulze V. and Löhe D. 2008. Relationships between process, microstructure and properties of molded zirconia micro specimens. *Microsystems Technologies*, **14**(12), 1831–1837.

[48] Yao M., He Y., Wang D. and Gao W. 2005. Nano–laminated ZrO2–Al2O3 films prepared by electrochemical deposition. *Electrochemical Solid State Letters*, **8**, C89–C90.

[49] Werner H.W. and Garten R.P.H. 1984. A comparative study of methods for thin film and surface analysis. Reports on Progress in Physics, **47**, 221.

[50] Fischmeister H.F. 1988. Applications of surface analysis in materials science and technology. *Fresenius' Journal of Analytical Chemistry*, **332**, 421–432.

[51] Kasanická B., Müller M., Auhorn M., Schulze V., Bauer W., Beck T., Ritzhaupt–Kleissl H.J. and Löhe D. 2006. Correlations between production process, states and mechanical properties of microspecimens made of zirconia. *Microsystem Technologies*, **12**(12), 1133–1141.

[52] Brandon D. and Kaplan W.Y. 1999. *Microstructural characterisation of Materials*, Wiley, New York, New York, USA.

[53] Schneider G., Postler I., Wührl I. and Böder H. 1994. Quantitative analyse der porenstruktur von sinterwerkstoffen. *Sonderband der Praktsche Metallographie*, **26**, 117–126.

[54] Yi T. and Kim C.J. 1999. Measurement of mechanical properties for MEMS materials. *Measurement Science and Technology*, **10**, 706–716.

[55] Okolo B., Rögner J., Kerscher E., Beck T., Schulze V., Wanner A. and Löhe A.D. 2007. Size effects in aluminium bronze cast specimens. In Proceedings of Micromaterials and Nanomaterials, First World Congress on MicroNanoReliability, 2–5 September 2007, Berlin, Germany. **6**, 179.

[56] Allameh S.M. 2003. An introduction to mechanical–properties–related issues in MEMS structures. *Journal of Material Sciences*, **38**, 4115–4123.

[57] Jadaan O., Nemeth N., Bagdahn J. and Sharpe W. 2003. Probabilistic Weibull behavior and mechanical properties of MEMS brittle materials. *Journal of Material Sciences*, **38**, 4087–4113.

[58] Obergfell K., Beck T., Schulze V. and Löhe D. 2000. Mikroproben quasistatisch und zyklisch prüfen. *MP Materialprüfung*, **42**, 391–395.

[59] Ruther P., Bacher W. and Feit K. 1997. Entwicklung eines LIGA–Mikrosystems zur Messung mechanischer Eigenschaften von Mikroproben, Wissenschaftliche Berichte Forschungszentrum Karlsruhe, 5986.

[60] Sharpe Jr. W.N. 2002. Mechanical properties of MEMS materials. In *The MEMS Handbook*, Gad-el-Hak M.(Ed), CRC Press, Boca Raton, Florida, USA.

[61] Sharpe Jr. W.N. and Bagdahn J. 2002. Fatigue of materials used in microelectro–mechanical systems (MEMS). In Proceedings of the 8th International Fatigue Congress, 2–6 June 2002, Stockholm, Sweden, Blom A.F.(Ed), 2197–2212.

[62] Sharpe Jr. W.N. 2003. Murray Lecture, Tensile testing at the micrometer scale: opportunities in experimental mechanics. In the *Proceedings of the Society for Experimental Mechanics*, **43**, 228–237.

[63] Weiss B. and Hadrboletz A. 2002. Fatigue of micromaterials. In the Proceedings of the 8th International Fatigue Congress, 2–6 June 2002, Stockholm, Sweden, Blom A.F.(Ed), 2233–2244.

[64] Connally J. and Brown S. 1993. Micromechanical Fatigue testing. *Experimental Mechanics*, **33**, 81–90.

[65] Bertsch A., Jiguet S. and Renaud P. 2004. Microfabrication of ceramic components by microstereolithography. *Journal of Micromechanics and Microengineering*, **14**, 197–203.

[66] Volkert C.A. and Lilleodden E.T. 2006. Size effects in the deformation of sub–micron Au columns. *Philosophical Magazine*, **86**, 5567–5579.

[67] Auhorn M., Beck T., Schulze V. and Löhe D. 2002. Quasi–static and cyclic testing of specimens with high aspect ratios produced by micro–casting and micro–powder–injection–moulding. *Microsystem Technologies*, **8**, 109–112.

[68] Auhorn M., Kasanická B., Beck T., Schulze V. and Löhe D. 2005. Microstructure, surface topography and mechanical properties of molded ZrO2 micro specimens, (Chapter 20). in: *Special Edition of Advanced Micro and Nanosystems: Micro–Engineering in Metals and Ceramics*, Löhe D. and Haußelt J.(Eds.), Wiley–VCH, Weinheim, Germany.

[69] Auhorn M., Beck T., Schulze V. and Löhe D. 2004. Determination of mechanical properties of slip cast, micro powder injection moulded and microcast high aspect ratio microspecimens. *Microsystem Technologies*, **10**, 489–492.

[70] Auhorn M., Kasanická B., Beck T., Schulze V. and Löhe D. 2006. Mechanical strength and microstructure of Stabilor–G and ZrO2 microspecimens. *Microsystem Technologies*, **12**, 713–716.

[71] Gronych D., Auhorn M., Beck T., Schulze V. and Löhe D. 2007. Influence of surface defects and edge geometry on the bending strength of slip–cast ZrO2 micro–specimens. *Zeitschrift fur Metallkunde*, **95**, 551–558.

[72] Rögner J., Schwind T., Kerscher E. and Schulze V. Investigations on the influence of size effect on the strength of silicon nitride. *Microsystem Technologies*, (submitted for publication).

[73] Tosoh Corporation. 2009. Zirconia powders easy sintering grades. http://www.tosoh.com/ Products/e+basic2_grades (accessed 2009).

[74] Petrovic J., Milewski J., Rohr D. and Gac F. 1985. Tensile mechanical properties of SiC whiskers. *Journal of Materials Science*, **20**, 1167–1177.

[75] Alcalá J. 2000. Instrumented micro–indentation of zirconia ceramics. *Journal of the American Ceramic Society*, **83**, 1977–1784.

[76] Li X. and Bhushan B. 1999. Micro/nanomechanical characterization of ceramic films for microdevices. *Thin Solid Films*, **340**(1–2), 210–217.

[77] Albers A., Burkardt N., Deigendesch T., Ellmer C. and Hauser S. 2008. Validation of micromechanical systems. *Microsystem Technologies*, **14**(9–11), 1481–1485

[78] Bhushan B. 2007. Nanotribology and nanomechanics of MEMS/NEMS and BioMEMS/ BioNEMS materials and devices. *Microelectronic Engineering*, **84**(3), 387–412.

[79] Scherge M. 2001. Scale dependence of friction. In *Proceedings of the World Tribology Conference*, 3–7 September 2001, Vienna, Austria. 489/1–489/7.

[80] Williams J. and Le H. 2006. Tribology of MEMS. *Journal of Physics D–Applied Physics*, **39**(12), R201–R214.

[81] Zum Gahr K.H. 1998. Tribological aspects of Microsystems. In *Micro Mechanical Systems – Principles and Technology*, Fukuda T., Menz W. (Eds.), Elsevier, Amsterdam, the Netherlands. 83–113.

[82] Zum Gahr K.H., Blattner R., Hwang D.H. and Pöhlmann K. 2001. Micro–and macro–tribological properties of SiC ceramics in sliding contact. *Wear*, **250**(1–12), 299–310.

[83] Bieger T. and Wallrabe U. 1996. Tribological investigations of LIGA–microstructures. *Microsystem Technologies*, **2**(2), 63–70.

[84] Thürigen C., Ehrfeld W., Hagemann B., Lehr H. and Michel F. 1998. Development, fabrication and testing of a multi–stage micro gear system. In *Tribology Issues and Opportunities in MEMS*, Bhushan B. (Ed.), Kluwer, Dordrecht, the Netherlands, 397–402.

[85] Stark K.C, Mehregany M. and Phillips S.M. 1998. Analysis of gear tooth performance of mechanically–coupled, outer–rotor polysilicon micromotors. In *Tribology Issues and Opportunities in MEMS*, Bhushan B. (Ed.), Kluwer, Dordrecht, the Netherlands, 403–406

[86] Flater E., Corwin A., de Boer M. and Carpick R. 2006. In situ wear studies of surface micromachined interfaces subject to controlled loading. *Wear*, **260**(6), 580–593.

[87] Kim S., Asay D. and Dugger M. 2007. Nanotribology and MEMS. *Nanotoday*, **2**(5), 22–29.

[88] Liu H. and Bhushan B. 2003. Adhesion and friction studies of microelectromechanical systems/nanoelectromechanical systems materials using a novel microtriboapparatus. *Journal of Vacuum Science and Technology*, **A21**(4), 1528–1538.

[89] Maboudian R., Ashurst W. and Carraro C. 2002. Tribological challenges in micromechanical systems. *Tribology Letters*, **12**(2), 95–100.

[90] Maboudian R. and Carraro C. 2004. Surface chemistry and tribology of MEMS. *Annual Review of Physical Chemistry*, **55**(1), 35–54.

[91] Tambe N.S. and Bhushan B. 2005. Durability studies of micro/nanoelectromechanical systems materials, coatings and lubricants at high sliding velocities (up to 10 mm/s) using a modified atomic force microscope. *Journal of Vacuum Science and Technology*, **A23**(4), 830–835.

[92] Bieger T. 1997. Fabrication, metallographic and tribologic investigations of LIGA–microstructures made of nickelphosphorous alloys. *Microsystem Technologies*, **3**(4), 155–163.

[93] Gee M. and Gee A. 2007. A cost effective test system for micro–tribology experiments. *Wear*, **263**(7–12), 1484–1491.

[94] Herz J., Schneider J. and Zum Gahr K.H. 2003. Microtribological characterisation of ceramic materials under sliding and slip–rolling conditions. In *Proceedings of MICRO.tec 2003*, 13–15 October 2003, Munich, Germany, VDE Verlag Berlin, 375–380.

[95] Scherge M., Ahmed S.I., Mollenhauer O. and Spiller F. 2000. Detection of Micronewton Forces in Tribology. *tm–Technisches Messen*, **67**(7–8), 324–327.

[96] Gitis N. 2005. Effective tribology testing of lubricating oils. In *Proceedings of the World Tribology Conference*, 12–16 September 2005, Washington, D.C. USA., paper 63078.

[97] Mollenhauer O., Ahmed S.I., Spiller F. and Haefke H. 2006. High–precision positioning and measurement systems for microtribotesting. *Tribotest*, **12**(3), 189–199.

[98] Drees D., Celis J.P. and Achanta S. 2004. Friction of thin coatings on three length scales under reciprocating sliding. *Surface and Coatings Technology*, **188–189**(1–2), 511–518.

[99] Hegadekatte V., Huber N. and Kraft O. 2005. 2005. Finite element based simulation of dry sliding wear. *Modelling and Simulation in Materials Science and Engineering*, **13**(1), 57–75.

[100] Hegadekatte V., Huber N. and Kraft O. 2006. Modeling and simulation of wear in a pin on disc tribometer. *Tribology Letters*, **24**(1), 51–60.

[101] Hegadekatte V., Kurzenhäuser S., Huber N. and Kraft O. 2008. A predictive modeling scheme for wear in pin-on-disc and twin-disc tribometers. *Tribology International*, **41**(11), 1020–1031.

[102] Herz J., Schneider J. and Zum Gahr K.H. 2002. Tribologische Untersuchungen an Gleitpaarungen aus ZrO_2, POM und Stahl für den Einsatz als wassergeschmierte mikromechanische Bauteile. *Materialwissenschaft und Werkstofftechnik*, **33**(7), 415–424.

[103] Herz J., Schneider J. and Zum Gahr K.H. 2004. Tribologische Charakterisierung von Werkstoffen für mikrotechnische Anwendungen. In *Proceedings of Tribologie Fachtagung 2004*, GfT, Gesellschaft für Tribologie e.V., Göttingen, (on CD).

[104] Schneider J., Zum Gahr K.H. and Herz J. 2005. Tribological characterization of mold inserts and micro components. *Advanced Micro and Nano Systems*, **3**, 579–604.

[105] Kurzenhäuser S., Schneider J. and Zum Gahr K.H. 2006. Untersuchungen zum Haft–und Gleitverhalten von Tribopaarungen für die Mikrotechnik. In Proceedings of *Tribologie Fachtagung 2006*, GfT, Gesellschaft für Tribologie e.V., Göttingen, (on CD).

[106] Kurzenhäuser S., Schneider J. and Zum Gahr K.H. 2007. Microtribological characterization of engineering ceramics and cemented carbide with respect to their application in high–performance Microsystems. In Proceedings of *Minat Congress 2007*, 12–14 June 2007, Stuttgart, Germany. (on CD).

[107] Rögner J., Kerscher E. and Löhe D. 2007. unpublished results.

Index